Multiantenna Channels: Capacity, Coding and Signal Processing

DIMACS
Series in Discrete Mathematics and Theoretical Computer Science

Volume 62

Multiantenna Channels: Capacity, Coding and Signal Processing

DIMACS Workshop
Signal Processing for Wireless Transmission
October 7–9, 2002
DIMACS Center

Gerard J. Foschini
Sergio Verdú
Editors

Center for Discrete Mathematics
and Theoretical Computer Science
A consortium of Rutgers University, Princeton University,
AT&T Labs–Research, Bell Labs (Lucent Technologies),
NEC Laboratories America, and Telcordia Technologies
(with partners at Avaya Labs, IBM Research, and Microsoft Research)

American Mathematical Society

This DIMACS volume presents the proceedings from the workshop Signal Processing for Wireless Transmission held at the DIMACS Center, Rutgers University, October 7–9, 2002.

2000 *Mathematics Subject Classification.* Primary 94A05, 94A13, 94A14, 94A15, 94A24, 94A40, 94B99, 15A52.

Library of Congress Cataloging-in-Publication Data

DIMACS Workshop Signal Processing for Wireless Transmission (2002 : Rutgers University)
Multiantenna channels : capacity, coding and signal processing : DIMACS Workshop Signal Processing for Wireless Transmission, October 7–9, 2002 / Gerard J. Foschini, Sergio Verdú, editors.
p. cm. — (DIMACS series in discrete mathematics and theoretical computer science ; v. 62)
Includes bibliographical references.
ISBN 0-8218-3407-X (alk. paper)
1. Signal processing—Congresses. 2. Wireless communication systems—Congresses. I. Foschini, Gerard J., 1940– II. Verdú, Sergio, 1958– III. Title. IV. Series.

TK5102.9.D58 2002
621.382′2—dc22 2003057746

∞ The paper used in this book is acid-free and falls within the guidelines
established to ensure permanence and durability.
Visit the AMS home page at `http://www.ams.org/`

10 9 8 7 6 5 4 3 2 1 08 07 06 05 04 03

Contents

Part III. Signal Processing for Multiantenna Channels

Part IV. Networking Capacity

Foreword

A workshop on Signal Processing for Wireless Transmission was held at Rutgers University on October 7–9, 2002. We would like to express our appreciation to Gerard Foschini and Sergio Verdú for their efforts to organize and plan this successful conference.

The workshop was part of the Special Focus on Computational Information Theory and Coding. We extend our thanks to Robert Calderbank, Chris Rose, Amin Shokrollahi, Emina Soljanin, and Sergio Verdú for their work as special focus organizers.

The workshop brought together theoreticians and practitioners working on wireless communications, information, and coding theory from a variety of perspectives. The main themes of the workshop included capacity of multiantenna channels and of vector broadcast channels, "dirty-paper" coding, signal processing, and ad hoc networking in wideband channels. These are all major themes in current research in physical-layer design for wireless communication.

DIMACS gratefully acknowledges the generous support that makes these programs possible. The National Science Foundation; the New Jersey Commission on Science and Technology; DIMACS' partners at Rutgers, Princeton, AT&T Labs-Research, Bell Labs, NEC Laboratories America, and Telcordia Technologies; and its affiliated partners at Avaya Labs, IBM Research, and Microsoft Research have generously supported the special focus.

Fred S. Roberts
Director

Robert Tarjan
Co-Director for Princeton

Preface

The DIMACS Workshop on Signal Processing for Wireless Transmission was held at Rutgers University on October 7–9, 2002, under the auspices of the DIMACS Special Focus on Computational Information Theory and Coding. Twenty-six invited presentations were delivered (http://dimacs.rutgers.edu/Workshops/Wireless/) by leading researchers in the fields of wireless communications, information and coding theory. Workshop attendees included researchers from institutions in North America, Europe, the Middle East and Asia. Through the agency of DIMACS, we were able to support the participation of a substantial number of students. In the true spirit of a workshop, there was considerable interaction among participants, accentuated by a sense of excitement over the presentation of fast-breaking results.

Several forces are coalescing to make this an exciting time in the field of physical-layer design for wireless communication: the relentless advance in VLSI technology, the recent breakthroughs in signal processing for multiantenna and multiuser systems, and the increased pressure for efficient utilization of scarce spectral resources. These forces make information-theoretic limits and the means of approaching them increasingly relevant for emerging wireless communication systems. This volume gives a snapshot of the range of challenges that are currently garnering a great deal of attention from the research community.

The first group of contributions on "Capacity of Multiantenna Channels" follows in the line of information-theoretic research pioneered at Bell Labs in the mid 1990s. The contribution by Xu, Chizhik, Huang and Valenzuela adds realism to the modeling of multiantenna channels by incorporating a number of key effects not captured in the canonical model. Even though much of the early emphasis on the information-theoretic analysis of multiantenna channels was on high signal-to-noise ratio (SNR) contexts, many practical embodiments, particularly in cellular systems, operate at low received energy per bit. The paper by Tulino, Lozano and Verdú considers the bandwidth-power tradeoff in the low SNR regime. The surprising relevance of large random matrix methods to systems with small numbers of antennas is illustrated in the paper by Moustakas, Simon and Sengupta, even with correlated channel coefficients. In another departure from the canonical setting, Simon and Moustakas analyze the impact of channel information at the transmitter array and determine under what conditions beamforming is optimal.

The second group of papers is devoted to the capacity of vector broadcast channels and its relationship with the problem of "dirty-paper coding". In dirty-paper coding, introduced by Costa in 1983 building upon earlier work, in addition to the background noise, an interference component is present which is known at the transmitter but not at the receiver. Dirty-paper coding is especially relevant to downlink communication, where the base station knows the channels and therefore knows the multiuser interference affecting each transmission. Although the sum-capacity has been solved recently, determination of the entire capacity region of the vector broadcast channel remains an open problem. Important advances towards its solution were announced at the workshop and reported in the paper by Jindal, Vishwanath, Jafar and Goldsmith; the paper by Tse and Viswanath; as well as the one by Vishwanath, Kramer, Shamai, Jafar and Goldsmith. The remaining issue to be settled in order to claim that the region put forth in those papers is the actual capacity region is the question of optimality of Gaussian codebooks.

The constructive counterpart to the capacity of dirty-paper coding is attracting considerable research. Caire and Shamai consider the application of low-density parity check codes and pulse-amplitude modulation in the design of systems that approach the fundamental limits when the interference is known at the transmitter only causally. After a tutorial presentation of dirty-paper coding, Foschini and Diaz explore what can be achieved with both one-dimensional lattices and in the asymptote of high-dimensional lattices.

The third part of the volume is devoted to signal processing. The analysis of the average computational complexity by Hassibi and Vikalo of the near-optimal sphere-decoding algorithm shows the considerable promise of this technique, which has applications beyond combating multiuser/multiantenna interference. A method to lower the signal processing complexity in channels with rapid fluctuations is presented in the contribution by Chizhik. This makes the gains promised by information-theoretic results approachable even in situations previously thought to be problematic. Foschini and Sellathurai show how to arrange a signal in space-time so as to achieve capacity with one-dimensional codecs when there is only one receive antenna. The technique is also shown to be effective in approaching capacity when the number of transmit antennas is much larger than the number of receive antennas. The use of iterative reception in conjunction with space-time codes is explored in the paper by Biglieri, Nordio and Taricco. Space-time block codes for channels that incorporate multiuser, intersymbol and multiantenna interference are considered by Diggavi, Al-Dhahir and Calderbank, who derive a simple formula for the diversity order. Most of the papers in this field assume perfect channel knowledge at the receiver. Samardzija and Mandayam explore the effect of imperfect channel estimation on achievable data rates.

The last group of papers in the volume explores issues related to ad hoc networking in wideband channels. Souilmi and Knopp examine the rates achievable by flash signaling in ultrawideband channels when only channel statistics are known at the receiver. Maric and Yates put forth the "accumulative broadcast" strategy in multihop environments where the objective is to minimize total transmitted energy.

It is a pleasure to acknowledge the expert guidance of Professor Fred S. Roberts, director of DIMACS. He generously provided his help and advice during the organization of the workshop and the preparation of this volume. We are also grateful to Ms. Shirley Hill of the American Mathematical Society for her diligent processing of the contents of this book. Finally, it is a pleasure to thank all the authors for their outstanding efforts that made this volume a reality.

Gerard J. Foschini and Sergio Verdú

Part I

Capacity of Multiantenna Channels

DIMACS Series in Discrete Mathematics
and Theoretical Computer Science
Volume **62**, 2003

Broadband MIMO Channel Model for Realistic Capacity and System Performance Evaluations

Hao Xu, Dmitry Chizhik, Howard Huang and Reinaldo Valenzuela

ABSTRACT. Realistic capacity evaluation requires accurate modeling of channel characteristics. This paper presents a multiple- input- multiple- output (MIMO) channel modeling approach for both single user and multiuser systems. Contrasting to the simplified complex Gaussian *iid* channel assumption, this model includes channel effects such as spatial interference, polarization, channel transitions, uplink and downlink channels, and key-hole/pin-hole channels. The spatial, temporal and frequency dispersions of the MIMO channels are implicitly modeled based on any given statistics. The modeling approach is demonstrated through an example with practical system setup and typical channel statistics.

1. Introduction

The channel capacity of a communications link can be significantly increased by deploying multiple transmit and multiple receive antennas [**11, 12**]. Early results on MIMO capacity were developed based on complex Gaussian *iid* channel matrices with additive white Gaussian noise (AWGN). The theoretical capacity, as well as performance of practical receivers, depends on the statistical properties of the channel. For realistic evaluation of MIMO capacity and algorithms, channel effects such as frequency selectivity, time variation, spatial and temporal correlations, and spatially non-white interference need to be taken into account.

Many MIMO channel modeling techniques have been developed, and can be generally divided into three classes: ray-tracing [**4, 6, 26**], scattering [**1, 5, 21, 22**], and correlation models [**3, 13, 17, 18**]. However, these models suffer either from inaccurate modeling of the temporal correlations, or computational complexity.

The goal of this work is twofold: 1) to develop a model that can accurately produce MIMO channel realizations based on any given statistics; 2) to reduce the computational complexity so that the model can be used for multiuser and multicell system evaluations.

Key words and phrases. MIMO, channel model, correlation, capacity, PDP, PAS.

2. Structure of the Channel Model

The discrete-time wideband MIMO channel can be described by a four dimensional channel transfer matrix $\mathbf{H}$ with dimensions $N_r \times N_t \times N_q \times N_s$, where N_r, N_t, N_q, N_s are the numbers of transmit antennas, receive antennas, resolved multipath components and time samples of the channel, respectively. The main task of the MIMO channel model is to construct $\mathbf{H}$ according to the desired statistical properties.

The entries of $\mathbf{H}$ can be correlated in space, time and frequency in realistic propagation environments. Therefore, a full statistical description of the channel is desirable in order to generate these $\mathbf{H}$ matrices. One such description is the joint distribution in angle , delay and Doppler domains such as in [**9**]. The distribution in angle can include angle of arrival (AOA) and angle of departure (AOD) in both azimuth and elevation in the most general case. Unfortunately, little channel measurements are available to fully describe these joint statistics for MIMO channels in different propagation environments. On the other hand, there is much of information available for single input single output (SISO), multiple input single output (MISO), or single input multiple output (SIMO) channel measurements [**5, 20, 8**]. Typically, these measurements report the spatial dispersion in terms of power azimuth spectrum (PAS) at the mobile or at the base, and the time dispersion in terms of power delay profiles (PDP). A PAS describes the power distribution in the azimuthal angle, while a PDP describes the power distribution in the delay time. In order to take advantage of these available measurement results, our channel model takes well-established PDP and PAS as input parameters.

The basic procedure for our modeling technique is summarized in Figure 1, which includes both large-scale and small-scale fading effects. The outer loop and inner loop are used to model the large-scale channel variation and small scale fading, respectively. The large-scale channel parameters (PDP, PAS and lognormal fading) are updated at a low speed of every few seconds per simulation run and remain fixed during the run according to the *quasistatic* assumption. For each choice of the large-scale parameters and mobile-basestation pair, the inner loop simulates the rapid small-scale fading due to the constructive and destructive superposition of the waves. The core of the channel generation consists of a basic module for typical MIMO channels and many optional modules, which address additional channel aspects depending on the propagation environments and system requirements. Details of each module are explained in the following sections.

3. Basic Module

We adopt the classical channel modeling approach, where the electromagnetic field in space is generated as a superposition of plane waves [**14**]. However, unlike in ray-tracing or scattering models where the wireless propagation medium is explicitly modeled to affect the wave components, we will treat the wireless medium as a "black box" so that the received signal's characteristics are derived entirely from statistical properties of the channel. This allows efficient channel modeling while capturing all important channel properties.

The propagation channel for the MIMO system is shown in Figure 2. The effects of the propagation medium to each of the wave component can be summarized as amplitude attenuation, phase shift, direction change, and time delay.

For the ith simulation run ***(outer loop for large-scale fading)***

Generate large-scale parameters (PAS, PDP, lognormal fading and velocity)

For the jth base-mobile pair ***(inner loop for small-scale fading)***

1. **Input: large-scale parameters for each base-mobile pair**
2. **MIMO Space-Time Model:**
   - **Basic Modules**
     - **For joint PAS of AOA and AOD (Section 3.1)**
     - **For separable PAS of AOA and AOD (Section 3.2)**
   - **Optional Modules**
     - **Model LOS component (Section 4.1)**
     - **Model uplink & downlink channels (Section 4.2)**
     - **Model polarization (Section 4.3)**
     - **Model transition (Section 4.5)**
     - **Model keyhole/pinhole channel (Section 4.6)**
3. **Output: four dimensional channel matrices H**

End

End

FIGURE 1. Simulation Procedure for General MIMO Channel Model

The modeling of direction change requires knowledge of the joint statistics of AOA and AOD, i.e. the joint PAS for AOA and AOD. Such information can be obtained from double directional measurements such as the one proposed in [**10**]. On the other hand, in rich scattering environments, it is reasonable to assume that the AOA and AOD are separable. In this case, large amounts of measurement results on PAS at the mobile or the base station can be readily implemented in the MIMO channel models. The following two subsections address the modeling issue for the cases of a) known joint statistics of AOD and AOA, and b) separable statistics of AOD and AOA.

3.1. Model Based on Joint PAS of AOA and AOD. The proposed channel model assumes that each of the entries in $\mathbf{H}$ is a superposition of the L wave components. The channel transfer function between receiver n and transmitter m at delay q and time s is determined by the sum of a large number of sinusoidal wave components:

$$h_{n,m,q,s} = \sqrt{P_q(\tau_q)} \sum_{l=1}^{L} A_{l,q} \sqrt{P_q^{BM}(\phi_l^B, \phi_l^M)} \underbrace{G_m^B(\phi_l^B) \exp\{i\vec{k}_l^B \cdot \vec{d}_m^B\}}_{base} \underbrace{G_n^M(\phi_l^M) \exp\{i\vec{k}_l^M \cdot \vec{d}_{n,s}^M\}}_{mobile} \tag{3.1}$$

where

L: is the number of wave components for each of the paths (assumed equal for different paths for simplicity);

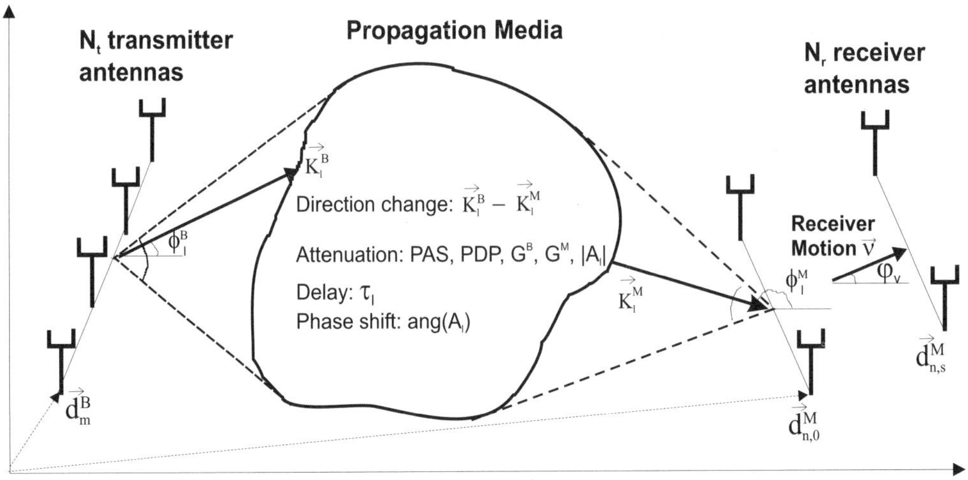

FIGURE 2. MIMO Channel Model

P_q: is the average power of the qth path given by the PDP, i.e. relative power as a function of τ_q;

τ_q: is the delay for the qth path;

P_q^{BM}: is the average power per wave component determined by the joint PAS at the base and the mobile, i.e. relative power as a function of AOD ϕ_l^B and AOA ϕ_l^M (the PAS can be a function of multipath delay in general);

ϕ_l^B: is the angle of departure (AOD) at the base for the lth wave component;

ϕ_l^M: is the angle of arrival (AOA) at the mobile for the lth wave component;

$A_{l,q}$: is a complex Gaussian random variable with zero mean and unit variance;

G_m^B: is the antenna pattern of the mth antenna at the base as a function of ϕ_l^B;

G_n^M: is the antenna pattern of the nth antenna at the mobile as a function of ϕ_l^M;

$\vec{k}_l^B$: is the directional vector at the base with amplitude $2\pi/\lambda$ and direction of the AOD of the lth component;

$\vec{k}_l^M$: is the directional vector at the mobile with amplitude $2\pi/\lambda$ and direction of the AOA of the lth component;

$\vec{d}_m^B$: is the position vector of the mth antenna of the base station antenna array;

$\vec{d}_{n,s}^M$: is the position vector of the nth antenna of the mobile antenna array at time s;

$\vec{x} \cdot \vec{y}$: represents the scalar product between two vectors.

Equation (3.1) provides a simple expression to generate four dimensional $\mathbf{H}$ for the case of joint PAS of AOA and AOD.

To be more general, the distribution of power vs. elevational angle can also be modeled. Measurements have shown that the elevational spread is much less than the azimuthal spread (e.g. [**25**]). Therefore, this dependency is omitted here for simplicity. As we mentioned before, all channel parameters in (3.1) are time-varying at different time scales. The antenna gains are fixed for all the simulation runs. The large scale parameters include PAS, PDP, AOD, AOA and $A_{l,q}$ are chosen in each run of the outer loop and remain fixed for the simulation run. The positional vector of the mobile is time-varying at the speed of the mobile. This leads to the small-scale fading due to the rapid phase changes in the L wave components. It is also worth noting that the model is flexible to include joint distribution of PDP and PAS. In fact, the PAS in (3.1), as well as the AOA and AOD can be functions of time delay.

3.2. Model Based on Separable PAS of AOA and AOD. In rich scattering environments, it is reasonable to assume that the PAS at the base and at the mobile are separable. One way to generate such separable PASs is to have L_1 AOAs and L_2 AODs, and let the total of $L = L1 * L2$ waves have all the possible combination of AOA and AOD. By doing this, power shaping of PAS at the base will have no effects on the PAS at the mobile [**24**]. Alternatively, as a result of separable PASs, one can model the correlations at the mobile and at the base separately by using different modeling techniques. We propose a hybrid technique of using a correlation matrix to model the spatial correlation at the base, while using the wave superposition to capture both the spatial and temporal correlations at the mobile. The main reduction in complexity comes from the fact that the base station antennas are stationary, therefore, the spatial correlation matrix at the base does not depend on the small scale fading that the mobile experiences. This spatial correlation can be calculated once for all of the small-scale fading realizations within each set of large-scale parameters. This two-step modeling approach is explained in Sections 3.2.1 and 3.2.2.

3.2.1. *Correlation at the Mobile.* The spatial and temporal correlations, that are caused by the motion of the mobile antenna array, are explicitly modeled by tracking the phase variations in each of the wave component.

The channel transfer function between receiver n and transmitter m at delay q and time s is given by

$$h^M_{n,m,q,s} = \sqrt{P_q}\sum_{l=1}^{L} A_{l,m,q}\sqrt{P^M_q(\phi^M_l)G^M_n(\phi^M_l)}\exp\{i\vec{k}^M_l \cdot \vec{d}^M_{n,s}\} \tag{3.2}$$

where index M represents mobile. The terms in (3.2) are identically defined as in (3.1) except for the P^M_q, which is now only the PAS at the mobile. As a result, the channel matrix $\mathbf{H}^M$ generated by (3.2) only includes the correlation at the mobile.

3.2.2. *Correlation at the Base.* Let $\mathbf{\Phi}^B$ denote the correlation matrix due to array geometry and the PAS at the base station, then the channel transfer matrix for any given time s and delay q are [**3, 21**]

$$\mathbf{H}_{s,q} = \mathbf{H}^M_{s,q}\sqrt{\mathbf{\Phi}^B} \tag{3.3}$$

where

- $\mathbf{H}^M_{s,q}$**:** is the $N_r \times N_t$ two dimensional matrix generated by (3.2), i.e. with correlations only at the mobile, for any given time s and delay q;
- $\mathbf{H}_{s,q}$**:** is the $N_r \times N_t$ two dimensional matrix with correlations at both the mobile and the base station for any given time s and delay q;

Each element of the correlation matrix, ϕ^B, is a function of power azimuth spectrum $P(\phi)$ and antenna separation d as [**14**]

$$R_d(\vec{d}) = \int_0^{2\pi} \exp\{-j\vec{k}\cdot\vec{d}\}P(\phi)d\phi \tag{3.4}$$

Note that the correlation is dependent on the direction of $\vec{d}$, i.e. the orientation of the antenna array.

Combining (3.2) and (3.3), the generated **H** matrices have correlated entries due to both spatial and temporal correlations from the transmitter and receiver arrays.

4. Optional Modules

4.1. LOS vs. NLOS. So far we have considered the basic propagation scenario, where the received signal is a superposition of many small wave components. Since all the wave components have time-varying phases and comparable amplitude, the amplitude of the resulting received signal obeys Rayleigh distribution for a large L. If it is desired to include a LOS component with a given Rician K factor, a strong plane wave signal from the base to the mobile will be added to the first multipath component with the shortest delay.

$$h^{LOS}_{n,m,1,s} = \sqrt{K}\exp\{i\vec{k}^B_{LOS}\cdot\vec{d}^B_m\}\exp\{i\vec{k}^M_{LOS}\cdot\vec{d}^M_{n,s}\} \tag{4.1}$$

$$h_{n,m,1,s} = h^{NLOS}_{n,m,1,s} + h^{LOS}_{n,m,1,s} \tag{4.2}$$

The elements of **H** at other delays remain the same. The AOA and AOD of the Rician component can be exactly determined from the coordinates of the base and the mobile. Note that in our model, the LOS component introduces additional power to the received signal. This reflects the increase of signal power when the receiver moves from a non-LOS location to a LOS location. If it is desirable to keep the received signal power constant, normalization can be performed by dividing the total power by K+1 for the Rician case.

4.2. Uplink vs. Downlink. The modeling technique can be applied both for the uplink and downlink. In adaptive beamforming applications, one implementation method is to obtain downlink beamforming information from the uplink measured channel. Therefore, uplink and downlink channels need to be generated jointly [**1**].

For a typical frequency division duplex (FDD) system, the frequency separation between uplink and downlink is small, so that the large-scale channel statistics such as PAS, PDP can be assumed statistically identical for both uplink and downlink. The only difference is the relative phases of the wave components due to different frequencies used at uplink and downlink. Thus, the channel matrices for the uplink and downlink with correlations at the mobile can be generated from

$$h^{u,M}_{n,m,q,s} = \sqrt{P_q}\sum_{l=1}^{L} A^u_{l,m,q}\sqrt{P^M_q(\phi^M_l)G^M_n(\phi^M_l)}\exp\{i\vec{k}^M_l\cdot\vec{d}^M_{n,s}\} \tag{4.3}$$

$$h^{d,M}_{n,m,q,s} = \sqrt{P_q}\sum_{l=1}^{L} A^d_{l,m,q}\sqrt{P^M_q(\phi^M_l)G^M_n(\phi^M_l)}\exp\{i\vec{k}^M_l\cdot\vec{d}^M_{n,s}\} \tag{4.4}$$

which results in channel transfer matrices $\mathbf{H}_{s,q}^{u,M}$ for uplink and $\mathbf{H}_{s,q}^{d,M}$ for the downlink. Note that two independent sets of random variables $A_{l,m,q}^{u}$ and $A_{l,m,q}^{d}$ are used to model the difference between uplink and downlink channels. Further inclusion of the spatial correlation at the base station leads to

$$\mathbf{H}_{s,q}^{u} = \mathbf{H}_{s,q}^{u,M}\sqrt{\mathbf{\Phi}^{B}} \tag{4.5}$$

$$\mathbf{H}_{s,q}^{d} = \mathbf{H}_{s,q}^{d,M}\sqrt{\mathbf{\Phi}^{B}} \tag{4.6}$$

with $\mathbf{H}_{s,q}^{u}$ for uplink and $\mathbf{H}_{s,q}^{d}$ for downlink.

4.3. Polarization Modeling. Polarized antennas are frequently used in MIMO systems to obtain decorrelation and reduce antenna spacing (for example dual $\pm 45^\circ$ slant antennas). This section briefly describes how polarization effects can be addressed in the context of our proposed model. For simplicity in notation, we restrict the discussion to one multipath component at a given time, thus the subindices of q and s for $\mathbf{H}$, which are used to show dependency on delay and time, are omitted.

When different polarizations are used, matrix $\mathbf{H}$ can be decomposed into four submatrices as

$$\mathbf{H} = \begin{pmatrix} \mathbf{H}_{v,v} & \sqrt{\eta}\mathbf{H}_{v,h} \\ \sqrt{\eta}\mathbf{H}_{h,v} & \mathbf{H}_{h,h} \end{pmatrix} \tag{4.7}$$

with $\mathbf{H}_{v(h),v(h)}$ denoting the channel matrix from v(h)-polarized transmitter antennas to v(h)-polarized receiver antennas, respectively, and η denoting cross-polarization discrimination (XPD) factor.

Measurements by Lee and Yeh [**15**] show that the received signals from the vertically and horizontally polarized antennas are uncorrelated, and the measured XPD is about 6 dB. Assuming zero cross correlation between different polarizations, the submatrices $\mathbf{H}_{v(h),v(h)}$ can be generated using the techniques described in the main module. The PAS and PDP for these four subchannels are identical as supported by measurements in [**7**] and [**19**]. The difference in the four subchannels results from the independent realizations of the random variables $A_{l,m,q}$ in (3.2).

For arbitrary orientations of the antenna arrays, for example in a system with dual $\pm 45^\circ$ slant antennas, the transmitted and received signals can be projected into the vertical and horizontal orientations.

4.4. System Model. In Sections 3 through 4.3, we have discussed generation of time varying $\mathbf{H}$ matrices for a single user single base station link. This section describes the extension from the link level to the system level channel model.

Consider the active sector with N_{usr} users and N_{base} closest base stations. For each simulation run, we will randomly distribute the users within the active sector. The initial positions of the users are random and evolve with time according to the array orientation and mobile velocity. The spatial vector $\vec{d_n}$ for the n_{th} receiver antenna for the i_{th} user can be determined from his location coordinates x_i and y_i. The mean AOA at the mobile is dictated by the local scatterers, thus can be assumed uniformly distributed in $[0, 2\pi]$. The mean AOD from the base stations can be either determined from the LOS direction to the mobile or uniformly distributed in $[-\pi/3, \pi/3]$ for a three-sector system. Finally, different PDPs and PASs can be assigned to different users from serving and interfering base stations according to the probability of occurrences of the standard set of PDPs and PASs, or chosen based on measurements.

Methodology described in Sections 3 to 4.3 can be used to generate **H** matrices from both the desired base station and interfering base stations.

$$\mathbf{r}(s) = \sum_{b=1}^{N_{base}} \sqrt{P_b \gamma_b} \sum_{q=1}^{N_{mp}} \mathbf{H}_{b,q}(s)\mathbf{X}_b(s - \tau_{b,q}) + \mathbf{n}(s) \tag{4.8}$$

where

- $\mathbf{r}(s)$**:** is an N_r dimensional received signal vector at time s;
- P_b**:** is the transmitted power from the bth base;
- γ_b**:** is the large scale path gain from the bth base, which includes lognormal fading;
- $\mathbf{H}_{b,q}(s)$**:** is an N_r by N_t matrix for the bth base and q^{th} delayed path at time s generated by (3.1);
- $\tau_{b,q}$**:** is the qth multipath delay given by the PDP selected for the bth base;
- $\mathbf{X}_b(s)$**:** is the N_t dimensional transmitted signal vector from the bth base at time s;
- $\mathbf{n}(s)$**:** is the N_r dimensional noise vector, which includes thermal noise and other interferences.

Pulse shaping can also be included in the transmitted signal $\mathbf{X}_b$ in (4.8).

The total system simulation requires at least hundreds of simulation runs in order to capture the large scale channel effects. The duration of each simulation run should be long enough to produce reliable statistics (e.g. bit error rate) but without violating the quasistationary assumption. One can use 1-10 s for each simulation run, for example, for UMTS applications at moderate vehicle speed of 3 km/hr to 30 km/hr. After each run, the users are redistributed in the cell, new PDP, PAP and γ_b are chosen, and the simulation is repeated for the next run. Since the proposed channel model does not require specific modeling of scatterers, it is simple and flexible to simulate MIMO channels with various statistical properties.

4.5. Model the Transition. In the system simulation procedure described in Section 4.4, the simulation runs are independent realizations from the large scale channel parameters. In applications where it is desirable to simulate the smooth transition from one simulation run to another run, we can adopt the two techniques developed for the scattering models. A transitional function can be introduced such as the one proposed in [**5**]. For the i_{th} and j_{th} simulation runs, independent channel transfer matrices $\mathbf{H}^i(s)$ and $\mathbf{H}^j(s)$ are generated according to (3.3). Then the channel transfer matrix during the transition from the ith to the jth simulation run can be modeled as a weighted sum of these two matrices

$$\mathbf{H_q}(s) = \sqrt{\alpha(s)}\mathbf{H}_q^i(s) + \sqrt{1-\alpha(s)}\mathbf{H}_q^j(s) \tag{4.9}$$

where the transitional function $\alpha(s) \in [0, 1]$ can be a linear function with $\alpha(0) = 1$ in the beginning of the transition and $\alpha(s_t) = 0$ in the end of the transition.

An alternative technique, which models the "birth-death" process of scatterers, can also be applied to our general channel model [**1**]. The main idea is that certain wave components can be replaced with new wave components within each simulation run. By doing so, the channel gradually evolves from one environment to another. Furthermore, additional randomness in the large-scale parameters can be introduced within each simulation run if desired.

4.6. Model Keyhole Channels. In particular propagation environments, it is possible to have degenerated channel phenomena so called "keyholes" [**2**] or "pinholes"[**13**]. In a keyhole channel, the **H** matrix can have zero correlation among the elements, yet have only one degree of freedom. However, the keyhole channel does not occur often in realistic propagation environments. In measurements performed in Manhattan, the only keyhole channel was observed when the car was driving through a tunnel [**16**]. Therefore, in our system model, we propose to keep keyhole channel as an optional module. The keyhole channel can be generated as a dyad:

$$\mathbf{H}_{kh} = \vec{B}^T \vec{A} = (b_1, b_2, \cdots, b_{N_r})^T (a_1, a_2, \cdots, a_{N_t}) \tag{4.10}$$

with T denoting the matrix transpose, vectors B and A are complex Gaussian variables describing the scattering at the receiver and the transmitter, respectively. Even for completely uncorrelated A and B, the rank of the matrix **H** is only one.

5. Example of Implementations

This section demonstrates the implementation of the channel modeling approach for a typical MIMO system setup and some realistic channel statistics. We use a 4 transmitter 4 receiver MIMO system with uniform linear arrays at both ends. The transmitter array has a separation of 4λ, and the receiver array has a separation of 0.5λ, where λ is the wave length. Table 1 summarizes the simulation assumptions.

TABLE 1. Simulation Assumptions.

	Base	Mobile
Antenna type	omni	omni
Array type	ULA	ULA
Antenna spacing	4λ	0.5λ
PAS shape	Laplacian	Laplacian
rms AS	5°	35°
PDP	Pedestrian A	
Pulse shaping	None	
Number of waves	30	
Mobile speed	10 km/hr	
Sampling interval	2/3 μs (1 sample/slot in UMTS)	
Samples per run	15000 (10 s duration)	

Based on these channel parameters, the realizations of the MIMO channel matrices can be generated from (3.2) and (3.3) assuming separable PASs at the transmitter and the receiver. The channel matrices will be correlated both spatially and temporally according to the PASs at the transmitter and the receiver. Next, we will verify the effects of these correlations on the theoretical MIMO channel capacity.

For wideband MIMO channels, the channel transfer matrix can be expressed as $\mathbf{H}(\omega)$ in frequency domain, or $\mathbf{H}(\tau)$ in the equivalent delay domain.

$$\mathbf{H}(\omega) = \int_{\tau} \mathbf{H}(\tau) \exp\{j\omega\tau\} d\tau = \sum_{q=1}^{N_{mp}} \mathbf{H}_q \exp\{j\omega\tau_q\} \tag{5.1}$$

where $\mathbf{H_q}$ is the channel response at the qth delay.

The open-loop link capacity per unit bandwidth for the wideband channel is then

$$C = \frac{1}{B} \int_{B} \log_2 \det[I + \frac{\rho}{N_t} H(f) H^{\dagger}(f)] df \tag{5.2}$$

where ρ is the average signal to noise ratio.

For narrowband communications systems, i.e. the signal bandwidth is smaller than the channel coherence bandwidth, the receiver narrowband filter effectively integrates over all the multipath components. The equivalent narrowband channel matrix is

$$\mathbf{H} = \sum_{q=1}^{N_{mp}} \mathbf{H}_q \tag{5.3}$$

The link capacity for the narrowband system is

$$C = \log_2 \det[I + \frac{\rho}{N_t} \mathbf{H}\mathbf{H}^{\dagger}] \tag{5.4}$$

In [**12**], this equation is derived for the MIMO channel matrix with complex Gaussian *iid* entries.

Here we apply (5.4) to the generated $\mathbf{H}$ matrics with correlated entries. Figure 3 shows the capacity results for the simulated channel matrices. The $\mathbf{H}$ matrices are generated based on our plane-wave based model and also the well-known correlation model. These two models produce essentially the same capacity CDFs. The link capacity and the effective rank of $\mathbf{H}$ depend on the spatial correlations due to the PASs. As a result, the capacity is noticeably lower than the capacity of the complex Gaussian *iid* channels. Further theoretical justification and experimental verification of the model are presented in a more detailed publication [**23**].

For this system setup, even for the realizations of 15000 channel matrices, the run time for each user-basestation pair is only about 30 s per simulation run. This is simulated by using Matlab in a laptop computer with a Pendium III processor. Further reduction in simulation time can be achieved by implementing the model in C.

6. Summary

This paper presented a MIMO space-time channel modeling approach that includes important channel effects such as polarization, key-hole/pin-hole channel, spatial and temporal correlations, and uplink-downlink modeling. The implementation of the MIMO channel model is demonstrated with a practical setup of a MIMO system and typical channel parameters. Realistic capacity as well as system performance can be evaluated based on the generated channels.

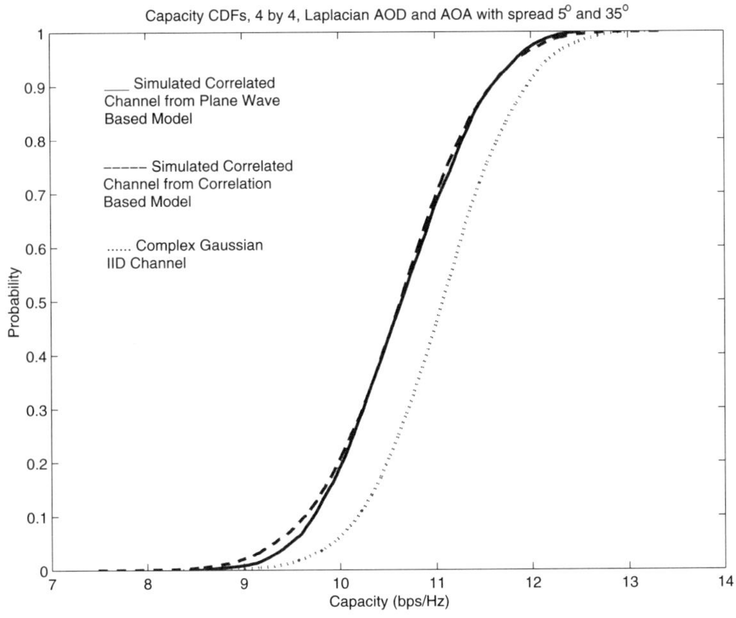

FIGURE 3. Capacity CDFs of the Simulated Channels, 4×4, SNR=10 dB.

References

[1] R. M. Buehrer, S. Arunachalam, K. H. Wu, and A. Tonello. Spatial Channel Model and Measurements for IMT-2000 Systems. In *IEEE VTC-2001*, Rhodes, Greece, May 6-9 2001.

[2] D. Chizhik, G. J. Foschini, and R. A. Valenzuela. Capacities of Multi-Element Transmit and Receive Antennas: Correlations and Keyholes. *Electronics Letters*, 36(13):1099–1100, June 2000.

[3] D. Chizhik, F. Rashid-Farrokhi, J. Ling, and A. Lozano. Effects of Antenna Separation on the Capacity of BLAST in Correlated Channels. *Communications Letters*, 4(11):337–339, November 2000.

[4] C. N. Chuah, D.N.C. Tse, J. M. Kahn, and R. Valenzuela. Capacity Scaling in MIMO Wireless Systems Under Correlated Fading. *IEEE Transactions on Information Theory*, 48(3):637–650, March 2002.

[5] L. Correia. *Wireless Flexible Personalized Communications*. Wiley, John & Sons Inc., 2001.

[6] P. Driessen and G. J. Foschini. On the Capacity Formula for Multiple Input-Multiple Output Wireless Channels: A Geometric Interpretation. *IEEE Transactions on Communications*, 47(2):173–176, Feb. 1999.

[7] P. C. Eggers, J. Toftgard, and A. Oprea. Antenna systems for base station diversity in urban small and micro cells. *IEEE Journal on Selected Areas in Communications*, 11(7), Sept 1993.

[8] R. B. Ertel, P. Cardieri, K. W. Sowerby, T. S. Rappaport, and J. H. Reed. Overview of Spatial Channel Models for Antenna Array Communication Systems. In *IEEE Personal Communications Magazine*, volume 5, pages 10–12, Feb 1998.

[9] B. H. Fleury. First- and Second-Order Characterization of Direction Dispersion and Space Selectivity in the Radio Channel. *IEEE Transactions on Information Theory*, 46(6):2027–2044, Sept. 2000.

[10] B. H. Fleury, M. Tschudin, R. Heddergott, D. Dahlhaus, and K.I. Pedersen. Channel parameter estimation in mobile radio environments using the sage algorithm. *IEEE Journal on Selected Areas in Communications*, 17(3), March 1999.

[11] G. J. Foschini. Layered Space-Time Architecture for Wireless Communication in a Fading Environment When Using Multiple Antennas. *Bell Labs Technical Journal*, 1(2):41–59, Autumn 1996.
[12] G. J. Foschini and M. J. Gans. On Limits of Wireless Communications in a Fading Environment when Using Multiple Antennas. *Wireless Personal Communications*, 6:311–335, 1998.
[13] D. Gesbert, H. Bolcskei, D. Gore, and A. Paulraj. MIMO Wireless Channels: Capacity and Performance Prediction. In *IEEE Globecom-2000*, San Fran., CA., Nov. 27-Dec. 1 2000.
[14] W. C. Jakes(Ed). *Microwave Mobile Communications.* IEEE Press, New York, 1974.
[15] W. C. Lee and Y. S. Yeh. Polarization divesity system for mobile radio. *IEEE Transactions on Communications*, COM-20(5), Oct 1972.
[16] J. Ling, D. Chizhik, P. Wolniansky, R. Valenzuela, N. Costa, and K. Huber. Multiple Transmitter Multiple Receiver (MTMR) Capacity Survey in Manhattan. *Electronic Letters*, 37(16), Aug. 2001.
[17] A. L. Moustakas. Communication through a diffusive medium: coherence and capacity. *Science*, 287:287–290, Jan. 2000.
[18] K. I. Pedersen, J. B. Andersen, J. P. Kermoal, and P. Modensen. A stochastic multiple-input-multiple-output radio channel model for evaluation of space-time coding algorithms. In *IEEE VTC-2000*, Boston, Sept. 24-28 2000.
[19] K. I. Pedersen, P. E. Mogensen, and B. H. Fleury. Dual-polarized model of outdoor propagation environments for adaptive antennas. In *IEEE VTC'99*, 1999.
[20] T. S. Rappaport. *Wireless Communications: Principles and Practice.* Prentice-Hall Inc., New Jersey, 1996.
[21] D. Shiu, G. J. Foschini, M. J. Gans, and J. M. Kahn. Fading Correlation and Its Effect on the Capacity of Multielement Antenna Systems. *IEEE Trans. Commun.*, 48(3):502–513, March 2000.
[22] M. Steinbauer, A. Molisch, and E. Bonek. The double-directional radio channel. *IEEE Antenna and Propagat. Mag.*, pages 51–63, Aug. 2001.
[23] H. Xu, D. Chizhik, H. Huang, and R. Valenzuela. A General Space-Time Multiple Input Multiple Output (MIMO) Channel Model. *submitted to IEEE Transactions on Wireless Communications*, 2002.
[24] H. Xu, D. Chizhik, H. Huang, and R. Valenzuela. A Wave-Based Wideband MIMO Channel Modeling Technique. *IEEE PIMRC*, Sept 2002.
[25] H. Xu, M. Gans, N. Amitay, R.A. Valenzuela, T. Sizer, R. Storz, D. Taylor, M. McDonald, and C. Tran. Measurement Results and Models for MTMR Capacity at 2.44 GHz in Suburban Outdoor Environments. In *IEEE VTC-2001*, Atlantic City, Oct. 7-11 2001.
[26] H. Xu, M. J. Gans, N. Amitay, and R. A. Valenzuela. Experimental verification of MTMR system capacity in a controlled propagation environment. *Electronics Letters*, 37(15), July 2001.

Wireless Communications Research Department, Bell-Laboratories/Lucent Technologies, 791 Holmdel Keyport Rd, Holmdel, NJ 07733
E-mail address: haoxu@ieee.org

DIMACS Series in Discrete Mathematics
and Theoretical Computer Science
Volume **62**, 2003

Bandwidth-Power Tradeoff of Multi-Antenna Systems in the Low-Power Regime

Antonia M. Tulino, Angel Lozano, and Sergio Verdú

ABSTRACT. This article provides an analytical characterization of the interplay between rate, power and bandwidth in multi-antenna channels being operated in the low-power regime. The analysis is conducted—both non-asymptotically and asymptotically in the number of antennas—in the region of energy per bit close to its minimum value. Throughout the derivations, several moments of the trace of the product of certain random matrices are evaluated and put forth as stand-alone results.

1. Introduction

Wireless communication through multiple transmit and receive antennas enables very large increases in the data rate that can be conveyed reliably per unit of bandwidth and power. In this article, we analyze the tradeoffs between these fundamental quantities (rate, bandwidth and power) in single-user multi-antenna channels. Such tradeoffs have been widely studied [**4, 16, 17, 21, 25**] but often only in the high-power regime and mostly with highly idealized canonical channel and noise models, respectively (i) uncorrelated transfer coefficients from each transmit to each receive antenna, and (ii) additive white Gaussian noise (AWGN). Our analysis, in contrast, is conducted in the low-power regime—a most important one in modern wireless systems—and it accounts for important features that are not captured by those canonical models.

Because of aggressive frequency reuse across neighboring cells and sectors, modern wireless systems are—by design—limited by their own interference. This has two direct consequences:

- Dominated by out-of-cell interference, the additive noise is spatially colored and subject to fading. Thus, it can no longer be modeled as AWGN.
- Since a majority of locations lie in the periphery of their cells, users must operate very often at low signal-to-noise ratio (SNR) while only rarely at high SNR.

1991 *Mathematics Subject Classification.* Primary 54C40, 14E20; Secondary 46E25, 20C20.
Key words and phrases. Channel capacity, Multi-antenna arrays, Low-power, Fading channels, Rayleigh, Ricean, Antenna correlation, Interference, Asymptotic analysis.

Denoting by $\mathsf{C}(\frac{E_b}{N_0})$ the capacity (in bits/s/Hz) as function of the normalized energy per bit, a system designed to achieve rate R (in bits/s) with power P reliably requires bandwidth

$$B = \frac{R}{\mathsf{C}\left(\frac{P}{RN_0}\right)}. \tag{1.1}$$

Unfortunately, $\mathsf{C}(\frac{E_b}{N_0})$ is known in closed form only for the unfaded AWGN channel and hence, in general, the tradeoffs governed by (1.1) cannot be characterized analytically. At low SNR, however, the two key performance measures that capture the behavior of $\mathsf{C}(\frac{E_b}{N_0})$, namely the $\frac{E_b}{N_0}$ at which the capacity becomes zero and the slope therein, can be expressed in closed form very generally [**24**]. This enables an analytical characterization of these tradeoffs and, in particular, of how they are impacted by important features that are not captured by the canonical models. Specifically, the multi-antenna features we consider are (i) antenna correlation, (ii) noise color, and (iii) time-varying interference.

Section 2 discusses the models we use to incorporate these non-canonical features. Section 3 presents general expressions for the key low-SNR performance measures. In Section 4, several moments of the trace of the product of some random matrices are evaluated. Besides required to proof the main results in the article, these moments are of independent interest. Then, Section 5 and Section 6 analyze the low-power capacity and the corresponding power-bandwidth tradeoffs in the presence of AWGN and out-of-cell interference, respectively. A summary follows in Section 7.

2. Definitions and Models

2.1. Definitions. Given a matrix $\mathbf{A}$, we shall use $(\mathbf{A})_{i,j}$ to denote its (i, j)-th element and $(\mathbf{A})_i$ to denote its i-th row.

With n_T transmit and n_R receive antennas, a baseband discrete-time complex-valued model for the multi-antenna channel with frequency-flat fading is

$$\mathbf{y} = \sqrt{\mathsf{g}}\,\mathbf{H}\,\mathbf{x} + \mathbf{n}$$

where $\mathbf{x}$ is the n_T-dimensional transmit signal while $\mathbf{y}$ and $\mathbf{n}$ are the received signal and the Gaussian noise, both of them n_R-dimensional. The channel is represented by the $(n_\mathrm{R} \times n_\mathrm{T})$ random matrix $\sqrt{\mathsf{g}}\,\mathbf{H}$, zero-mean and independent of both $\mathbf{x}$ and $\mathbf{n}$. For convenience, we factor out the scalar $\sqrt{\mathsf{g}}$ (average channel gain) so as to yield a normalized matrix $\mathbf{H}$ with unit-variance entries.

When $n_\mathrm{R} = n_\mathrm{T}$ or when a particular equation applies to both transmitter and receiver, we may use n to refer to the number of antennas generically. Also, for the purpose of asymptotic analysis in number of antennas, we define

$$\beta \stackrel{\text{def}}{=} \frac{n_\mathrm{T}}{n_\mathrm{R}}.$$

The single-sided spectral density of the noise is denoted by

$$N_0 = \frac{E[\|\mathbf{n}\|^2]}{n_\mathrm{R}}.$$

and its normalized spatial covariance is defined as

$$\mathbf{\Phi}_\mathbf{n} \stackrel{\text{def}}{=} \frac{E[\mathbf{n}\mathbf{n}^\dagger]}{N_0}.$$

Throughout the article, we focus on those scenarios where $\mathbf{H}$ and $\mathbf{\Phi_n}$ are known to the receiver. At the transmitter, however, we shall presume that no information about either the channel or the noise is available. In such conditions, the transmitter must spread its power evenly on all dimensions and thus [**21, 22**]

$$E[\mathbf{x}\mathbf{x}^\dagger] = \frac{E[\|\mathbf{x}\|^2]}{n_\mathrm{T}}\mathbf{I} \tag{2.1}$$

with $\mathbf{I}$ the identity matrix. We shall consider two distinct types of noise:

- Thermal AWGN, characterized by $\mathbf{\Phi_n}=\mathbf{I}$ and $N_0=\gamma$ with γ the thermal spectral density per receive antenna.
- Out-of-cell interference, spatially colored and subject to fading. Conditioned on its fading, this interference is still modeled as Gaussian.

When both types are present, the aggregate noise adopts the form

$$\mathbf{n} = \sum_{\ell=1}^{L} \sqrt{\mathsf{g}_\ell}\,\mathbf{H}_\ell \mathbf{x}_\ell + \mathbf{n}_{\mathsf{th}} \tag{2.2}$$

with L the number of interferers, $\mathbf{x}_\ell$ the signal transmitted by the ℓ-th interferer, $\sqrt{\mathsf{g}_\ell}\,\mathbf{H}_\ell$ the channel from such interferer, and $\mathbf{n}_{\mathrm{th}}$ the thermal AWGN. The number of transmit antennas at the ℓ-th interferer is denoted by m_ℓ. Since the contribution of each interferer is subject to fading, the short-term covariance of $\mathbf{n}$ depends on $\mathbf{H}_\ell$, $\ell{=}1\ldots L$, which we assemble into $\overline{\mathbf{H}}{=}[\mathbf{H}_1|\ldots|\mathbf{H}_L]$. With the various interferers mutually independent and signaling as in (2.1), the normalized covariance matrix of (2.2) conditioned on $\overline{\mathbf{H}}$ becomes

$$\begin{aligned}\mathbf{\Phi_n}(\overline{\mathbf{H}}) &= \frac{E[\mathbf{n}\mathbf{n}^\dagger|\overline{\mathbf{H}}]}{N_0}\\ &= \frac{1}{N_0}\left(\sum_{\ell=1}^{L}\mathsf{g}_\ell\frac{E[\|\mathbf{x}_\ell\|^2]}{m_\ell}\mathbf{H}_\ell\mathbf{H}_\ell^\dagger + \gamma\,\mathbf{I}\right).\end{aligned} \tag{2.3}$$

It is important to stress that the receiver does not have access to $\overline{\mathbf{H}}$, but only to $\mathbf{\Phi_n}(\overline{\mathbf{H}})$. By conveniently defining

$$\mathcal{I}_\ell \stackrel{\text{def}}{=} \mathsf{g}_\ell\frac{E[\|\mathbf{x}_\ell\|^2]}{m_\ell}\,\frac{E[\mathrm{Tr}\{\mathbf{H}_\ell\mathbf{H}_\ell^\dagger\}]}{n_\mathrm{R}} \tag{2.4}$$

as the average energy per antenna received from the ℓ-th interferer, we can write

$$N_0 = \sum_{\ell=1}^{L}\mathcal{I}_\ell + \gamma \tag{2.5}$$

noting that, in mature systems, γ tends to be small with respect to N_0.

Given a received per-symbol SNR, the corresponding normalized received energy per bit relates to it through the number of bits per dimension, that is,

$$\frac{E_b^r}{N_0} = \frac{\mathsf{SNR}}{(R/B)}n_\mathrm{R}.$$

The SNR, in turn, is computed as as the average of the signal-to-noise ratios along each of the principal directions of the noise space, which yields

$$\mathsf{SNR} = \mathsf{g}\frac{E[\|\mathbf{x}\|^2]}{N_0}\,\frac{E[\mathrm{Tr}\{\mathbf{H}\mathbf{H}^\dagger\mathbf{\Phi_n}^{-1}(\overline{\mathbf{H}})\}]}{n_\mathrm{T}n_\mathrm{R}} \tag{2.6}$$

with expectation over $\mathbf{x}$, $\mathbf{H}$ and $\overline{\mathbf{H}}$.[1] Since the largest rate per unit bandwidth achievable with arbitrary reliability is precisely the capacity $C(\mathsf{SNR})$, the $\frac{E_b}{N_0}$ required at a receiver operating at capacity is

$$\frac{E_b^r}{N_0} = \frac{\mathsf{SNR}}{C(\mathsf{SNR})} n_{\mathrm{R}}.$$

Rather than $\frac{E_b^r}{N_0}$, though, we are interested in the *transmitted* $\frac{E_b}{N_0}$, given by

$$\begin{aligned} \frac{E_b}{N_0} &= \frac{E[\|\mathbf{x}\|^2]}{N_0 C(\mathsf{SNR})} \\ &= \frac{n_{\mathrm{T}} n_{\mathrm{R}}}{\mathsf{g} E[\mathrm{Tr}\{\mathbf{H}\mathbf{H}^\dagger \mathbf{\Phi}_{\mathbf{n}}^{-1}(\overline{\mathbf{H}})\}]} \frac{\mathsf{SNR}}{C(\mathsf{SNR})}. \end{aligned} \tag{2.7}$$

To conclude this section, a final definition: given a $(n \times n)$ matrix $\mathbf{A}$, in general random, we define its *dispersion* as

$$\zeta(\mathbf{A}) = n \frac{E[\mathrm{Tr}\{\mathbf{A}^2\}]}{E^2[\mathrm{Tr}\{\mathbf{A}\}]}. \tag{2.8}$$

Note that $\zeta(\mathbf{A})$ achieves its minimum value (equal to 1) when $\mathbf{A}$ is equal to the identity matrix. As we shall see, the dispersion serves to succinctly quantify how much the channel and noise depart from those in the canonical model.

2.2. Channel Model. The fading encountered by wireless systems is typically Rayleigh-distributed and thus we model the entries of $\mathbf{H}$ as jointly Gaussian and zero-mean. With that, the characterization of $\mathbf{H}$ entails simply determining the correlation between its entries. Such correlation can be obtained, in most cases, from the local correlation between transmit antennas and the local correlation between receive antennas, separately.[2] With that, the standard model for Rayleigh channel matrices is given by [**1, 15, 18**]

$$\mathbf{H} = \mathbf{\Theta}_{\mathrm{R}}^{1/2} \mathbf{H}_w \mathbf{\Theta}_{\mathrm{T}}^{1/2} \tag{2.9}$$

where the entries of $\mathbf{H}_w$ are independent unit-variance complex Gaussian random variables whereas $\mathbf{\Theta}_{\mathrm{T}}$ and $\mathbf{\Theta}_{\mathrm{R}}$ are $(n_{\mathrm{T}} \times n_{\mathrm{T}})$ and $(n_{\mathrm{R}} \times n_{\mathrm{R}})$ correlation matrices, deterministic and Hermitian. Specifically, $\mathbf{\Theta}_{\mathrm{T}}$ contains the correlation coefficients between transmit antennas while $\mathbf{\Theta}_{\mathrm{R}}$ contains the correlation coefficients between receive antennas.

Since a correlation matrix—either transmit or receive—has unit diagonal elements, its dispersion (2.8) particularizes to

$$\zeta(\mathbf{\Theta}) = \frac{\mathrm{Tr}\{\mathbf{\Theta}^2\}}{n}$$

which we shall refer to as *correlation number*. For example, if $n=2$,

$$\zeta\left(\begin{bmatrix} 1 & \rho \\ \rho & 1 \end{bmatrix}\right) = 1 + \rho^2.$$

[1]The SNR in (2.6) can also be interpreted as the average signal power divided by the average noise power, both measured at the output of a noise whitening filter.

[2]This separation applies if the immediate surroundings to each array are responsible for the correlation between its antennas but have no impact on the correlation between the antennas at the other end of the link [**5, 14**].

Some immediate properties of such correlation number are:

PROPERTY 2.1. *The correlation number is bounded by*

$$1 \leq \zeta(\mathbf{\Theta}) \leq n \tag{2.10}$$

with the lower bound achieved if and only if the antennas are uncorrelated and the upper bound achieved if and only if the antennas are fully correlated.

PROPERTY 2.2. *If* $\mathbf{\Theta}$ *is Toeplitz, then* $(\mathbf{\Theta})_{k\ell}=\rho(k-\ell)$ *where* $\rho(i)$ *denotes the correlation coefficient between antennas whose indices differ by* i. *The correlation number is then given by*

$$\zeta(\mathbf{\Theta}) = 1 + \frac{2}{n}\sum_{i=1}^{n-1}(n-i)\,|\rho(i)|^2 \tag{2.11}$$

and the eigenvalues of $\mathbf{\Theta}$ *become—as* n *goes to infinity—samples of the discrete Fourier transform of any of its rows* [**8**]. *Furthermore, if the correlation decays faster than* $1/i$ *across the array, (2.11) converges to a finite quantity*

$$\lim_{n\to\infty} \zeta(\mathbf{\Theta}) = 1 + 2\sum_{i=1}^{\infty}|\rho(i)|^2 < \infty$$

whereas if the correlation decays no faster than $1/i$ *(as in the classic Jakes model* [**11**], *for example) the above sum diverges.*

If the discrete random process described by each row (resp. each column) of $\mathbf{H}$ is wide-sense stationary, then $\mathbf{\Theta}_{\rm T}$ (resp. $\mathbf{\Theta}_{\rm R}$) is Toeplitz and it satisfies (2.11). This is always the case when the corresponding array has a regular structure. The eigenvalues of $\mathbf{\Theta}_{\rm T}$ (resp. $\mathbf{\Theta}_{\rm R}$) can also be interpreted, as $n_{\rm T}$ (resp. $n_{\rm R}$) goes to infinity, as samples of the power spectral density of the random process represented by any row (resp. any column) of $\mathbf{H}$.

PROPERTY 2.3. *Let* $\mathbf{\Theta}_{\rm B}$ *be a* $(n_{\rm B}\times n_{\rm B})$ *Toeplitz correlation matrix and let* $\mathbf{\Theta}_{\rm A}$ *be a* $(n_{\rm A}\times n_{\rm A})$ *principal submatrix of* $\mathbf{\Theta}_{\rm B}$. *(These may be the correlation matrices of an array whose number of antennas has grown from* $n_{\rm A}$ *to* $n_{\rm B}$ *preserving its original structure and antenna spacing.) Then,*

$$\zeta(\mathbf{\Theta}_{\rm B}) \geq \zeta(\mathbf{\Theta}_{\rm A})$$

with equality only if $\rho(i)=0\ \forall i$. *Hence, given a regular array and a certain antenna spacing, the correlation number increases with the number of antennas.*

3. Capacity in the Low-Power Regime

The ergodic capacity can be expressed as function of the SNR via

$$C(\mathsf{SNR}) = E\left[\log_2\det\left(\mathbf{I} + \frac{\mathsf{SNR}\, n_{\rm R}}{E[\mathrm{Tr}\{\mathbf{H}\mathbf{H}^\dagger\mathbf{\Phi}_{\mathbf{n}}^{-1}(\overline{\mathbf{H}})\}]}\,\mathbf{H}\mathbf{H}^\dagger\mathbf{\Phi}_{\mathbf{n}}^{-1}(\overline{\mathbf{H}})\right)\right] \tag{3.1}$$

where the expectations are over the distributions of $\mathbf{H}$ and $\overline{\mathbf{H}}$.[3] However, we are interested in the function $\mathsf{C}(\frac{E_b}{N_0})$, which at low SNR can be expressed as [**24**]

$$\mathsf{C}(\tfrac{E_b}{N_0}) = \frac{S_0}{3\,\text{dB}}\left(\frac{E_b}{N_0}|_{\text{dB}} - \frac{E_b}{N_0}_{\min}|_{\text{dB}}\right) + \epsilon$$

with $\frac{E_b}{N_0}_{\min}$ representing the minimum energy per information bit required to convey any positive rate reliably, S_0 the capacity slope therein in bits/s/Hz/(3 dB), and with ϵ vanishing faster than the main term as $\frac{E_b}{N_0} \downarrow \frac{E_b}{N_0}_{\min}$. Both $\frac{E_b}{N_0}_{\min}$ and S_0, the low-SNR performance measures, can be obtained from the first and second derivatives of $C(\text{SNR})$ at SNR=0, specifically [**24**]

$$\begin{aligned}\frac{E_b}{N_0}_{\min} &= \lim_{\text{SNR}\to 0}\frac{E_b}{N_0} \\ &= \frac{n_{\mathrm{T}} n_{\mathrm{R}}}{\mathsf{g}\, E[\operatorname{Tr}\{\mathbf{H}\mathbf{H}^\dagger \mathbf{\Phi}_{\mathbf{n}}^{-1}(\overline{\mathbf{H}})\}]}\frac{1}{\dot{C}(0)}\end{aligned}$$

and

$$S_0 = \frac{2\,[\dot{C}(0)]^2}{-\ddot{C}(0)}.$$

Computing the above derivatives, $\frac{E_b}{N_0}_{\min}$ can be found to be

$$\frac{E_b}{N_0}_{\min} = \frac{\log_e 2}{\mathsf{g}}\frac{n_{\mathrm{T}}}{E[\operatorname{Tr}\{\mathbf{H}\mathbf{H}^\dagger \mathbf{\Phi}_{\mathbf{n}}^{-1}(\overline{\mathbf{H}})\}]} \tag{3.2}$$

while S_0 becomes

$$S_0 = \frac{2\,n_{\mathrm{R}}}{\zeta(\mathbf{H}\mathbf{H}^\dagger \mathbf{\Phi}_{\mathbf{n}}^{-1}(\overline{\mathbf{H}}))}. \tag{3.3}$$

The received $\frac{E_b}{N_0}_{\min}$, in turn, equals

$$\frac{E_b^r}{N_0}_{\min} = \log_e 2$$

which is approximately -1.59 dB. This value represents a fundamental property of Gaussian noise [**24**, Theorem 1].

In the canonical case (uncorrelated channel impaired by AWGN), (3.2) and (3.3) particularize to

$$\frac{E_b}{N_0}_{\min} = \frac{\log_e 2}{\mathsf{g}}\frac{1}{n_{\mathrm{R}}}$$

and

$$S_0 = 2\frac{n_{\mathrm{T}} n_{\mathrm{R}}}{n_{\mathrm{T}} + n_{\mathrm{R}}} \tag{3.4}$$

from which we observe that

- The $\frac{E_b}{N_0}_{\min}$ depends on n_{R}, but not on n_{T}. This is a direct consequence of the total transmit power being constrained while the total captured power increases with the number of receive antennas.

[3]Notice that our fading model is purposely frequency-flat in nature. In frequency-selective environments, the channel can always be decomposed into a number of parallel non-interacting subchannels, each experiencing frequency-flat fading and having the same ergodic capacity as the overall channel.

- The slope is symmetric with respect to n_{T} and n_{R}.
- The low-SNR slope is at least as large as the high-SNR slope [**3**], given by $\min(n_{\mathrm{T}},n_{\mathrm{R}})$, with equality only if $n_{\mathrm{T}}=n_{\mathrm{R}}$.

4. Auxiliary Results

From (2.3) and (3.1), it is clear that the capacity depends on the set of random channel matrices $\mathbf{H}$ and $\mathbf{H}_\ell$, $\ell=1,\ldots,L$. The low-SNR performance measures, in particular, depend on some moments of the trace of products of these matrices as indicated by (3.2) and (3.3).

All of the aforementioned matrices are modeled as Gaussian with separable transmit and receive antenna correlations. In the case of $\mathbf{H}$, for example, recall from (2.9) that

$$\mathbf{H} = \mathbf{\Theta}_{\mathrm{R}}^{1/2}\mathbf{H}_w\mathbf{\Theta}_{\mathrm{T}}^{1/2}$$

where $\mathbf{\Theta}_{\mathrm{T}}$ and $\mathbf{\Theta}_{\mathrm{R}}$ are deterministic and the randomness is thus contained within $\mathbf{H}_w$, whose entries are independent zero-mean unit-variance complex Gaussian. As a result of this structure, there are two classes of matrices that permeate the computation of the required trace moments:

Wishart matrices [**27**]: Defined as $(\mathbf{H}_w\mathbf{H}_w^\dagger)$ with $\mathbf{H}_w$ as above.

Haar unitary matrices: A random unitary matrix is said to be Haar if it is uniformly distributed over the submanifold of all unitary matrices [**10**]. In our context, Haar unitary matrices emerge from the singular-value decomposition of $\mathbf{H}_w$

$$\mathbf{H}_w = \mathbf{U}\mathbf{\Lambda}^{1/2}\mathbf{V}^\dagger$$

where $\mathbf{\Lambda}$ is a diagonal matrix whose elements are the eigenvalues of a Wishart matrix while $\mathbf{U}$ and $\mathbf{V}$ are Haar unitary. The most important property of the Haar probability measure is that it is invariant to left and right unitary transformations. (For additional properties of the Haar measure, the reader is referred to Appendix A.)

In this section, we present some results that emerge from the computation of the moments required to evaluate $\frac{E_b}{N_0}_{\min}$ and S_0 and that are independently interesting.

LEMMA 4.1. *Let* $\mathbf{A}$ *and* $\mathbf{B}$ *be independent Hermitian* $(n\times n)$ *random matrices. Let* $\mathbf{U}$ *be a Haar unitary matrix independent of* $\mathbf{A}$ *and* $\mathbf{B}$ *with* $\mathbf{V}=\mathbf{U}\mathbf{B}\mathbf{U}^\dagger$. *Then,*

$$E[\mathrm{Tr}\{\mathbf{AV}\}] = \frac{1}{n}E[\mathrm{Tr}\{\mathbf{A}\}]\,E[\mathrm{Tr}\{\mathbf{V}\}] \tag{4.1}$$

$$\begin{aligned} E[\mathrm{Tr}\{(\mathbf{AV})^2\}] &= \frac{E[\mathrm{Tr}\{\mathbf{A}^2\}]}{n^2-1}\left(E[\mathrm{Tr}^2\{\mathbf{V}\}]-\frac{1}{n}E[\mathrm{Tr}\{\mathbf{V}^2\}]\right)\\ &\quad+\frac{E[\mathrm{Tr}^2\{\mathbf{A}\}]}{n^2-1}\left(E[\mathrm{Tr}\{\mathbf{V}^2\}]-\frac{1}{n}E[\mathrm{Tr}^2\{\mathbf{V}\}]\right) \end{aligned} \tag{4.2}$$

$$\begin{aligned} E[\mathrm{Tr}^2\{\mathbf{AV}\}] &= \frac{E[\mathrm{Tr}\{\mathbf{A}^2\}]}{n^2-1}\left(E[\mathrm{Tr}\{\mathbf{V}^2\}]-\frac{1}{n}E[\mathrm{Tr}^2\{\mathbf{V}\}]\right)\\ &\quad+\frac{E[\mathrm{Tr}^2\{\mathbf{A}\}]}{n^2-1}\left(E[\mathrm{Tr}^2\{\mathbf{V}\}]-\frac{1}{n}E[\mathrm{Tr}\{\mathbf{V}^2\}]\right). \end{aligned} \tag{4.3}$$

Proof: See Appendix B.

LEMMA 4.2. *An* $(n \times m)$ *matrix* $\mathbf{H}_w$ *with independent zero-mean unit-variance complex Gaussian random entries satisifes*

$$\begin{aligned} E[\mathrm{Tr}\{\mathbf{H}_w \mathbf{H}_w^\dagger\}] &= m\,n \\ E[\mathrm{Tr}\{(\mathbf{H}_w \mathbf{H}_w^\dagger)^2\}] &= m\,n\,(m+n) \\ E[\mathrm{Tr}^2\{\mathbf{H}_w \mathbf{H}_w^\dagger\}] &= m\,n\,(m\,n+1). \end{aligned}$$

LEMMA 4.3. *An* $(n \times m)$ *matrix* $\mathbf{H}_w$ *with independent zero-mean unit-variance complex Gaussian random entries and* $m>n$ *satisfies*

$$E\left[\mathrm{Tr}\left\{(\mathbf{H}_w \mathbf{H}_w^\dagger)^{-1}\right\}\right] = \frac{n}{m-n} \tag{4.4}$$

and, for $m>n+1$,

$$E\left[\mathrm{Tr}\left\{(\mathbf{H}_w \mathbf{H}_w^\dagger)^{-2}\right\}\right] = \frac{m\,n}{(m-n)^3-(m-n)} \tag{4.5}$$

$$E\left[\mathrm{Tr}^2\left\{(\mathbf{H}_w \mathbf{H}_w^\dagger)^{-1}\right\}\right] = \frac{n}{m-n}\left(\frac{m}{(m-n)^2-1}+\frac{n-1}{m-n+1}\right). \tag{4.6}$$

Proof: See Appendix C.

We point out that (4.4) and (4.5) correspond to the first and second moment of an eigenvalue of an inverse Wishart matrix, respectively, whereas (4.6) gives the mixed moment of two distinct such eigenvalues.

5. Bandwidth-Power Tradeoffs in AWGN

When the noise is white and unfaded, we have $\mathbf{\Phi_n}=\mathbf{I}$ and the following result applies.

PROPOSITION 5.1. *Consider a correlated Rayleigh channel* $\mathbf{H}=\mathbf{\Theta}_{\mathrm{R}}^{1/2}\mathbf{H}_w\mathbf{\Theta}_{\mathrm{T}}^{1/2}$. *In AWGN, the* $\frac{E_b}{N_0}_{\min}$ *and the capacity slope therein in bits/s/Hz/(3 dB) are*

$$\frac{E_b}{N_0}_{\min} = \frac{\log_e 2}{\mathrm{g}}\frac{1}{n_{\mathrm{R}}} \tag{5.1}$$

and

$$S_0 = \frac{2 n_{\mathrm{T}} n_{\mathrm{R}}}{n_{\mathrm{T}}\,\zeta(\mathbf{\Theta}_{\mathrm{R}}) + n_{\mathrm{R}}\,\zeta(\mathbf{\Theta}_{\mathrm{T}})} \tag{5.2}$$

Proof: See Appendix D.

The $\frac{E_b}{N_0}_{\min}$ is unaffected by the existence of transmit and receive antenna correlations, which are only reflected in S_0. Furthermore, the effects of such correlations appear only through the correlation numbers of the transmit and receive arrays. Thus, any two correlation matrices mapping onto the same correlation number are equivalent in terms of low-SNR capacity. Using (2.10), it is straightforward to see that the slope in (5.2) satisfies

$$1 \le S_0 \le \frac{2 n_{\mathrm{T}} n_{\mathrm{R}}}{n_{\mathrm{T}} + n_{\mathrm{R}}}$$

confirming the intuitive result that antenna correlation can only reduce the capacity.

Fixing the rate, R, and the power, P, the bandwidth B required with correlation matrices $\mathbf{\Theta}_{\mathrm{T}}$ and $\mathbf{\Theta}_{\mathrm{R}}$ relative to the canonical bandwidth B_c required in the absence of correlation is

$$\frac{B}{B_{\mathrm{c}}} = \frac{n_{\mathrm{T}}\,\zeta(\mathbf{\Theta}_{\mathrm{R}}) + n_{\mathrm{R}}\,\zeta(\mathbf{\Theta}_{\mathrm{T}})}{n_{\mathrm{T}} + n_{\mathrm{R}}}. \tag{5.3}$$

Conversely, if we fix the rate and the bandwidth, the power penalty that results from correlation is

$$\frac{P|_{\mathrm{dB}} - P_{\mathrm{c}}|_{\mathrm{dB}}}{3\,\mathrm{dB}} = \frac{R}{2B}\left(\frac{\zeta(\mathbf{\Theta}_{\mathrm{T}}) - 1}{n_{\mathrm{T}}} + \frac{\zeta(\mathbf{\Theta}_{\mathrm{R}}) - 1}{n_{\mathrm{R}}}\right) \tag{5.4}$$

which, in contrast with (5.3), depends on the operating point (R/B).

EXAMPLE 5.2. Let $n_{\mathrm{T}}{=}2$ and $n_{\mathrm{R}}{=}1$. The low-SNR capacity in a Rayleigh channel is

$$\mathsf{C}\left(\tfrac{E_b}{N_0}\right) \approx \frac{4}{3 + |\rho|^2}\log_2\left(\frac{\mathsf{g}}{\log_e 2}\frac{E_b}{N_0}\right)$$

with ρ the correlation coefficient between the transmit antennas. With fixed power and rate, the bandwidth expansion factor caused by such correlation is $1{+}\frac{1}{3}|\rho|^2$ while the power penalty with fixed rate and bandwidth, measured in 3-dB units, is $(R/B)\frac{1}{4}|\rho|^2$.

The slope in the presence of correlation is equivalent to that of $n_{\mathrm{T}}^{\mathrm{eq}}$ and $n_{\mathrm{R}}^{\mathrm{eq}}$ uncorrelated antennas, with

$$n_{\mathrm{T}}^{\mathrm{eq}} = \frac{n_{\mathrm{T}}}{\zeta(\mathbf{\Theta}_{\mathrm{T}})} \qquad\qquad n_{\mathrm{R}}^{\mathrm{eq}} = \frac{n_{\mathrm{R}}}{\zeta(\mathbf{\Theta}_{\mathrm{R}})}$$

and, therefore, the reciprocals of the transmit and receive correlation numbers determine the fraction of n_{T} and n_{R} that would result in the same low-SNR slope if the antennas were uncorrelated.

If the correlation matrices are Toeplitz, it follows from Property 2.3 that the bandwidth expansion factor increases monotonically with the number of antennas when their spacing is preserved. Hence, even though the canonical capacity increases linearly with the number of antennas, the penalty associated with antenna correlation compounds and thus only a diminishing fraction of such capacity can be attained. Experimental data [**13**] confirms that this progressive decrease in slope occurs even with non-Toeplitz correlation matrices and beyond the low-SNR regime.

We conclude this subsection by analyzing the limiting behavior as the number of antennas is driven to infinity with ratio $\beta{=}\frac{n_{\mathrm{T}}}{n_{\mathrm{R}}}$ with the correlation matrices being Toeplitz. Property 2.2 indicates that, given a rate R and some power P, the bandwidth expansion factor due to correlation converges to

$$\begin{aligned} \lim_{n_{\mathrm{R}}\to\infty}\frac{B}{B_{\mathrm{c}}} &= 1 + \frac{2}{\beta+1}\sum_{i=1}^{\infty}\left(\beta\,|\rho_{\mathrm{R}}(i)|^2 + |\rho_{\mathrm{T}}(i)|^2\right) \\ &= \frac{1}{\beta+1}\int_0^1\left(\beta\,\mathcal{S}_{\mathrm{R}}^2(\nu) + \mathcal{S}_{\mathrm{T}}^2(\nu)\right)d\nu \end{aligned} \tag{5.5}$$

with $\rho_{\mathrm{T}}(i)$ and $\rho_{\mathrm{R}}(i)$ the transmit and receive correlation between antennas whose indices differ by i and with $\mathcal{S}_{\mathrm{T}}(\nu)$ and $\mathcal{S}_{\mathrm{R}}(\nu)$ the power spectral densities of the random processes represented by any of the rows and any of the columns, respectively, of $\mathbf{H}$. Since the variance of the entries of $\mathbf{H}$ is unity, these power spectral densities satisfy $\int_0^1\mathcal{S}_{\mathrm{T}}(\nu){=}\int_0^1\mathcal{S}_{\mathrm{R}}(\nu){=}1$.

If correlation at either array does not decay faster than $1/i$, (5.5) is infinite and thus the intended rate R cannot be achieved with power P. Otherwise, (5.5) provides a limiting value for the bandwidth expansion factor, a value that is approached as the number of antennas grows large. In that case, the achievable fraction of the canonical capacity, which—as we have seen—decreases monotonically with the number of antennas, approaches a non-zero asymptote [**2**].

6. Bandwidth-Power Tradeoffs in the Presence of Interference

Recall, from (2.3) and (2.5), that in the presence of L out-of-cell interferers each equipped with m_ℓ antennas, $\ell=1,\dots,L$, the normalized conditional covariance of the noise is given by

$$\mathbf{\Phi_n}(\overline{\mathbf{H}}) = \frac{1}{\sum_{\ell=1}^{L} \mathcal{I}_\ell + \gamma} \Big(\sum_{\ell=1}^{L} \mathcal{I}_\ell \frac{\mathbf{H}_\ell \mathbf{H}_\ell^\dagger}{m_\ell} + \gamma \mathbf{I}\Big) \tag{6.1}$$

which satisfies

$$E[\mathrm{Tr}\{\mathbf{\Phi_n}\}] = n_{\mathrm{R}}. \tag{6.2}$$

Recall also that $\beta_\ell = \frac{m_\ell}{n_{\mathrm{R}}}$ and define

$$\bar{\beta} \stackrel{\mathrm{def}}{=} \sum_{\ell=1}^{L} \beta_\ell.$$

PROPOSITION 6.1. *Consider a correlated Rayleigh channel* $\mathbf{H}=\mathbf{\Theta}_{\mathrm{R}}^{1/2}\mathbf{H}_w\mathbf{\Theta}_{\mathrm{T}}^{1/2}$. *If the normalized conditional covariance of the noise is* $\mathbf{\Phi_n}(\overline{\mathbf{H}})$, *then*

$$\frac{E_b}{N_0}_{\min} = \frac{\log_e 2}{\mathsf{g}} \frac{1}{E[\mathrm{Tr}\{\mathbf{\Theta}_{\mathrm{R}} \mathbf{\Phi_n}^{-1}(\overline{\mathbf{H}})\}]} \tag{6.3}$$

and

$$S_0 = \frac{2\, n_{\mathrm{T}} n_{\mathrm{R}}}{n_{\mathrm{T}}\, \zeta(\mathbf{\Theta}_{\mathrm{R}} \mathbf{\Phi_n}^{-1}(\overline{\mathbf{H}})) + n_{\mathrm{R}}\, \zeta(\mathbf{\Theta}_{\mathrm{T}})\, \varphi(\mathbf{\Theta}_{\mathrm{R}} \mathbf{\Phi_n}^{-1}(\overline{\mathbf{H}}))} \tag{6.4}$$

with expectation over the fading of the interferers, $\overline{\mathbf{H}}$, *and with*

$$\varphi(\mathbf{A}) \stackrel{\mathrm{def}}{=} \frac{E[\mathrm{Tr}^2\{\mathbf{A}\}]}{E^2[\mathrm{Tr}\{\mathbf{A}\}]}.$$

Proof: See Appendix D.

Using (6.2), it is easy to show via Jensen's inequality that

$$E[\mathrm{Tr}\{\mathbf{\Theta}_{\mathrm{R}} \mathbf{\Phi_n}^{-1}\}] \geq n_{\mathrm{R}}$$

and, therefore, (6.3) is upper bounded by the value it takes when $\mathbf{\Phi_n}=\mathbf{I}$. Consequently, noise fading and color can only diminish the minimum energy necessary at the transmitter to convey information reliably.

If we let the number of transmit, receive, and interfering antennas go to infinity with ratios $\beta=\frac{n_{\mathrm{T}}}{n_{\mathrm{R}}}$ and $\beta_\ell=\frac{m_\ell}{n_{\mathrm{R}}}$, $\ell=1,\dots,L$, we can elaborate on Proposition 6.1 as follows.

PROPOSITION 6.2. *Consider a channel* $\mathbf{H}=\mathbf{\Theta}_{\mathrm{R}}^{1/2}\mathbf{H}_w\mathbf{\Theta}_{\mathrm{T}}^{1/2}$ *with* $\mathbf{\Theta}_{\mathrm{R}}$ *Toeplitz. Consider also AWGN with spectral density* γ *as well as* L *out-of-cell interferers, each with* m_ℓ *transmit antennas, average energy per receive antenna* $\mathcal{I}_\ell$*, and fading channel*

$$\mathbf{H}_\ell = \mathbf{\Theta}_{\mathrm{R}}^{1/2}(\ell)\,\mathbf{H}_w(\ell)\,\mathbf{\Theta}_{\mathrm{T}}^{1/2}(\ell)$$

where $\mathbf{\Theta}_{\mathrm{R}}(\ell)$, $\ell=1,\ldots,L$, *are Toeplitz. Asymptotically in the number of antennas,*

$$\lim_{n_{\mathrm{R}}\to\infty} n_{\mathrm{R}}\,\frac{E_b}{N_0}_{\min} = \frac{\log_e 2}{\mathsf{g}}\,\frac{1}{(\gamma+\sum_{\ell=1}^{L}\mathcal{I}_\ell)\int_0^1 a(x,0,\gamma)\,dx}$$

and

$$\lim_{n_{\mathrm{R}}\to\infty}\frac{S_0}{n_{\mathrm{R}}} = \frac{2}{\frac{1}{\beta}\zeta(\mathbf{\Theta}_{\mathrm{T}}) + \frac{\int_0^1 \mathcal{R}(x)\dot{a}(x,0,\gamma)dx}{\left(\int_0^1 \mathcal{R}(x)a(x,0,\gamma)dx\right)^2}}$$

with

$$\dot{a}(x,0,\gamma) = \lim_{z\to 0}\frac{d}{dz}a(x,z,\gamma)$$

and with $a(x,z,\gamma)$ *the solution to the following fixed-point equation*

$$a(x,z,\gamma) = \frac{\mathcal{R}(x)}{-z+\gamma+\displaystyle\int_0^{\bar{\beta}}\frac{\mathcal{V}_{\mathrm{R}}(x,y)\mathcal{V}_{\mathrm{T}}(y)}{1+\mathcal{V}_{\mathrm{T}}(y)\int_0^1 a(\tau,z,\gamma)\mathcal{V}_{\mathrm{R}}(\tau,y)\,d\tau}dy}$$

where $\mathcal{R}(x)$, $\mathcal{V}_{\mathrm{T}}(y)$ *and* $\mathcal{V}_{\mathrm{R}}(x,y)$, *with* $x\in(0,1)$ *and* $y\in(0,\bar{\beta})$, *represent the limits, as* $n_{\mathrm{R}}\to\infty$, *of the power spectral densities*

$$\begin{aligned}
\mathcal{R}^{(n_{\mathrm{R}})}(x) &= \lambda_{\mathrm{R}}(i) && \tfrac{i}{n_{\mathrm{R}}}\le x<\tfrac{i+1}{n_{\mathrm{R}}},\\
\mathcal{V}_{\mathrm{T}}^{(n_{\mathrm{R}})}(y) &= \lambda_{\mathrm{T}}(j,\ell) && \tfrac{j+\sum_{k=1}^{\ell-1}m_k}{n_{\mathrm{R}}}\le y<\tfrac{j+1+\sum_{k=1}^{\ell-1}m_k}{n_{\mathrm{R}}}\\
\mathcal{V}_{\mathrm{R}}^{(n_{\mathrm{R}})}(x,y) &= \lambda_{\mathrm{R}}(i,\ell) && \tfrac{i}{n_{\mathrm{R}}}\le x<\tfrac{i+1}{n_{\mathrm{R}}},\quad \tfrac{\sum_{k=1}^{\ell-1}m_k}{n_{\mathrm{R}}}\le y<\tfrac{\sum_{k=1}^{\ell}m_k}{n_{\mathrm{R}}}
\end{aligned}$$

with $1\le i\le n_{\mathrm{R}}$, $1\le\ell\le L$, *and* $1\le j\le m_\ell$. *Here,* $\lambda_{\mathrm{R}}(n)$, $\lambda_{\mathrm{T}}(n,\ell)$, *and* $\lambda_{\mathrm{R}}(n,\ell)$, *represent the* n*-th eigenvalue of* $\mathbf{\Theta}_{\mathrm{R}}$, $\mathbf{\Theta}_{\mathrm{T}}(\ell)$, *and* $(\mathcal{I}_\ell\,\mathbf{\Theta}_{\mathrm{R}}(\ell))$, *respectively.*

Proof: See Appendix E.

6.1. Single Interferer. In the presence of a single interferer, the noise covariance in (6.1) simplifies to

$$\mathbf{\Phi}_{\mathbf{n}}(\mathbf{H}_1) = \tfrac{\mathcal{I}_1}{\mathcal{I}_1+\gamma}\,\frac{\mathbf{H}_1\mathbf{H}_1^\dagger}{m_1} + \tfrac{\gamma}{\mathcal{I}_1+\gamma}\,\mathbf{I}$$

facilitating the analysis. In the non-asymptotic regime, in particular, we find explicit expressions for $\frac{E_b}{N_0}_{\min}$ and S_0 in interference-limited conditions, i.e. $\frac{\mathcal{I}_1}{\gamma}\to\infty$. We start by assuming the interfering antennas exhibit no transmit correlation but, later in the section, such correlation is brought in.

PROPOSITION 6.3. *Consider a correlated Rayleigh channel* $\mathbf{H}=\mathbf{\Theta}_{\mathrm{R}}^{1/2}\mathbf{H}_w\mathbf{\Theta}_{\mathrm{T}}^{1/2}$ *and a single interferer transmitting from* m_1 *uncorrelated antennas. Let the fading of such interferer be Rayleigh with arbitrary correlation at the receiver*

$$\mathbf{H}_1 = \mathbf{\Theta}_1^{1/2}\mathbf{H}_w(1).$$

The interference-limited $\frac{E_b}{N_0}_{\min}$ *satisfies*

$$\lim_{(\mathcal{I}_1/\gamma)\to\infty} \frac{E_b}{N_0}_{\min} = \frac{\log_e 2}{\mathsf{g}} \frac{n_\mathrm{R}}{\mathrm{Tr}\{\boldsymbol{\Theta}_\mathrm{R}\boldsymbol{\Theta}_1^{-1}\}} \left[\frac{1}{n_\mathrm{R}} - \frac{1}{m_1}\right]^+ \tag{6.5}$$

with $[x]^+ = x$ *if* $x \geq 0$ *and* $[x]^+ = 0$ *otherwise. The corresponding slope is given by*

$$\lim_{(\mathcal{I}_1/\gamma)\to\infty} \mathcal{S}_0 = \frac{2}{\left(\frac{\Psi}{n_\mathrm{R}} + \frac{\zeta(\boldsymbol{\Theta}_\mathrm{T})\,\Upsilon}{n_\mathrm{T}}\right)\frac{n_\mathrm{R}^2 - \zeta(\boldsymbol{\Theta}_\mathrm{R}\boldsymbol{\Theta}_1^{-1})}{n_\mathrm{R}^2-1} + \frac{\zeta(\boldsymbol{\Theta}_\mathrm{R}\boldsymbol{\Theta}_1^{-1})-1}{n_\mathrm{R}^2-1}\left(n_\mathrm{R}\,\Upsilon + \frac{\zeta(\boldsymbol{\Theta}_\mathrm{T})\,\Psi}{n_\mathrm{T}}\right)}$$

where

$$\Psi = \begin{cases} \dfrac{m_1(m_1 - n_\mathrm{R})}{(m_1 - n_\mathrm{R})^2 - 1} & m_1 > n_\mathrm{R} + 1 \\[2ex] \dfrac{n_\mathrm{R}}{n_\mathrm{R} - m_1} & m_1 \leq n_\mathrm{R} \end{cases} \tag{6.6}$$

$$\Upsilon = \begin{cases} \dfrac{\Psi}{n_\mathrm{R}} + \dfrac{(n_\mathrm{R} - 1)(m_1 - n_\mathrm{R})}{n_\mathrm{R}(m_1 - n_\mathrm{R} + 1)} & m_1 > n_\mathrm{R} + 1 \\[2ex] 1 & m_1 \leq n_\mathrm{R}. \end{cases} \tag{6.7}$$

Proof: See Appendix F.

The results in Proposition 6.3 reveal that, in the presence of a strongly dominant interferer:

- The receive antenna correlations experienced by signal and interference, respectively $\boldsymbol{\Theta}_\mathrm{R}$ and $\boldsymbol{\Theta}_1$, compound into an *effective* correlation given by $\boldsymbol{\Theta}_\mathrm{R}\boldsymbol{\Theta}_1^{-1}$. If the receiver correlations experienced by signal and interference coincide, it is as if both were uncorrelated.
- For $m_1 < n_\mathrm{R}$, the $\frac{E_b}{N_0}_{\min}$ vanishes as $\frac{\mathcal{I}_1}{\gamma}$ grows. This is clearly a direct consequence of the fact that interference with dimensionality inferior to the number of receive antennas can be completely suppressed, through mere linear processing, if only $\boldsymbol{\Phi}_\mathbf{n}$ is known [**12, 26**]. Such interference-limited conditions enable reliable communication at negligible levels of normalized energy. More precisely, when noise and interference experience identical correlation at the receiver,

$$\lim_{(\mathcal{I}_1/\gamma)\to\infty} \frac{E_b}{\gamma} = \frac{\log_e 2}{\mathsf{g}} \frac{1}{n_\mathrm{R}}$$

  and thus, as $\frac{\mathcal{I}_1}{\gamma}$ grows, reliable communication becomes feasible if only the received energy per bit is -1.59 dB above the thermal noise, irrespective of the strength of the interference. The corresponding slope, in turn, is

$$\lim_{(\mathcal{I}_1/\gamma)\to\infty} \mathcal{S}_0 = \frac{2n_\mathrm{T}(n_\mathrm{R} - m_1)}{n_\mathrm{T} + \zeta(\boldsymbol{\Theta}_\mathrm{T})(n_\mathrm{R} - m_1)}$$

  which, contrasted with (5.2), indicates that the equivalent of m_1 receive antennas have been spent suppressing the m_1-dimensional interference and thus the slope equals that which would be experienced with the remaining $(n_\mathrm{R} - m_1)$ receive antennas and the n_T transmit antennas in the presence of only the underlying thermal noise, γ.

- When $m_1{=}n_{\mathrm{R}}$, both the $\frac{E_b}{N_0}_{\min}$ and S_0 vanish as $\frac{\mathcal{I}_1}{\gamma}$ grows without bound. In this case, however, the fact that the slope also approaches zero as the $\frac{E_b}{N_0}_{\min}$ decreases warns that non-negligible capacity will require non-negligible energy per bit.

If the antennas at the interferer exhibit transmit correlation, the number of degrees of freedom within $\mathbf{\Phi_n}$ will diminish. In the limit of full correlation the interference exhibits a single degree of freedom and, therefore, it is as if the interferer was equipped with a single transmit antenna:

PROPOSITION 6.4. *Consider the scenario of Proposition 6.3 but with the m_1 transmit antennas at the interferer fully correlated, i.e. $\mathbf{H}_1{=}\mathbf{\Theta}_1^{1/2}\mathbf{H}_w\,\mathbf{1}^{1/2}$, with $\mathbf{1}$ an $(m_1{\times}m_1)$ matrix whose entries are all 1. The interference-limited $\frac{E_b}{N_0}_{\min}$ and S_0 are given by their expressions in Proposition 6.3 with $m_1{=}1$.*

Proof: See Appendix G.

More generally, partial transmit correlation at the interferer results in an *equivalent* number of transmit antennas therein that ranges between 1 and m_1.

In the asymptotic regime, the interference-limited constraint can be relaxed and explicit expressions for $\frac{E_b}{N_0}_{\min}$ and S_0 can be found for arbitrary levels of $\frac{\mathcal{I}_1}{\gamma}$ in the presence of a single interferer. Recalling that $\beta{=}\frac{n_{\mathrm{T}}}{n_{\mathrm{R}}}$ and $\beta_1{=}\frac{m_1}{n_{\mathrm{R}}}$, we present the following result.

PROPOSITION 6.5. *Consider a correlated Rayleigh channel $\mathbf{H}{=}\mathbf{\Theta}_{\mathrm{R}}^{1/2}\mathbf{H}_w\mathbf{\Theta}_{\mathrm{T}}^{1/2}$ and a single interferer transmitting from m_1 uncorrelated antennas. If the fading of such interferer is Rayleigh and uncorrelated also at the receiver, i.e. $\mathbf{H}_1{=}\mathbf{H}_w(1)$,*

$$\lim_{n_{\mathrm{R}}\to\infty} n_{\mathrm{R}}\frac{E_b}{N_0}_{\min} = \frac{\log_e 2}{\mathsf{g}}\;\frac{\frac{2}{\beta_1}}{\left(1+\frac{\mathcal{I}_1}{\gamma}\right)\mathcal{D}(\beta_1,\frac{\mathcal{I}_1}{\gamma})}$$

and

$$\lim_{n_{\mathrm{R}}\to\infty}\frac{S_0}{n_{\mathrm{R}}} = \frac{2}{\frac{\frac{2}{\beta_1}}{\mathcal{D}^2(\beta_1,\frac{\mathcal{I}_1}{\gamma})}\left(\mathcal{D}(\beta_1,\frac{\mathcal{I}_1}{\gamma})+\frac{\mathcal{D}(\beta_1,\frac{\mathcal{I}_1}{\gamma})-\frac{2}{\beta_1}}{1+\frac{\mathcal{I}_1}{\gamma}\left(\mathcal{D}(\beta_1,\frac{\mathcal{I}_1}{\gamma})+1-\frac{1}{\beta_1}\right)}\right)+\frac{\zeta(\mathbf{\Theta}_{\mathrm{T}})}{\beta}+\zeta(\mathbf{\Theta}_{\mathrm{R}})-1}$$

given the auxiliary function

$$\mathcal{D}(x,y)=\sqrt{\left(\frac{x-1}{x}+\frac{1}{y}\right)^2+\frac{4}{x\,y}}-\left(\frac{x-1}{x}+\frac{1}{y}\right). \tag{6.8}$$

Proof: See appendix H.

6.2. Multiple Interferers. In the presence of multiple equal-energy interferers, the computation of non-asymptotic interference-limited $\frac{E_b}{N_0}_{\min}$ and S_0 boils down to a scaled version of the single-interferer case analyzed in the previous section. Specifically, the capacity in the presence of L equal-energy interferers experiencing identical receive antenna correlation equals the capacity in the presence of a single *equivalent* interferer with that same receive correlation. The energy of this equivalent interferer is equal to the sum of energies of the L interferers and, if the transmit antennas within each of the L actual interferers are uncorrelated (resp. fully correlated), the equivalent interferer has $\sum_{\ell=1}^{L} m_\ell$ (resp. L) uncorrelated antennas.

When the multiple interferers exhibit unequal levels of energy, non-asymptotic analysis becomes very involved and it is convenient to resort to computer simulations (see [**13**] for some numerical examples). In the asymptotic regime, the general solution presented in Proposition 6.2 can be specialized to fit any particular scenario.

7. Summary

Throughout this paper we have derived—focusing on those scenarios where the transmitter has no knowledge about the channel or its statistics—expressions for the two performance measures that characterize the capacity function $\mathsf{C}(\frac{E_b}{N_0})$ in the low-SNR regime, namely the normalized energy at which such capacity becomes zero and the slope therein. From these expressions, we have learned how the tradeoffs between rate, power, and bandwidth, are affected by the existence of antenna correlation as well as fading and correlation within the noise.

Our proof of the formulas presented in the paper require the evaluation of several moments of the trace of the product of certain random matrices. To compute these moments, which we have put forth as results of independent interest, we have invoked recent results from the theory of free probability and some properties related to the concept of freeness with amalgamation.

Appendix A. The Haar Probability Measure

Let $\mathbf{U}$ be an $(n\times n)$ matrix such that $\mathbf{U}\mathbf{U}^\dagger=\mathbf{I}$. $\mathbf{U}$ is uniquely specified by n^2 real parameters. As a subspace of $\mathbb{R}^{2n^2}$, these matrices form a submanifold $\mathbf{V}_{n,n}$ of dimension n^2. The uniform distribution over the submanifold $\mathbf{V}_{n,n}$ is called *Haar* probability measure. The most important property of the Haar probability measure is that it is left and right invariant under unitary transformation. This invariance implies that most mixed moments of the entries of a Haar unitary matrix are zero. In particular,

LEMMA A.1. [**10**] *Consider the product of the k_p-th power of $(\mathbf{U})_{i_p,j_p}$ and the m_p-th power of $(\mathbf{U})^*_{i_p,j_p}$ with $\mathbf{U}$ a Haar matrix. Whenever $k_1,\dots,k_\ell,m_1,\dots,m_\ell>0$ and there exist some i or j such that*

$$\sum_{p:i_p=i}(k_p-m_p)\neq 0 \qquad \text{or} \qquad \sum_{p:j_p=j}(k_p-m_p)\neq 0$$

then

$$E\left[\prod_{p=1}^{\ell}\left((\mathbf{U})_{i_p,j_p}\right)^{k_p}\left((\mathbf{U}^\dagger)_{j_p,i_p}\right)^{m_p}\right]=0. \tag{A.1}$$

As a special case, (A.1) holds whenever the sum $\sum_{p=1}^{\ell}(k_p+m_p)$ is odd.

LEMMA A.2. [**10**] *If $1\leq i_1,j_1,i_2,j_2\leq n$, $i_1\neq i_2$, $j_1\neq j_2$, and $\mathbf{U}$ is a Haar matrix, then*

$$E[|(\mathbf{U})_{i_1,j_1}|^2] = \frac{1}{n}$$

$$E[|(\mathbf{U})_{i_1,j_1}|^4] = \frac{2}{n(n+1)}$$

$$
\begin{aligned}
E[|(\mathbf{U})_{i_1,j_1}|^2|(\mathbf{U})_{i_2,j_1}|^2] = E[|(\mathbf{U})_{i_1,j_1}|^2|(\mathbf{U})_{i_1,j_2}|^2] &= \frac{1}{n(n+1)} \\
E[|(\mathbf{U})_{i_1,j_1}|^2|(\mathbf{U})_{i_2,j_2}|^2] &= \frac{1}{n^2-1} \\
E[(\mathbf{U})_{i_1,j_1}(\mathbf{U})_{i_2,j_2}(\mathbf{U})_{i_1,j_2}(\mathbf{U})_{i_2,j_1}] &= -\frac{1}{n(n^2-1)}.
\end{aligned}
$$

In addition, we shall make use of the following result.

LEMMA A.3. *Given a $(n\times m)$ matrix $\mathbf{H}_w$ with independent zero-mean unit-variance complex Gaussian random entries as well as a $(m\times m)$ matrix $\mathbf{D}$ and a $(n\times n)$ matrix $\mathbf{S}$, the matrices $\mathbf{H}_w\mathbf{D}\mathbf{H}_w^\dagger$ and $\mathbf{H}_w^\dagger\mathbf{S}\mathbf{H}_w$ admit the following eigenvalue decomposition*

$$
\begin{aligned}
\mathbf{H}_w\mathbf{D}\mathbf{H}_w^\dagger &= \mathbf{U}\mathbf{M}\mathbf{U}^\dagger \\
\mathbf{H}_w^\dagger\mathbf{S}\mathbf{H}_w &= \widetilde{\mathbf{U}}\mathbf{N}\widetilde{\mathbf{U}}^\dagger
\end{aligned}
$$

where $\mathbf{U}$ and $\widetilde{\mathbf{U}}$ are $(n\times n)$ and $(m\times m)$ Haar matrices and $\mathbf{M}$ and $\mathbf{N}$ are $(n\times n)$ and $(m\times m)$ random matrices independent of $\mathbf{U}$ and $\widetilde{\mathbf{U}}$, respectively.

Proof: This lemma is derived as part of the proof of [**10**, Theorem 3.2].

Appendix B. Proof of Lemma 4.1

This Lemma can be proved using the properties of the Haar measure presented in the previous section. Since $\mathbf{A}$ is independent of $\mathbf{B}$ and $\mathbf{U}$,

$$
E[\mathrm{Tr}\{\mathbf{A}\mathbf{V}\}] = \mathrm{Tr}\{E[\mathbf{A}]\,E[\mathbf{U}\mathbf{B}\mathbf{U}^\dagger]\}. \tag{B.1}
$$

$\mathbf{U}$ is Haar and thus the (i,j)-th element of $E[\mathbf{U}\mathbf{B}\mathbf{U}^\dagger]$ is given by

$$
\begin{aligned}
(E[\mathbf{U}\mathbf{B}\mathbf{U}^\dagger])_{i,j} &= E[(\mathbf{U})_i\mathbf{B}(\mathbf{U})_j^\dagger] \\
&= \sum_{m=1}^{n}\sum_{p=1}^{n} E[(\mathbf{U})_{i,m}(\mathbf{B})_{m,p}(\mathbf{U})_{j,p}^*] \\
&= \delta_{i,j}\sum_{m=1}^{n} E[|(\mathbf{U})_{i,m}|^2]E[(\mathbf{B})_{m,m}] \qquad \text{(B.2)} \\
&= \delta_{i,j}\frac{1}{n}\mathrm{Tr}\{E[\mathbf{B}]\}
\end{aligned}
$$

where (B.2) follows from Lemma A.2 and where $\delta_{i,j}$ is the Kronecker delta. Thus, $E[\mathbf{U}\mathbf{B}\mathbf{U}^\dagger]$ is a scalar matrix, i.e.

$$
\begin{aligned}
E[\mathbf{U}\mathbf{B}\mathbf{U}^\dagger] &= \frac{1}{n}\mathrm{Tr}\{E[\mathbf{B}]\}\,\mathbf{I} \\
&= \frac{1}{n}E[\mathrm{Tr}\{\mathbf{V}\}]\,\mathbf{I}
\end{aligned}
$$

which, plugged into (B.1), proves (4.1).

Now, let $\mathbf{A}=\mathbf{Q_A}\Lambda\mathbf{Q_A}^\dagger$ be the eigenvalue decomposition of $\mathbf{A}$. From the invariance of the Haar probability measure, we have that

$$\begin{aligned} E[\mathrm{Tr}\{(\mathbf{AV})^2\}] &= E[\mathrm{Tr}\{\mathbf{\Lambda V \Lambda V}\}] \\ &= E[\mathrm{Tr}\{\mathbf{\Lambda}^{1/2}\mathbf{V\Lambda V\Lambda}^{1/2}\}] \\ &= \sum_{ij} E[(\mathbf{\Lambda}^{1/2}\mathbf{V}\mathbf{\Lambda}^{1/2})_{i,j}^2]. \end{aligned}$$

With some algebraic manipulations and neglecting those terms whose expected value is zero, we find that, for $i\neq j$

$$\begin{aligned} E[(\mathbf{\Lambda}^{1/2}\mathbf{V}\mathbf{\Lambda}^{1/2})_{i,j}^2] &= E[(\mathbf{\Lambda})_{i,i}(\mathbf{\Lambda})_{j,j}(\mathbf{V})_{i,j}^2] \\ &= E[(\mathbf{\Lambda})_{i,i}(\mathbf{\Lambda})_{j,j}]\; E\left[\left(\sum_{m\,p}(\mathbf{U})_{i,m}(\mathbf{B})_{m,p}(\mathbf{U})_{j,p}^*\right)^2\right] \\ &= E[(\mathbf{\Lambda})_{i,i}(\mathbf{\Lambda})_{j,j}]\left(E\left[\sum_m |(\mathbf{B})_{m,m}|^2|(\mathbf{U})_{i,m}|^2|(\mathbf{U})_{j,m}|^2\right]\right. \\ &\quad +E\left[\sum_{m\neq p}|(\mathbf{B})_{m,p}|^2|(\mathbf{U})_{i,m}|^2|(\mathbf{U})_{j,p}|^2\right] \\ &\quad \left.+\,E\left[\sum_{m\neq \ell}(\mathbf{B})_{m,m}(\mathbf{B})_{\ell,\ell}^*(\mathbf{U})_{i,m}(\mathbf{U})_{j,m}^*(\mathbf{U})_{i,\ell}^*(\mathbf{U})_{j,\ell}\right]\right) \\ &= E[(\mathbf{\Lambda})_{i,i}(\mathbf{\Lambda})_{j,j}]\left(\frac{E[\sum_m|(\mathbf{B})_{m,m}|^2]}{n(n+1)}+\frac{E[\sum_{m\neq p}|(\mathbf{B})_{m,p}|^2]}{n^2-1}\right. \\ &\quad \left.-\frac{E[\sum_{m\neq\ell}(\mathbf{B})_{m,m}(\mathbf{B})_{\ell,\ell}^*]}{n(n^2-1)}\right) \\ &= E[(\mathbf{\Lambda})_{i,i}(\mathbf{\Lambda})_{j,j}]\left(\frac{E[\mathrm{Tr}\{\mathbf{B}^2\}]}{n^2-1}-\frac{E[\mathrm{Tr}^2\{\mathbf{B}\}]}{n(n^2-1)}\right) \end{aligned}$$

while, for $i=j$

$$\begin{aligned} E[(\mathbf{\Lambda}^{1/2}\mathbf{V}\mathbf{\Lambda}^{1/2})_{i,i}^2] &= E[(\mathbf{\Lambda})_{i,i}^2(\mathbf{V})_{i,i}^2] \\ &= E[(\mathbf{\Lambda})_{i,i}^2]\; E\left[\left(\sum_{m\,p}(\mathbf{U})_{i,m}(\mathbf{B})_{m,p}(\mathbf{U})_{i,p}^*\right)^2\right] \\ &= E[(\mathbf{\Lambda})_{i,i}^2]\; E\left[\sum_{m\,p}\sum_{\ell\,q}(\mathbf{B})_{m,p}(\mathbf{B})_{\ell,q}^*(\mathbf{U})_{i,m}(\mathbf{U})_{i,p}^*(\mathbf{U})_{i,\ell}^*(\mathbf{U})_{i,q}\right] \end{aligned}$$

$$
\begin{aligned}
&= E[(\mathbf{\Lambda})_{i,i}^2]\, E\left[\sum_m |(\mathbf{B})_{m,m}|^2 |(\mathbf{U})_{i,m}|^4\right] \\
&\quad + E[(\mathbf{\Lambda})_{i,i}^2]\, E\left[\sum_{m\neq p} |(\mathbf{B})_{m,p}|^2 |(\mathbf{U})_{i,m}|^2 |(\mathbf{U})_{i,p}|^2\right] \\
&\quad + E[(\mathbf{\Lambda})_{i,i}^2]\, E\left[\sum_{m\neq \ell} (\mathbf{B})_{m,m} (\mathbf{B})_{\ell,\ell}^* |(\mathbf{U})_{i,m}|^2 |(\mathbf{U})_{i,\ell}|^2\right] \\
&= E[(\mathbf{\Lambda})_{i,i}^2] \left(\frac{2E[\sum_m |(\mathbf{B})_{m,m}|^2]}{n(n+1)} + \frac{E[\sum_{m\neq p} |(\mathbf{B})_{m,p}|^2]}{n(n+1)} \right. \\
&\quad \left. + \frac{E[\sum_{m\neq \ell} (\mathbf{B})_{m,m} (\mathbf{B})_{\ell,\ell}^*]}{n(n+1)} \right) \\
&= \frac{E[(\mathbf{\Lambda})_{i,i}^2]}{n(n+1)} \left(E[\mathrm{Tr}\{\mathbf{B}^2\}] + E[\mathrm{Tr}^2\{\mathbf{B}\}] \right).
\end{aligned}
$$

Altogether,

$$
\begin{aligned}
E[\mathrm{Tr}\{(\mathbf{AV})^2\}] &= \sum_{ij} E[(\mathbf{\Lambda}^{1/2}\mathbf{V}\mathbf{\Lambda}^{1/2})_{i,j}^2] \\
&= \sum_i E[(\mathbf{\Lambda})_{i,i}^2] \left(\frac{E[\mathrm{Tr}\{\mathbf{B}^2\}] + E[\mathrm{Tr}^2\{\mathbf{B}\}]}{n(n+1)} \right) \\
&\quad + \sum_{i\neq j} E[(\mathbf{\Lambda})_{i,i}(\mathbf{\Lambda})_{j,j}] \left(\frac{E[\mathrm{Tr}\{\mathbf{B}^2\}]}{n^2-1} - \frac{E[\mathrm{Tr}^2\{\mathbf{B}\}]}{n(n^2-1)} \right) \\
&= E[\mathrm{Tr}\{\mathbf{B}^2\}] \left(\sum_i \frac{E[(\mathbf{\Lambda})_{i,i}^2]}{n(n+1)} + \sum_{i\neq j} \frac{E[(\mathbf{\Lambda})_{i,i}(\mathbf{\Lambda})_{j,j}]}{n^2-1} \right) \\
&\quad + E[\mathrm{Tr}^2\{\mathbf{B}\}] \left(\sum_i \frac{E[(\mathbf{\Lambda})_{i,i}^2]}{n(n+1)} - \sum_{i\neq j} \frac{E[(\mathbf{\Lambda})_{i,i}(\mathbf{\Lambda})_{j,j}]}{n(n^2-1)} \right) \\
&= \frac{E[\mathrm{Tr}\{\mathbf{B}^2\}\, E[\mathrm{Tr}^2\{\mathbf{A}\}]}{n^2-1} + \frac{E[\mathrm{Tr}^2\{\mathbf{B}\}]\, E[\mathrm{Tr}\{\mathbf{A}^2\}]}{n(n+1)} \\
&\quad + \frac{E[\mathrm{Tr}\{\mathbf{A}^2\}] \left(E[\mathrm{Tr}^2\{\mathbf{B}\}] - E[\mathrm{Tr}\{\mathbf{B}^2\}] \right)}{n(n^2-1)} \\
&\quad - \frac{E[\mathrm{Tr}^2\{\mathbf{B}\}]\, E[\mathrm{Tr}^2\{\mathbf{A}\}]}{n(n^2-1)}
\end{aligned}
$$

which, given that $\mathrm{Tr}\{\mathbf{B}\}=\mathrm{Tr}\{\mathbf{V}\}$ and $\mathrm{Tr}\{\mathbf{B}^2\}=\mathrm{Tr}\{\mathbf{V}^2\}$, proves (4.2). Finally,

$$
\begin{aligned}
E[\mathrm{Tr}^2\{\mathbf{AV}\}] &= E[\mathrm{Tr}^2\{\mathbf{Q_A}\Lambda\mathbf{Q_A}^\dagger\mathbf{V}\}] \\
&= E[\mathrm{Tr}^2\{\Lambda\mathbf{V}\}] \\
&= E[\mathrm{Tr}^2\{\Lambda\mathbf{U}\Lambda_{\mathbf{B}}\mathbf{U}^\dagger\}]
\end{aligned}
$$

with the eigenvalue decomposition $\mathbf{B}=\mathbf{Q_B}\Lambda_\mathbf{B}\mathbf{Q}_\mathbf{B}^\dagger$. Further algebra yields

$$\begin{aligned}
E[\mathrm{Tr}^2\{\mathbf{AV}\}] &= E\left[\left(\sum_i (\Lambda)_{i,i}(\mathbf{U})_i\Lambda_\mathbf{B}(\mathbf{U})_i^\dagger\right)^2\right] \\
&= E\left[\sum_i\sum_j (\Lambda)_{i,i}(\Lambda)_{j,j}(\mathbf{U})_j\Lambda_\mathbf{B}(\mathbf{U})_j^\dagger(\mathbf{U})_i\Lambda_\mathbf{B}(\mathbf{U})_i^\dagger)\right] \\
&= \sum_i E[(\Lambda)_{i,i}^2]\, E[(\mathbf{U})_i\Lambda_\mathbf{B}(\mathbf{U})_i^\dagger(\mathbf{U})_i\Lambda_\mathbf{B}(\mathbf{U})_i^\dagger)] \\
&\quad + \sum_{i\neq j} E[(\Lambda)_{i,i}(\Lambda)_{j,j}]\, E[(\mathbf{U})_j\Lambda_\mathbf{B}(\mathbf{U})_j^\dagger(\mathbf{U})_i\Lambda_\mathbf{B}(\mathbf{U})_i^\dagger)].
\end{aligned}$$

For $i\neq j$,

$$\begin{aligned}
E[(\mathbf{U})_j\Lambda_\mathbf{B}(\mathbf{U})_j^\dagger(\mathbf{U})_i\Lambda_\mathbf{B}(\mathbf{U})_i^\dagger)] &= E\left[\sum_p\sum_\ell (\Lambda_\mathbf{B})_{p,p}(\Lambda_\mathbf{B})_{\ell,\ell}|(\mathbf{U})_{j,p}|^2|(\mathbf{U})_{i,\ell}|^2\right] \\
&= \frac{E\left[\sum_p(\Lambda_\mathbf{B})_{p,p}^2\right]}{n(n+1)} + \frac{E\left[\sum_{p\neq\ell}(\Lambda_\mathbf{B})_{p,p}(\Lambda_\mathbf{B})_{\ell,\ell}\right]}{n^2-1} \\
&= -\frac{E[\mathrm{Tr}\{\mathbf{B}^2\}]}{n(n^2-1)} + \frac{E[\mathrm{Tr}^2\{\mathbf{B}\}]}{n^2-1}
\end{aligned}$$

while, for $i=j$,

$$\begin{aligned}
E[(\mathbf{U})_j\Lambda_\mathbf{B}(\mathbf{U})_j^\dagger(\mathbf{U})_i\Lambda_\mathbf{B}(\mathbf{U})_i^\dagger)] &= E\left[\sum_p\sum_\ell (\Lambda_\mathbf{B})_{p,p}(\Lambda_\mathbf{B})_{\ell,\ell}|(\mathbf{U})_{i,p}|^2|(\mathbf{U})_{i,\ell}|^2\right] \\
&= 2\,\frac{E\left[\sum_p(\Lambda_\mathbf{B})_{p,p}^2\right]}{n(n+1)} + \frac{E\left[\sum_{p\neq\ell}(\Lambda_\mathbf{B})_{p,p}(\Lambda_\mathbf{B})_{\ell,\ell}\right]}{n(n+1)} \\
&= \frac{E[\mathrm{Tr}\{\mathbf{B}^2\}]}{n(n+1)} + \frac{E[\mathrm{Tr}^2\{\mathbf{B}\}]}{n(n+1)}.
\end{aligned}$$

Altogether,

$$\begin{aligned}
E[\mathrm{Tr}^2\{\mathbf{AV}\}] &= E[\mathrm{Tr}\{\mathbf{A}^2\}]\left(\frac{E[\mathrm{Tr}\{\mathbf{B}^2\}+\mathrm{Tr}^2\{\mathbf{B}\}]}{n(n+1)} + \frac{E[\mathrm{Tr}\{\mathbf{B}^2\}]}{n(n^2-1)} - \frac{E[\mathrm{Tr}^2\{\mathbf{B}\}]}{n^2-1}\right) \\
&\quad + E[\mathrm{Tr}^2\{\mathbf{A}\}]\left(\frac{E[\mathrm{Tr}^2\{\mathbf{B}\}]}{n^2-1} - \frac{E[\mathrm{Tr}\{\mathbf{B}^2\}]}{n(n^2-1)}\right)
\end{aligned}$$

from which (4.3) follows easily.

Appendix C. Proof of Lemma 4.3

$$E\left[\mathrm{Tr}\left\{(\mathbf{H}_w\mathbf{H}_w^\dagger)^{-1}\right\}\right] = n\,E\left[\frac{1}{\lambda}\right]$$

with λ an arbitrary eigenvalue of $(\mathbf{H}_w\mathbf{H}_w^\dagger)$, whose marginal density distribution is given by **[21]**

$$f_\lambda(z) = \frac{1}{n}\sum_{k=0}^{n-1}\frac{k!\,(L_k^{m-n}(z))^2}{(k+m-n)!}\,z^{m-n}e^{-z}$$

with L_k^{m-n} the corresponding Laguerre polynomial of order k [**7**]. The expectation can be computed to yield

$$E\left[\frac{1}{\lambda}\right] = \int_0^\infty \frac{f_\lambda(z)}{z}\,\mathrm{d}z = \frac{1}{m-n}$$

which leads to (4.4). At the same time,

$$\begin{aligned} E\left[\mathrm{Tr}\left\{(\mathbf{H}_w\mathbf{H}_w^\dagger)^{-2}\right\}\right] &= n\,E\left[\frac{1}{\lambda^2}\right] \\ &= n\int_0^\infty \frac{f_\lambda(z)}{z^2}\,\mathrm{d}z \\ &= \frac{mn}{(m-n)^3-(m-n)} \end{aligned}$$

and, from the joint density of any two different eigenvalues of $(\mathbf{H}_w\mathbf{H}_w^\dagger)$ given by

$$\begin{aligned} f_{\lambda\lambda'} = {} & \frac{(\lambda\lambda')^{m-n}}{n\,(n-1)} e^{-(\lambda+\lambda')}\left[-\left(\sum_{k=0}^{n-1}\frac{k!\,L_k^{m-n}(\lambda)L_k^{m-n}(\lambda')}{(k+m-n)!}\right)^2\right. \\ & \left. +\sum_{k=0}^{n-1}\frac{k!\,\left(L_k^{m-n}(\lambda)\right)^2}{(k+m-n)!}\sum_{k=0}^{n-1}\frac{k!\,\left(L_k^{m-n}(\lambda')\right)^2}{(k+m-n)!}\right] \end{aligned}$$

we find that

$$\begin{aligned} E\left[\mathrm{Tr}^2\left\{(\mathbf{H}_w\mathbf{H}_1^\dagger)^{-1}\right\}\right] &= \sum_{i=1}^{n}\sum_{j=1}^{n} E\left[\frac{1}{\lambda_i\lambda_j}\right] \\ &= \left(n\,E\left[\frac{1}{\lambda^2}\right]+(n^2-n)\,E\left[\frac{1}{\lambda\lambda'}\right]\right) \\ &= n\left(\int_0^\infty \frac{f_\lambda(z)}{z^2}\,\mathrm{d}z+(n-1)\iint_0^\infty \frac{f_{\lambda\lambda'}(z,z')}{zz'}\,\mathrm{d}z\mathrm{d}z'\right) \\ &= \frac{n}{m-n}\left(\frac{m}{(m-n)^2-1}+\frac{n-1}{m-n+1}\right). \end{aligned}$$

Appendix D. Proof of Propositions 5.1 and 6.1

To prove both these results, we first derive the more general expressions for $\frac{E_b}{N_0}_{\min}$ and S_0 claimed in Proposition 6.1 and then particularize them to Proposition 5.1. Notice from (3.2) and (3.3) that, in its most general form, expressing $\frac{E_b}{N_0}_{\min}$ and S_0 entails characterizing the terms

(D.1) $$E[\mathrm{Tr}\{\mathbf{H}\mathbf{H}^\dagger\mathbf{\Phi}_\mathbf{n}^{-1}(\overline{\mathbf{H}})\}]$$

and

(D.2) $$E[\mathrm{Tr}\{(\mathbf{H}\mathbf{H}^\dagger\mathbf{\Phi}_\mathbf{n}^{-1}(\overline{\mathbf{H}}))^2\}].$$

Expression (D.1) can be expanded, using the channel model, as

(D.3) $$E[\mathrm{Tr}\{\mathbf{H}\mathbf{H}^\dagger\mathbf{\Phi}_\mathbf{n}^{-1}(\overline{\mathbf{H}})\}] = E[\mathrm{Tr}\{\mathbf{\Theta}_\mathrm{R}^{1/2}\mathbf{H}_w\mathbf{\Theta}_\mathrm{T}\mathbf{H}_w^\dagger\mathbf{\Theta}_\mathrm{R}^{1/2}\mathbf{\Phi}_\mathbf{n}^{-1}(\overline{\mathbf{H}})\}].$$

From Lemma A.3, $\mathbf{H}_w\mathbf{\Theta}_{\mathrm{T}}\mathbf{H}_w^\dagger$ admits the eigenvalue decomposition

$$\mathbf{H}_w\mathbf{\Theta}_{\mathrm{T}}\mathbf{H}_w^\dagger = \mathbf{U}\mathbf{B}\mathbf{U}^\dagger$$

where $\mathbf{U}$ is an $(n_{\mathrm{R}}\times n_{\mathrm{R}})$ Haar random matrix and $\mathbf{B}$ is the $(n_{\mathrm{R}}\times n_{\mathrm{R}})$ random eigenvalue matrix independent of $\mathbf{U}$. Furthermore, $\mathbf{H}$ and $\mathbf{\Phi}_{\mathbf{n}}^{-1}(\overline{\mathbf{H}})$ are statistically independent. Thus, $\mathbf{U}$ and $\mathbf{B}$ are mutually independent and both are independent of $\mathbf{\Phi}_{\mathbf{n}}^{-1}(\overline{\mathbf{H}})$. Lemma 4.1 applies and we can expand (D.3) as

$$\begin{aligned} E[\mathrm{Tr}\{\mathbf{\Theta}_{\mathrm{R}}^{1/2}\mathbf{H}_w\mathbf{\Theta}_{\mathrm{T}}\mathbf{H}_w^\dagger\mathbf{\Theta}_{\mathrm{R}}^{1/2}\mathbf{\Phi}_{\mathbf{n}}^{-1}(\overline{\mathbf{H}})\}] &= E[\mathrm{Tr}\{\mathbf{H}_w\mathbf{\Theta}_{\mathrm{T}}\mathbf{H}_w^\dagger\mathbf{\Theta}_{\mathrm{R}}^{1/2}\mathbf{\Phi}_{\mathbf{n}}^{-1}(\overline{\mathbf{H}})\mathbf{\Theta}_{\mathrm{R}}^{1/2}\}] \\ &= E[\mathrm{Tr}\{\mathbf{U}\mathbf{B}\mathbf{U}^\dagger\mathbf{\Theta}_{\mathrm{R}}^{1/2}\mathbf{\Phi}_{\mathbf{n}}^{-1}(\overline{\mathbf{H}})\mathbf{\Theta}_{\mathrm{R}}^{1/2}\}] \\ &= \tfrac{1}{n_{\mathrm{R}}}E[\mathrm{Tr}\{\mathbf{B}\}]\,E[\mathrm{Tr}\{\mathbf{\Theta}_{\mathrm{R}}^{1/2}\mathbf{\Phi}_{\mathbf{n}}^{-1}(\overline{\mathbf{H}})\mathbf{\Theta}_{\mathrm{R}}^{1/2}\}] \\ &= \mathrm{Tr}\{\mathbf{\Theta}_{\mathrm{T}}\}E[\mathrm{Tr}\{\mathbf{\Theta}_{\mathrm{R}}\mathbf{\Phi}_{\mathbf{n}}^{-1}(\overline{\mathbf{H}})\}] \\ &= n_{\mathrm{T}}E[\mathrm{Tr}\{\mathbf{\Theta}_{\mathrm{R}}\mathbf{\Phi}_{\mathbf{n}}^{-1}(\overline{\mathbf{H}})\}] \end{aligned}$$

where we have used $E[\mathrm{Tr}\{\mathbf{B}\}]{=}E[\mathrm{Tr}\{\mathbf{\Theta}_{\mathrm{T}}\mathbf{H}_w^\dagger\mathbf{H}_w\}]{=}n_{\mathrm{R}}\mathrm{Tr}\{\mathbf{\Theta}_{\mathrm{T}}\}$. All in all, (D.1) is given by

$$E[\mathrm{Tr}\{\mathbf{H}\mathbf{H}^\dagger\mathbf{\Phi}_{\mathbf{n}}^{-1}(\overline{\mathbf{H}})\}] = n_{\mathrm{T}}E[\mathrm{Tr}\{\mathbf{\Theta}_{\mathrm{R}}\mathbf{\Phi}_{\mathbf{n}}^{-1}(\overline{\mathbf{H}})\}]. \tag{D.4}$$

Let us now turn our attention to (D.2), which expands—disregarding those terms whose mean is zero—into

$$\begin{aligned} E[\mathrm{Tr}\{(\mathbf{H}\mathbf{H}^\dagger\mathbf{\Phi}_{\mathbf{n}}^{-1}(\overline{\mathbf{H}}))^2\}] &= E[\mathrm{Tr}\{\mathbf{H}\mathbf{H}^\dagger\mathbf{\Phi}_{\mathbf{n}}^{-1}(\overline{\mathbf{H}})\mathbf{H}\mathbf{H}^\dagger\mathbf{\Phi}_{\mathbf{n}}^{-1}(\overline{\mathbf{H}})\}] \\ &= E[\mathrm{Tr}\{(\mathbf{\Theta}_{\mathrm{R}}^{1/2}\mathbf{H}_w\mathbf{\Theta}_{\mathrm{T}}\mathbf{H}_w^\dagger\mathbf{\Theta}_{\mathrm{R}}^{1/2}\mathbf{\Phi}_{\mathbf{n}}^{-1}(\overline{\mathbf{H}}))^2\}]. \end{aligned} \tag{D.5}$$

Define $\mathbf{A}\stackrel{\mathrm{def}}{=}\mathbf{\Theta}_{\mathrm{R}}^{1/2}\mathbf{\Phi}_{\mathbf{n}}^{-1}(\overline{\mathbf{H}})\mathbf{\Theta}_{\mathrm{R}}^{1/2}$ and notice that

$$\begin{aligned} E[\mathrm{Tr}\{(\mathbf{\Theta}_{\mathrm{R}}^{1/2}\mathbf{H}_w\mathbf{\Theta}_{\mathrm{T}}\mathbf{H}_w^\dagger\mathbf{\Theta}_{\mathrm{R}}^{1/2}\mathbf{\Phi}_{\mathbf{n}}^{-1}(\overline{\mathbf{H}}))^2\}] &= E[\mathrm{Tr}\{(\mathbf{H}_w\mathbf{\Theta}_{\mathrm{T}}\mathbf{H}_w^\dagger\mathbf{A})^2\}] \\ &= E[\mathrm{Tr}\{(\mathbf{U}\mathbf{B}\mathbf{U}^\dagger\mathbf{A})^2\}]. \end{aligned}$$

Applying Lemma 4.1 twice followed by Lemma A.3 and Lemma 4.2, we find that

$$\begin{aligned} E[\mathrm{Tr}\{(\mathbf{\Theta}_{\mathrm{R}}^{1/2}\mathbf{H}_w\mathbf{\Theta}_{\mathrm{T}}\mathbf{H}_w^\dagger\mathbf{\Theta}_{\mathrm{R}}^{1/2}\mathbf{\Phi}_{\mathbf{n}}^{-1}(\overline{\mathbf{H}}))^2\}] &= E[\mathrm{Tr}\{(\mathbf{\Theta}_{\mathrm{R}}\mathbf{\Phi}_{\mathbf{n}}^{-1}(\overline{\mathbf{H}}))^2\}]\,\mathrm{Tr}^2\{\mathbf{\Theta}_{\mathrm{T}}\} \\ &\quad + E[\mathrm{Tr}^2\{\mathbf{\Theta}_{\mathrm{R}}\mathbf{\Phi}_{\mathbf{n}}^{-1}(\overline{\mathbf{H}})\}]\,\mathrm{Tr}\{\mathbf{\Theta}_{\mathrm{T}}^2\}. \end{aligned}$$

Putting the pieces together, (D.2) is given by

$$\begin{aligned} E[\mathrm{Tr}\{(\mathbf{H}\mathbf{H}^\dagger\mathbf{\Phi}_{\mathbf{n}}^{-1}(\overline{\mathbf{H}}))^2\}] &= n_{\mathrm{T}}^2E[\mathrm{Tr}\{(\mathbf{\Theta}_{\mathrm{R}}\mathbf{\Phi}_{\mathbf{n}}^{-1}(\overline{\mathbf{H}}))^2\}] \\ &\quad + \mathrm{Tr}\{\mathbf{\Theta}_{\mathrm{T}}^2\}E[\mathrm{Tr}^2\{\mathbf{\Theta}_{\mathrm{R}}\mathbf{\Phi}_{\mathbf{n}}^{-1}(\overline{\mathbf{H}})\}]. \end{aligned} \tag{D.6}$$

Using (D.4) and (D.6), the $\frac{E_b}{N_0}_{\min}$ and $\mathcal{S}_0$ claimed in Proposition 6.1 are readily obtained. These, in turn, specialize to Proposition 5.1 by setting $\mathbf{\Phi}_{\mathbf{n}}{=}\mathbf{I}$.

Appendix E. Proof of Proposition 6.2

Since $\boldsymbol{\Theta}_{\rm R}$ and $\boldsymbol{\Theta}_{\rm R}(\ell)$ are Toeplitz, asymptotically they admit decompositions which involve the same eigenvector matrices [**8**] and thus

$$\text{(E.1)} \qquad \lim_{n_{\rm R}\to\infty} \frac{E[\mathrm{Tr}\{\boldsymbol{\Theta}_{\rm R}\boldsymbol{\Phi}_{\mathbf{n}}^{-1}\}]}{n_{\rm R}} = \left(\sum_{\ell=1}^{L} \mathcal{I}_\ell + \gamma\right) \lim_{n_{\rm R}\to\infty} \frac{E[\mathrm{Tr}\{\boldsymbol{\Xi}^{-1}\}]}{n_{\rm R}}$$

with[4]

$$\boldsymbol{\Xi} = \boldsymbol{\Lambda}_{\rm R}^{-1/2}\left(\sum_{\ell=1}^{L} \boldsymbol{\Lambda}_{\rm R}(\ell)^{1/2}\frac{\mathbf{H}_w(\ell)}{\sqrt{m_\ell}}\boldsymbol{\Lambda}_{\rm T}(\ell)\frac{\mathbf{H}_w(\ell)^\dagger}{\sqrt{m_\ell}}\boldsymbol{\Lambda}_{\rm R}(\ell)^{1/2}\right)\boldsymbol{\Lambda}_{\rm R}^{-1/2} + \gamma\boldsymbol{\Lambda}_{\rm R}^{-1}$$

and with $\boldsymbol{\Lambda}_{\rm R}$, $\boldsymbol{\Lambda}_{\rm T}(\ell)$ and $\boldsymbol{\Lambda}_{\rm R}(\ell)$ diagonal matrices containing the eigenvalues of $\boldsymbol{\Theta}_{\rm R}$, $\boldsymbol{\Theta}_{\rm T}(\ell)$ and $(\mathcal{I}_\ell\,\boldsymbol{\Theta}_{\rm R}(\ell))$. From [**9**], the empirical eigenvalue distribution of $\boldsymbol{\Xi}$ converges almost surely to a non-random limit as $n_{\rm R}\to\infty$. Consequently, we have

$$\lim_{n_{\rm R}\to\infty} \varphi(\boldsymbol{\Phi}_{\mathbf{n}}^{-1}) = 1.$$

Furthermore, (E.1) can be interpreted as the Stieltjes transform of $\boldsymbol{\Xi}$, $S_{\boldsymbol{\Xi}}(z)$, evaluated for $z\downarrow 0$

$$\begin{aligned}\lim_{n_{\rm R}\to\infty} \frac{E[\mathrm{Tr}\{(\boldsymbol{\Xi})^{-1}\}]}{n_{\rm R}} &= \lim_{z\to 0}\lim_{n_{\rm R}\to\infty} \frac{E[\mathrm{Tr}\{(\boldsymbol{\Xi}-z\mathbf{I})^{-1}\}]}{n_{\rm R}} \\ &= \lim_{z\to 0} S_{\boldsymbol{\Xi}}(z).\end{aligned}$$

Using [**19**], it can be shown that $\boldsymbol{\Lambda}_{\rm R}^{-1}$ and

$$\boldsymbol{\Lambda}_{\rm R}^{-1/2}\left(\sum_{\ell=1}^{L} \boldsymbol{\Lambda}_{\rm R}(\ell)^{1/2}\frac{\mathbf{H}_w(\ell)}{\sqrt{m_\ell}}\boldsymbol{\Lambda}_{\rm T}(\ell)\frac{\mathbf{H}_w(\ell)^\dagger}{\sqrt{m_\ell}}\boldsymbol{\Lambda}_{\rm R}(\ell)^{1/2}\right)\boldsymbol{\Lambda}_{\rm R}^{-1/2}$$

are asymptotically free with amalgamation. This, in conjunction with [**6**], yields

$$\lim_{n_{\rm R}\to\infty} \frac{E[\mathrm{Tr}\{(\boldsymbol{\Xi}-z\mathbf{I})^{-1}\}]}{n_{\rm R}} = \int_0^1 \eta(x,z,\gamma)\,dx$$

with $\eta(x,z,\gamma)$ given by the fixed-point equation

$$\eta(x,z,\gamma) = \frac{1}{-z+\gamma\,\vartheta_{\rm R}(x)+\vartheta_{\rm R}(x)\displaystyle\int_0^{\bar\beta}\frac{\mathcal{V}_{\rm R}(x,y)\mathcal{V}_{\rm T}(y)}{1+\mathcal{V}_{\rm T}(y)\int_0^1\eta(\tau,z,\gamma)\vartheta_{\rm R}(\tau)\mathcal{V}_{\rm R}(\tau,y)\,d\tau}dy}$$

where $\mathcal{V}_{\rm T}(y)$, $\mathcal{V}_{\rm R}(x,y)$ and $\vartheta_{\rm R}(x)$, with $x\in(0,1)$ and $y\in(0,\bar\beta)$, represent the limits, as $n_{\rm R}\to\infty$, of the sequences of functions

$$\vartheta_{\rm R}^{(n_{\rm R})}(x) = \frac{1}{\lambda_{\rm R}(i)} \qquad \tfrac{i}{n_{\rm R}} \le x < \tfrac{i+1}{n_{\rm R}},$$

$$\mathcal{V}_{\rm T}^{(n_{\rm R})}(y) = \lambda_{\rm T}(j,\ell) \qquad \tfrac{j+\sum_{k=1}^{\ell-1}m_k}{n_{\rm R}} \le y < \tfrac{j+1+\sum_{k=1}^{\ell-1}m_k}{n_{\rm R}}$$

$$\mathcal{V}_{\rm R}^{(n_{\rm R})}(x,y) = \lambda_{\rm R}(i,\ell) \qquad \tfrac{i}{n_{\rm R}} \le x < \tfrac{i+1}{n_{\rm R}}, \quad \tfrac{\sum_{k=1}^{\ell-1}m_k}{n_{\rm R}} \le y < \tfrac{\sum_{k=1}^{\ell}m_k}{n_{\rm R}}$$

[4]The correlation matrices $\boldsymbol{\Theta}_{\rm R}$ and $\boldsymbol{\Theta}_{\rm R}(\ell)$ are herein assumed non-singular. The proof, however, can be extended to singular such matrices using a generalized version of [**9**, Lemma 14].

with $1 \le i \le n_{\mathrm{R}}$, $1 \le \ell \le L$, and $1 \le j \le m_\ell$. Here, $\lambda_{\mathrm{R}}(n)$, $\lambda_{\mathrm{T}}(n,\ell)$, and $\lambda_{\mathrm{R}}(n,\ell)$, represent the n-th eigenvalue of the matrices $\mathbf{\Theta}_{\mathrm{R}}$, $\mathbf{\Theta}_{\mathrm{T}}(\ell)$, and $(\mathcal{I}_\ell\,\mathbf{\Theta}_{\mathrm{R}}(\ell))$, respectively. Defining $\mathcal{R}(x) \stackrel{\text{def}}{=} 1/\vartheta_{\mathrm{R}}(x)$,

$$\lim_{n_{\mathrm{R}}\to\infty} \frac{E[\mathrm{Tr}\{\mathbf{\Theta}_{\mathrm{R}}\mathbf{\Phi}_{\mathbf{n}}^{-1}\}]}{n_{\mathrm{R}}} = \left(\sum_{\ell=1}^{L} \mathcal{I}_\ell + \gamma\right) \lim_{z\to 0} \int_0^1 \eta(x,z,\gamma)dx$$

with $\eta(x,z,\gamma)$ the solution to the fixed-point equation

$$\begin{aligned}\eta(x,z,\gamma) &= \frac{1}{-z+\gamma\dfrac{1}{\mathcal{R}(x)} + \dfrac{1}{\mathcal{R}(x)}\displaystyle\int_0^{\bar\beta} \frac{\mathcal{V}_{\mathrm{R}}(x,y)\mathcal{V}_{\mathrm{T}}(y)}{1+\mathcal{V}_{\mathrm{T}}(y)\int_0^1 \eta(\tau,z,\gamma)\frac{1}{\mathcal{R}(\tau)}\mathcal{V}_{\mathrm{R}}(\tau,y)\,d\tau}dy} \\ &= \frac{\mathcal{R}(x)}{-z+\gamma+\displaystyle\int_0^{\bar\beta} \frac{\mathcal{V}_{\mathrm{R}}(x,y)\mathcal{V}_{\mathrm{T}}(y)}{1+\mathcal{V}_{\mathrm{T}}(y)\int_0^1 \eta(\tau,z,\gamma)\frac{1}{\mathcal{R}(\tau)}\mathcal{V}_{\mathrm{R}}(\tau,y)\,d\tau}dy}.\end{aligned}$$

If we further define $a(x,z,\gamma) = \frac{\eta(x,z,\gamma)}{\mathcal{R}(x)}$, then $a(x,z,\gamma)$ must satisfy

$$a(x,z,\gamma) = \frac{\mathcal{R}(x)}{-z+\gamma+\displaystyle\int_0^{\bar\beta} \frac{\mathcal{V}_{\mathrm{R}}(x,y)\mathcal{V}_{\mathrm{T}}(y)}{1+\mathcal{V}_{\mathrm{T}}(y)\int_0^1 a(\tau,z,\gamma)\mathcal{V}_{\mathrm{R}}(\tau,y)\,d\tau}dy}$$

and thus

$$\lim_{n_{\mathrm{R}}\to\infty} \frac{E[\mathrm{Tr}\{\mathbf{\Theta}_{\mathrm{R}}\mathbf{\Phi}_{\mathbf{n}}^{-1}\}]}{n_{\mathrm{R}}} = \left(\sum_{\ell=1}^{L} \mathcal{I}_\ell + \gamma\right) \int_0^1 a(x,0,\gamma)\,\mathcal{R}(x)dx. \tag{E.2}$$

Next, we need to compute $\lim_{n_{\mathrm{R}}\to\infty} E[\mathrm{Tr}\{(\mathbf{\Theta}_{\mathrm{R}}\mathbf{\Phi}_{\mathbf{n}}^{-1})^{-2}\}]/n_{\mathrm{R}}$. Notice that

$$\begin{aligned}\lim_{n_{\mathrm{R}}\to\infty} \frac{E[\mathrm{Tr}\{(\mathbf{\Theta}_{\mathrm{R}}\mathbf{\Phi}_{\mathbf{n}}^{-1})^2\}]}{n_{\mathrm{R}}} &= \left(\sum_{\ell=1}^{L} \mathcal{I}_\ell + \gamma\right)^2 \lim_{z\to 0}\lim_{n_{\mathrm{R}}\to\infty} \frac{E[\mathrm{Tr}\{(\mathbf{\Xi}-z\mathbf{I})^{-2}\}]}{n_{\mathrm{R}}} \\ &= \left(\sum_{\ell=1}^{L} \mathcal{I}_\ell + \gamma\right)^2 \lim_{z\to 0} \int \frac{1}{(\lambda(\mathbf{\Xi})-z)^2} f_{\mathbf{\Xi}}(\lambda)d\lambda \\ &= \left(\sum_{\ell=1}^{L} \mathcal{I}_\ell + \gamma\right)^2 \lim_{z\to 0} \frac{d}{d\,\Re(z)} \int \frac{1}{(\lambda(\mathbf{\Xi})-z)} f_{\mathbf{\Xi}}(\lambda)d\lambda \\ &= \left(\sum_{\ell=1}^{L} \mathcal{I}_\ell + \gamma\right)^2 \lim_{z\to 0} \int_0^1 \dot\eta(x,z,\gamma)dx\end{aligned}$$

(E.3)

where $\dot\eta(x,z,\gamma) = \frac{d}{d\,\Re(z)}\dot\eta(x,z,\gamma)$ and $f_{\mathbf{\Xi}}(\lambda)$ represents the empirical eigenvalue density distribution of $\mathbf{\Xi}$. Therefore, from (E.2) and (E.3)

$$\zeta(\mathbf{\Theta}_{\mathrm{R}}\mathbf{\Phi}_{\mathbf{n}}^{-1}) = \frac{\int_0^1 \mathcal{R}(x)\dot a(x,0,\gamma)dx}{\left(\int_0^1 \mathcal{R}(x)a(x,0,\gamma)dx\right)^2}$$

with

$$\dot{a}(x,0,\gamma) = \lim_{z\to 0} \frac{d}{dz} a(x,z,\gamma).$$

Appendix F. Proof of Proposition 6.3

Starting from Proposition 6.1, we need to elaborate on the expressions for $E[\mathrm{Tr}\{\boldsymbol{\Theta}_{\mathrm{R}}\boldsymbol{\Phi}_{\mathbf{n}}^{-1}\}]$, $\zeta(\boldsymbol{\Theta}_{\mathrm{R}}\boldsymbol{\Phi}_{\mathbf{n}}^{-1})$ and $\varphi(\boldsymbol{\Theta}_{\mathrm{R}}\boldsymbol{\Phi}_{\mathbf{n}}^{-1})$. The last two require, in addition to the former, $E[\mathrm{Tr}\{(\boldsymbol{\Theta}_{\mathrm{R}}\boldsymbol{\Phi}_{\mathbf{n}}^{-1})^2\}]$ and $E[\mathrm{Tr}^2\{\boldsymbol{\Theta}_{\mathrm{R}}\boldsymbol{\Phi}_{\mathbf{n}}^{-1}\}]$. In the presence of a single interferer with uncorrelated transmit antennas,

$$\boldsymbol{\Phi}_{\mathbf{n}}^{-1}(\mathbf{H}_1) = (1+\tfrac{\gamma}{\mathcal{I}_1})\left(\frac{\boldsymbol{\Theta}_1^{1/2}\mathbf{H}_w\mathbf{H}_w^{\dagger}\boldsymbol{\Theta}_1^{1/2}}{m_1} + \tfrac{\gamma}{\mathcal{I}_1}\mathbf{I}\right)^{-1}.$$

In the interference-limited regime, we have

$$\text{(F.1)} \qquad \lim_{\mathcal{I}_1/\gamma\to\infty} \boldsymbol{\Phi}_{\mathbf{n}}^{-1}(\mathbf{H}_1) = m_1 \boldsymbol{\Theta}_1^{-1/2}\left(\mathbf{H}_w\mathbf{H}_w^{\dagger}\right)^{-1}\boldsymbol{\Theta}_1^{-1/2}.$$

From Lemma A.3, it follows that $(\mathbf{H}_w\mathbf{H}_w^{\dagger})$ admits an eigenvalue decomposition such that (F.1) conforms to the structure of $\mathbf{V}$ in Lemma 4.1, which can be applied to yield

$$\text{(F.2)} \qquad \lim_{\mathcal{I}_1/\gamma\to\infty} E[\mathrm{Tr}\{\boldsymbol{\Theta}_{\mathrm{R}}\boldsymbol{\Phi}_{\mathbf{n}}^{-1}\}] = m_1 \frac{\mathrm{Tr}\{\boldsymbol{\Theta}_{\mathrm{R}}\boldsymbol{\Theta}_1^{-1}\}}{n_{\mathrm{R}}} E[\mathrm{Tr}\{(\mathbf{H}_w\mathbf{H}_w^{\dagger})^{-1}\}]$$

and

$$\begin{aligned}
\lim_{\mathcal{I}_1/\gamma\to\infty} E[\mathrm{Tr}\{(\boldsymbol{\Theta}_{\mathrm{R}}\boldsymbol{\Phi}_{\mathbf{n}}^{-1})^2\}] &= \frac{m_1^2 E[\mathrm{Tr}\{(\mathbf{H}_w\mathbf{H}_w^{\dagger})^{-2}\}]}{n_{\mathrm{R}}^2-1}\left(\mathrm{Tr}^2\{\boldsymbol{\Theta}_{\mathrm{R}}\boldsymbol{\Theta}_1^{-1}\} - \frac{\mathrm{Tr}\{(\boldsymbol{\Theta}_{\mathrm{R}}\boldsymbol{\Theta}_1^{-1})^2\}}{n_{\mathrm{R}}}\right) \\
&+ \frac{m_1^2 E[\mathrm{Tr}^2\{(\mathbf{H}_w\mathbf{H}_w^{\dagger})^{-1}\}]}{n_{\mathrm{R}}^2-1}\left(\mathrm{Tr}\{(\boldsymbol{\Theta}_{\mathrm{R}}\boldsymbol{\Theta}_1^{-1})^2\} - \frac{\mathrm{Tr}^2\{\boldsymbol{\Theta}_{\mathrm{R}}\boldsymbol{\Theta}_1^{-1}\}}{n_{\mathrm{R}}}\right) \\
\lim_{\mathcal{I}_1/\gamma\to\infty} E[\mathrm{Tr}^2\{\boldsymbol{\Theta}_{\mathrm{R}}\boldsymbol{\Phi}_{\mathbf{n}}^{-1}\}] &= \frac{m_1^2 E[\mathrm{Tr}\{(\mathbf{H}_w\mathbf{H}_w^{\dagger})^{-2}\}]}{n_{\mathrm{R}}^2-1}\left(\mathrm{Tr}\{(\boldsymbol{\Theta}_{\mathrm{R}}\boldsymbol{\Theta}_1^{-1})^2\} - \frac{\mathrm{Tr}^2\{\boldsymbol{\Theta}_{\mathrm{R}}\boldsymbol{\Theta}_1^{-1})\}}{n_{\mathrm{R}}}\right) \\
&+ \frac{m_1^2 E[\mathrm{Tr}^2\{(\mathbf{H}_w\mathbf{H}_w^{\dagger})^{-1}\}]}{n_{\mathrm{R}}^2-1}\left(\mathrm{Tr}^2\{\boldsymbol{\Theta}_{\mathrm{R}}\boldsymbol{\Theta}_1^{-1}\} - \frac{\mathrm{Tr}\{(\boldsymbol{\Theta}_{\mathrm{R}}\boldsymbol{\Theta}_1^{-1})^2\}}{n_{\mathrm{R}}}\right).
\end{aligned}$$

Defining

$$\Psi \stackrel{\text{def}}{=} \zeta\left((\mathbf{H}_w\mathbf{H}_w^{\dagger})^{-1}\right) \qquad \Upsilon \stackrel{\text{def}}{=} \varphi\left((\mathbf{H}_w\mathbf{H}_w^{\dagger})^{-1}\right)$$

we can use the above expressions to obtain

$$\begin{aligned}
\lim_{(\mathcal{I}_1/\gamma)\to\infty} \zeta(\boldsymbol{\Theta}_{\mathrm{R}}\boldsymbol{\Phi}_{\mathbf{n}}^{-1}) &= \frac{n_{\mathrm{R}}^2 - \zeta(\boldsymbol{\Theta}_{\mathrm{R}}\boldsymbol{\Theta}_1^{-1})}{n_{\mathrm{R}}^2-1}\Psi + \frac{n_{\mathrm{R}}^2(\zeta(\boldsymbol{\Theta}_{\mathrm{R}}\boldsymbol{\Theta}_1^{-1})-1)}{n_{\mathrm{R}}^2-1}\Upsilon \\
\lim_{(\mathcal{I}_1/\gamma)\to\infty} \varphi(\boldsymbol{\Theta}_{\mathrm{R}}\boldsymbol{\Phi}_{\mathbf{n}}^{-1}) &= \frac{\zeta(\boldsymbol{\Theta}_{\mathrm{R}}\boldsymbol{\Theta}_1^{-1})-1}{n_{\mathrm{R}}^2-1}\Psi + \frac{n_{\mathrm{R}}^2 - \zeta(\boldsymbol{\Theta}_{\mathrm{R}}\boldsymbol{\Theta}_1^{-1})}{n_{\mathrm{R}}^2-1}\Upsilon
\end{aligned}$$

from which the slope in (6.4) becomes the one claimed in the Proposition. For $m_1 > n_{\mathrm{R}}$, Lemma 4.3 yields

$$E\left[\mathrm{Tr}\left\{(\mathbf{H}_w\mathbf{H}_w^{\dagger})^{-1}\right\}\right] = \frac{n_{\mathrm{R}}}{m_1 - n_{\mathrm{R}}}$$

while, for $m_1 \leq n_{\mathrm{R}}$,

$$\lim_{(\mathcal{I}_1/\gamma)\to\infty} E\left[\mathrm{Tr}\left\{\left(\frac{\mathbf{H}_1\mathbf{H}_1^{\dagger}}{m_1} + \tfrac{\gamma}{\mathcal{I}_1}\mathbf{I}\right)^{-1}\right\}\right] \to \infty$$

and thus, altogether,

$$E[\mathrm{Tr}\{(\mathbf{H}_w\mathbf{H}_w^\dagger)^{-1}\}] = \frac{1}{\left[\frac{1}{n_\mathrm{R}} - \frac{1}{m_1}\right]^+}$$

from which, given (F.2), the $\frac{E_b}{N_0}_{\min}$ in (6.5) follows.

For $m_1 > n_\mathrm{R}+1$, (6.6) and (6.7) are obtained through Lemma 4.3, whereas, for $m_1 \leq n_\mathrm{R}$, they result from

$$\lim_{(\mathcal{I}_1/\gamma)\to\infty} \frac{E\left[\mathrm{Tr}\left\{\left(\frac{\mathbf{H}_1\mathbf{H}_1^\dagger}{m_1} + \frac{\gamma}{\mathcal{I}_1}\mathbf{I}\right)^{-2}\right\}\right]}{E^2\left[\mathrm{Tr}\left\{\left(\frac{\mathbf{H}_1\mathbf{H}_1^\dagger}{m_1} + \frac{\gamma}{\mathcal{I}_1}\mathbf{I}\right)^{-1}\right\}\right]} = \frac{1}{n_\mathrm{R} - m_1}$$

and

$$\lim_{(\mathcal{I}_1/\gamma)\to\infty} \frac{E\left[\mathrm{Tr}^2\left\{\left(\frac{\mathbf{H}_1\mathbf{H}_1^\dagger}{m_1} + \frac{\gamma}{\mathcal{I}_1}\mathbf{I}\right)^{-1}\right\}\right]}{E^2\left[\mathrm{Tr}\left\{\left(\frac{\mathbf{H}_1\mathbf{H}_1^\dagger}{m_1} + \frac{\gamma}{\mathcal{I}_1}\mathbf{I}\right)^{-1}\right\}\right]} = 1.$$

Appendix G. Proof of Proposition 6.4

The channel for an interferer with fully correlated transmit antennas is given by

$$\begin{aligned} \mathbf{H}_1 &= \mathbf{\Theta}_1^{1/2}\mathbf{H}_w\,\mathbf{1}^{1/2} \\ &= \mathbf{\Theta}_1^{1/2}\frac{1}{\sqrt{n_\mathrm{T}}}\mathbf{H}_w\,\mathbf{1} \\ &= \mathbf{\Theta}_1^{1/2}\frac{1}{\sqrt{n_\mathrm{T}}}[\mathbf{h}\,\mathbf{h}\,\ldots\,\mathbf{h}] \end{aligned}$$

and thus, prior to applying receive correlations, the n_T columns of the channel matrix are identical and given by the vector $\mathbf{h}$. It is easy to see that the corresponding normalized conditional covariance $\mathbf{\Phi}_\mathbf{n}(\mathbf{H}_1)$ is identical to that of a single-antenna interferer with channel

$$\mathbf{H}_1 = \mathbf{\Theta}_1^{1/2}\mathbf{h}$$

and the same total power, $\mathcal{I}_1$. The vector $\mathbf{h}$, in turn, is given by

$$\mathbf{h} = \left[\sum_{j=1}^{n_\mathrm{T}}(\mathbf{H}_w)_{1,j} \quad \sum_{j=1}^{n_\mathrm{T}}(\mathbf{H}_w)_{2,j} \quad \cdots \quad \sum_{j=1}^{n_\mathrm{T}}(\mathbf{H}_w)_{n_\mathrm{T},j}\right]^T.$$

Since the entries of $\mathbf{H}_w$ are independent zero-mean unit-variance Gaussian random variables, the entries of $\mathbf{h}$ are likewise and Proposition 6.3 can be applied simply by setting $m_1=1$.

Appendix H. Proof of Proposition 6.5

Since in this case $\mathbf{\Phi}_{\rm n}$ is unitarily invariant, we can apply Lemma 4.1 to obtain

$$\begin{aligned}
E[\mathrm{Tr}\{\mathbf{\Theta}_{\rm R}\mathbf{\Phi}_{\mathbf{n}}^{-1}\}] &= E[\mathrm{Tr}\{\mathbf{\Phi}_{\mathbf{n}}^{-1}\}] \\
E[\mathrm{Tr}\{(\mathbf{\Theta}_{\rm R}\mathbf{\Phi}_{\mathbf{n}}^{-1})^2\}] &= \tfrac{n_{\rm R}^2-\zeta(\mathbf{\Theta}_{\rm R})}{n_{\rm R}^2-1}E[\mathrm{Tr}\{\mathbf{\Phi}_{\mathbf{n}}^{-2}\}] + n_{\rm R}\tfrac{\zeta(\mathbf{\Theta}_{\rm R})-1}{n_{\rm R}^2-1}E[\mathrm{Tr}^2\{\mathbf{\Phi}_{\mathbf{n}}^{-1}\}] \\
E[\mathrm{Tr}^2\{\mathbf{\Theta}_{\rm R}\mathbf{\Phi}_{\mathbf{n}}^{-1}\}] &= n_{\rm R}\tfrac{\zeta(\mathbf{\Theta}_{\rm R})-1}{n_{\rm R}^2-1}E[\mathrm{Tr}\{\mathbf{\Phi}_{\mathbf{n}}^{-2}\}] + \tfrac{n_{\rm R}^2-\zeta(\mathbf{\Theta}_{\rm R})}{n_{\rm R}^2-1}E[\mathrm{Tr}^2\{\mathbf{\Phi}_{\mathbf{n}}^{-1}\}]
\end{aligned}$$

and, in turn,

$$\begin{aligned}
\zeta(\mathbf{\Theta}_{\rm R}\mathbf{\Phi}_{\mathbf{n}}^{-1}) &= \tfrac{n_{\rm R}^2-\zeta(\mathbf{\Theta}_{\rm R})}{n_{\rm R}^2-1}\zeta\left(\mathbf{\Phi}_{\mathbf{n}}^{-1}\right) + \tfrac{n_{\rm R}^2(\zeta(\mathbf{\Theta}_{\rm R})-1)}{n_{\rm R}^2-1}\varphi\left(\mathbf{\Phi}_{\mathbf{n}}^{-1}\right) \\
\varphi(\mathbf{\Theta}_{\rm R}\mathbf{\Phi}_{\mathbf{n}}^{-1}) &= \tfrac{\zeta(\mathbf{\Theta}_{\rm R})-1}{n_{\rm R}^2-1}\zeta\left(\mathbf{\Phi}_{\mathbf{n}}^{-1}\right) + \tfrac{n_{\rm R}^2-\zeta(\mathbf{\Theta}_{\rm R})}{n_{\rm R}^2-1}\varphi\left(\mathbf{\Phi}_{\mathbf{n}}^{-1}\right)
\end{aligned}$$

from which the slope in (6.4) becomes

$$S_0 = \frac{2}{\left(\frac{\zeta(\mathbf{\Phi}_{\mathbf{n}}^{-1})}{n_{\rm R}} + \frac{\zeta(\mathbf{\Theta}_{\rm T})\,\varphi(\mathbf{\Phi}_{\mathbf{n}}^{-1})}{n_{\rm T}}\right)\frac{n_{\rm R}^2-\zeta(\mathbf{\Theta}_{\rm R})}{n_{\rm R}^2-1} + \frac{\zeta(\mathbf{\Theta}_{\rm R})-1}{n_{\rm R}^2-1}\left(n_{\rm R}\,\varphi(\mathbf{\Phi}_{\mathbf{n}}^{-1}) + \frac{\zeta(\mathbf{\Theta}_{\rm T})\,\zeta(\mathbf{\Phi}_{\mathbf{n}}^{-1})}{n_{\rm T}}\right)}.$$

The empirical eigenvalue distribution of $\mathbf{\Phi}_{\mathbf{n}}^{-1}$ converges almost surely to a non-random limit as $n_{\rm R}\to\infty$ [**20**] and thus

$$\lim_{n_{\rm R}\to\infty}\varphi(\mathbf{\Phi}_{\mathbf{n}}^{-1}) = 1.$$

Defining $\Psi_\infty = \lim_{n_{\rm R}\to\infty}\Psi^{(n_{\rm R})}$, the slope per antenna becomes

$$\lim_{n_{\rm R}\to\infty}\frac{S_0}{n_{\rm R}} = \frac{2}{\Psi_\infty + \frac{\zeta(\mathbf{\Theta}_{\rm T})}{\beta} + \zeta(\mathbf{\Theta}_{\rm R}) - 1}. \tag{H.1}$$

In order to evaluate Ψ_∞, recall that

$$\Psi_\infty = \lim_{n_{\rm R}\to\infty} n_{\rm R}\frac{E[\mathrm{Tr}\{\mathbf{\Phi}_{\mathbf{n}}^{-2}\}]}{E^2[\mathrm{Tr}\{\mathbf{\Phi}_{\mathbf{n}}^{-1}\}]}$$

On one hand,

$$\begin{aligned}
\lim_{n_{\rm R}\to\infty}\frac{E[\mathrm{Tr}\{\mathbf{\Phi}_{\mathbf{n}}^{-1}\}]}{n_{\rm R}} &= (\mathcal{I}_1+\gamma)\lim_{n_{\rm R}\to\infty}\frac{1}{n_{\rm R}}E\left[\mathrm{Tr}\left\{\left(\tfrac{\mathcal{I}_1}{m_1}\mathbf{H}_w(1)\mathbf{H}_w^\dagger(1)+\gamma\mathbf{I}\right)^{-1}\right\}\right] \\
&= (1+\frac{\gamma}{\mathcal{I}_1})\int\frac{1}{(\lambda+\frac{\gamma}{\mathcal{I}_1})}f(\lambda,\tfrac{1}{\beta})d\lambda \\
&= \frac{\beta_1}{2}(1+\tfrac{\mathcal{I}_1}{\gamma})\mathcal{D}(\beta_1,\tfrac{\mathcal{I}_1}{\gamma})
\end{aligned}\tag{H.2}$$

with the function $\mathcal{D}(\cdot,\cdot)$ defined in (6.8). In (H.2), $f(\lambda,\frac{1}{\beta})$ represents the limiting empirical eigenvalue density distribution of $\mathbf{H}_w(1)\mathbf{H}_w^\dagger(1)/m_1$ [**23**]. On the other hand,

$$\begin{aligned}\lim_{n_{\mathrm{R}}\to\infty} \frac{E[\mathrm{Tr}\{\mathbf{\Phi}_{\mathbf{n}}^{-2}\}]}{n_{\mathrm{R}}} &= (1+\tfrac{\gamma}{\mathcal{I}_1})\int \frac{1}{(\lambda+\frac{\gamma}{\mathcal{I}_1})^2} f(\lambda,\tfrac{1}{\beta})d\lambda \\ &= (1+\tfrac{\gamma}{\mathcal{I}_1})\frac{d}{d\gamma}\int_a^b \frac{1}{\lambda+\frac{\gamma}{\mathcal{I}_1}} f(\lambda,\tfrac{1}{\beta})d\lambda\end{aligned}$$

which can be solved to yield, in combination with (H.1) and (H.2), the claims of the proposition.

References

1. D. Chizhik, F. R. Farrokhi, J. Ling, and A. Lozano, *Effect of antenna separation on the capacity of BLAST in correlated channels*, IEEE Commun. Letters **4** (2000), no. 11.
2. C. Chuah, D. Tse, J. Kahn, and R. Valenzuela, *Capacity scaling in dual-antenna-array wireless systems*, IEEE Trans. on Inform. Theory **48** (2002), no. 3, 637–650.
3. G. J. Foschini, *Layered space-time architecture for wireless communications in a fading environment when using multi-element antennas*, Bell Labs Tech. J. (1996), 41–59.
4. G. J. Foschini and M. J. Gans, *On the limits of wireless communications in a fading environment when using multiple antennas*, Wireless Personal Communications (1998), 315–335.
5. D. Gesbert, H. Bolcskei, D. Gore, and A. J. Paulraj, *MIMO wireless channels: Capacity and performance prediction*, Proc. of IEEE GLOBECOM'2000, San Francisco, CA, USA (2000).
6. V. L. Girko, *Theory of random determinants*, Kluwer, 1990.
7. I. S. Gradshteyn and I. M. Ryzhik, *Table of integrals, series, and products*, New York: Academic Press (corrected and enlarged edition), 1980.
8. R. M. Gray, *On the asymptotic eigenvalue distribution of Toeplitz matrices*, IEEE Trans. Inform. Theory **18** (1972), 725–730.
9. S. V. Hanly and D. N. C. Tse, *Resource pooling and effective bandwidth in CDMA networks with multiuser receivers and spatial diversity*, IEEE Trans. Inform. Theory **47** (2001), no. 4, 1328–1351.
10. F. Hiai and D. Petz, *Asymptotic freeness almost everywhere for random matrices*, Acta Sci. Math. (Szeged) **66** (2000), 801–826.
11. W. C. Jakes, *Microwave mobile communications*, New York, IEEE Press, 1974.
12. A. Lozano and A. M. Tulino, *Capacity of multiple-transmit multiple-receive antenna architectures*, IEEE Trans. on Inform. Theory **48** (2002), no. 12.
13. A. Lozano, A. M. Tulino, and S. Verdú, *Multi-antenna capacity in the low-power regime*, Bell Labs Technical Memorandum 10009652-020916-13TM, (also submitted to IEEE Trans. Inform. Theory, Special Issue on Space-Time Coding) (2002).
14. A. Moustakas, H. Baranger, L. Balents, A. Sengupta, and S. Simon, *Communication through a diffusive medium: Coherence and capacity*, Science **287** (2000), 287–290.
15. K. I. Pedersen, J. B. Andersen, J. P. Kermoal, and P. E. Mogensen, *A stochastic multiple-input multiple-output radio channel model for evaluations of space-time coding algorithms*, Proc. IEEE Vehic. Tech. Conf. (VTC'2000 Fall), Boston, MA (2000).
16. G. Raleigh and J. M. Cioffi, *Spatio-temporal coding for wireless communications*, IEEE Trans. on Commun. **46** (1998), no. 3.
17. P. Rapajic and D. Popescu, *Information capacity of a random signature multiple-input multiple-output channel*, IEEE Trans. on Commmun. **48** (2000), no. 8, 1245–1248.
18. D.-S. Shiu, G. J. Foschini, M. J. Gans, and J. M. Kahn, *Fading correlation and its effects on the capacity of multielement antenna systems*, IEEE Trans. on Commun. **48** (2000), no. 3, 502–511.
19. D. Shlyakhtenko, *Random Gaussian band matrices and freeness with amalgamation*, Int. Math. Res. **20** (1996), 1013–1025.
20. J. W. Silverstein and Z. D. Bai, *On the empirical distribution of eigenvalues of a class of large dimensional random matrices*, Journal Multivariate Analysis **54** (1995), 175–192.
21. I. E. Telatar, *Capacity of multi-antenna Gaussian channels*, Eur. Trans. Telecom **10** (1999), 585–595.

22. S. Verdú, *Capacity region of Gaussian CDMA channel: The symbol-synchronous case*, Proc. of 24th Annual Allerton Conf. on Communication, Control, Computing, Monticello, IL (1986), 1025–1034.
23. S. Verdú, *Multiuser detection*, Cambridge University Press, 1998.
24. S. Verdú, *Spectral efficiency in the wideband regime*, IEEE Trans. on Inform. Theory **48** (2002), no. 6, 1319–1343.
25. S. Verdú and S. Shamai, *Spectral efficiency of CDMA with random spreading*, IEEE Trans. on Inform. Theory **45** (1999), no. 3, 622–640.
26. J. H. Winters, *Optimum combining in digital mobile radio with cochannel interference*, J. Sel. Areas in Commun. **2** (1984), no. 4, 528–539.
27. J. Wishart, *The generalized product moment distribution in samples from a normal multivariate population*, Biometrika **20 A** (1928), 32–52.

Universitá Degli Studi di Napoli, 81125 Napoli, Italy
Current address: Princeton University, Princeton, NJ 08540, USA
E-mail address: atulino@ee.princeton.edu

Bell Laboratories (Lucent Technologies), Holmdel, NJ 07733, USA
E-mail address: aloz@lucent.com

Princeton University, Princeton, NJ 08540, USA
E-mail address: verdu@ee.princeton.edu

DIMACS Series in Discrete Mathematics
and Theoretical Computer Science
Volume **62**, 2003

Distribution of MIMO capacity in the presence of correlated signals and interferers: A (not so) large N analysis

Aris L. Moustakas, Steven H. Simon, and Anirvan M. Sengupta

ABSTRACT. The use of multi-antenna arrays in both transmission and reception promises huge increases in the throughput of wireless communication systems. It is therefore important to analyze the capacities of such systems in realistic situations, such as correlated channels and the presence of interference. Here we present an approach that provides analytic expressions for the statistics, i.e. the moments of the distribution of the mutual information of multi-antenna systems with arbitrary correlations and interferers. Although this method is valid formally for large antenna numbers, it produces extremely accurate results even for arrays with as few as two or three antennas. We also develop a method to analytically optimize over the input signal covariance when the channel covariance is available at the receiver and thus to calculate analytic closed-loop capacities with covariance feedback. We apply this analytic approach to a number of important examples and also compare to simulations to establish its validity. This method provides a simple tool to analyze the statistics of throughput for arrays of any size.

1. Introduction

In recent years the use of multi-antenna arrays in both transmission and reception has attracted considerable interest. Theoretical work in [**1**] and [**2**] has shown that for sufficiently rich scattering environments the Shannon capacity of an n_T-element transmitting and an n_R-element receiving array is roughly proportional to $\min(n_R, n_T)$ for large numbers of antennas. To assess the capacity gains using such MIMO (Multiple Input Multiple Output) technologies in realistic situations, several factors have to be taken into account. First, the degree of spatial correlations between antennas at both transmitter and receiver determines the effective number of independent channels. This is especially important in the case of outdoor propagation where the base-station antennas are typically placed on high towers with a few nearby scatterers. In addition, the amount of channel information at both receiver and transmitter may be crucial in achieving increased throughputs. For example, for interference-limited systems, the accuracy of *instantaneous* channel information of the interference at the receiver will determine how effective the

2000 *Mathematics Subject Classification.* Primary 94A15, 94A17, 94A24; Secondary 94A05.
Key words and phrases. Multi-antenna systems, capacity, information theory.

interference suppression will be. Similarly, the amount of channel information at the transmitter may be crucial in achieving increased throughputs. Realistically, full channel information at the transmitter is impractical, due to the associated feedback overload. Instead, partial or statistical channel information, such as the channel covariance may be fed back more efficiently.

For large antenna numbers and Gaussian channel distributions the analysis of MIMO ergodic capacities (expectation value of mutual information over channel realizations) is greatly facilitated by the use of asymptotic techniques of random matrix theory (RMT) techniques. These methods were introduced in this context by various authors, starting with Foschini [**1**] and Telatar [**2**]. Verdu and Shamai [**3**] derived the infinite antenna open-loop capacity expression in the context of CDMA codes and more recently Rapajic and Popescu [**4**] applied to the context of multi-antenna systems. Lozano and Tulino [**5**] using methods developed by Tse and Hanley [**6**] calculated the infinite antenna capacity with uncorrelated channels and uncorrelated interferers. Very recently, Chuah *et al.* [**7**] extended the results of [**3**] to calculate the open-loop capacity of spatially correlated channels in the infinite antenna limit. In all previous studies, the scaling of the capacity with the number of antennas was studied as the number of antennas grows indefinitely. However, for finite antenna numbers or for slowly decaying spatial correlations this scaling is not adequate to describe the finite antenna array capacity. For example, as shown in Fig. 2 of Ref. [**8**], for square arrays with $\lambda/2$ spacing the capacity per antenna does not reach the limiting value even at $10,000$ antennas per array! In Ref. [**8**], using techniques developed by Sengupta and Mitra [**9**] in a different context, a method was developed to analytically calculate the capacity of spatially correlated channels for large but finite antenna numbers.

In this paper, extending work done in [**8, 10**] we analyze the distribution of the MIMO capacity and present analytic expressions for the first few moments of the capacity distribution in the presence of correlated channels and interferers. Specifically, we find that in the formal limit of large antenna numbers, the first, second and third moments of the capacity distribution are of order $O(n)$, $O(1)$ and $O(1/n)$, where n is the number of antennas, strongly suggesting that the capacity distribution is Gaussian. The mathematical framework and the derivation of these expressions is presented elsewhere.[**11**] Although formally these expressions are valid in the large antenna number limit, we verify numerically that they are quite accurate even for just a few (i.e. 2-3) antennas in each array. We also find by comparing to simulations that the capacity distribution is very close to a Gaussian with mean and variance the ergodic average and covariance of the capacity evaluated using the expressions presented in this paper. In addition, we develop a method to analytically optimize over the input signal covariance subject to the channel covariance. This method enables us to derive analytic expressions of the closed loop capacity, when partial knowledge of the channel is available at the transmitter (covariance feedback). This method, although formally valid for large antenna numbers, gives extremely accurate results for all multi-antenna systems, as discussed also in [**12**]. The analytic methods presented here provide a powerful tool for analyzing array systems with even few antennas. In [**11**] the general case of correlated channels, interferers and noise is considered in more detail. Here we focus on two specific situations: First, we analyze the case of correlated transmitting antennas with white noise and no other interference. Second we consider the situation of a multi-antenna interferer with channel matrix partially known at the receiver. In the latter case all channels are assumed i.i.d.

In the next section we provide definitions of important quantities and define the problem we will analyze. In section 3 we present results for the average mutual information, including its optimization over signal covariance to obtain the closed loop capacity with covariance feedback. In section 4 we discuss the two specific examples as mentioned above, presenting results for the mean of the mutual information distribution, its variance and its skewness and we compare our results to simulations.

2. Definitions

We consider the case of single-user transmission from n_T transmit antennas to n_R receive antennas over a narrow band fading channel. We also include a number n_I of interfering transmitters in addition to the noise at each receiver. The received complex signal vector y_α, where $\alpha = 1 \dots n_R$ can be written as

$$y_\alpha = \sqrt{\frac{\rho_S}{n_T}} \sum_{a=1}^{n_T} \mathbf{G}^S_{\alpha a} x^S_a + \sqrt{\frac{\rho_i}{n_I}} \sum_{a=1}^{n_I} \mathbf{G}^i_{\alpha a} x^I_a + z_\alpha \tag{2.1}$$

$\mathbf{G}^S$ is a complex matrix with the channel coefficients from the transmitting to the receiving arrays. Similarly, $\mathbf{G}^i$ describes the channel from the interfering antennas to the receiver array. x^S_a and x^I_a are the transmitted and interfering signals, assumed to be both Gaussian. The signal covariance $\mathbf{Q}$ with entries $Q_{ab} = E\left[x^S_a x^{S*}_b\right]$, is normalized so that $\mathrm{Tr}\left\{\mathbf{Q}\right\} = n_T$. The covariance of the interfering signal x^I_α is taken to be unity. ρ_S and ρ_i are the average signal-to-noise (SNR) and interference-to-noise ratios. The signal to interference and noise ratio can also be written as $SINR = \rho_S/(1+\rho_i)$. Finally, the Gaussian noise vector z_α is assumed to be i.i.d. with entries $E\left[z_\alpha z^*_\beta\right] = \delta_{\alpha\beta}$.

The associated mutual information can be expressed as[**5**] (the derivation follows directly from Appendix A in [**1**])

$$\begin{aligned} I &= I_S - I_I \\ I_S &= \log\det\left(\mathbf{1} + \frac{\rho_i}{n_i}\mathbf{G}^i\mathbf{G}^{i\dagger} + \frac{\rho_S}{n_T}\mathbf{G}^S\mathbf{Q}\mathbf{G}^{S\dagger}\right) \\ I_I &= \log\det\left(\mathbf{1} + \frac{\rho_i}{n_i}\mathbf{G}^i\mathbf{G}^{i\dagger}\right) \end{aligned} \tag{2.2}$$

where $\mathbf{1}$ is the unit matrix. The log above (and throughout the whole paper) represents the natural logarithm and thus I, I_S and I_I are all expressed in nats/sec/Hz. It should be noted that this equation for the mutual information is also applicable to multi-user detection. This case corresponds e.g. to an uplink scenario with the interferer(s) representing other uplink users.

Due to the underlying randomness of $\mathbf{G}^S$ and $\mathbf{G}^i$, the mutual information I is also a random quantity. To analyze its statistics we assume that $\mathbf{G}^S$, $\mathbf{G}^i$ are zero mean Gaussian matrices with covariance

$$\begin{aligned} \langle G^S_{\alpha a} G^{S\ *}_{\beta b} \rangle &= T_{ab}\, R_{\alpha\beta} \\ \langle G^i_{\alpha a} G^{i\ *}_{\beta b} \rangle &= \delta_{ab}\, \delta_{\alpha\beta} \end{aligned} \tag{2.3}$$

and $\langle G^s_{\alpha a} G^i_{\beta b}{}^* \rangle = 0$. The notation $\langle\cdot\rangle$ denotes the ensemble average over channel realizations. $\mathbf{R}$ is the $n_R \times n_R$ correlation matrix of the incoming signal at the receiver, respectively, normalized so that $\mathrm{Tr}\,\{\mathbf{R}\} = n_R$. $\mathbf{T}$ is the $n_T \times n_T$ correlation matrix describing the antenna correlations at the transmitter array with $\mathrm{Tr}\,\{\mathbf{T}\} = n_T$. The simple form of correlations of Eq. 2.3 has been shown[**8**] to be the leading term in a controlled approximation for diffusive environments. The dependence of these matrices on array properties, polarization and channel properties as angle spread have been described in various references[**7, 8, 13**].

In Eq. 2.2 it is assumed that the receiver knows the channel matrices $\mathbf{G}^S$, $\mathbf{G}^i$. The difference between $\mathbf{G}^i\mathbf{G}^{i\dagger}$ and the noise term (the "$\mathbf{1}$" term inside the log det term) is that in the former term the receiver has detailed knowledge of the interfering channel, while in the latter no such knowledge is available.

3. Ergodic Capacity

The purpose of this paper is to analyze the statistics of the mutual information I in Eq. 2.2 for Gaussian channels having correlations given by Eq. 2.3. In this section we analyze the average mutual information $\langle I\rangle_{\mathbf{Q}}$ where the average is taken for fixed $\mathbf{Q}$. We will use the general analytical expression derived in [**11**], which is valid for large antenna numbers. Two ergodic capacities are considered. For the first case no channel information is available at the transmitter. Here, by symmetry the optimal signal covariance $\mathbf{Q}$ is unity—there are no preferred antennas or directions at the transmitter. Thus the open loop ergodic capacity is simply $C_{OL} = \langle I\rangle$, where the subscript $\mathbf{Q}$ will be dropped for simplicity. In contrast, when channel information is available at the transmitter in the form of channel statistics (e.g. $\mathbf{R}$, $\mathbf{T}$), the optimal $\mathbf{Q}$ that maximizes $\langle I\rangle_{\mathbf{Q}}$ may in general be non-trivial. Thus the closed loop ergodic capacity will be $C_{CL} = \max_{\mathbf{Q}, Tr\mathbf{Q}=n_T} \langle I\rangle_{\mathbf{Q}}$. Note that maximization over $\mathbf{Q}$ is performed *after* the average, since only partial (statistical) information is available at the transmitter. This is a realistic model for incorporating closed loop feedback, since full and timely channel knowledge at the transmitter is not possible without overloading the reverse link.

3.1. Average Mutual Information $\langle I\rangle_{\mathbf{Q}}$.

The average mutual information $\langle I\rangle_{\mathbf{Q}}$ in the limit of large antenna numbers is derived in [**11**] and can be expressed as

$$\langle I\rangle_{\mathbf{Q}} = \langle I_S\rangle_{\mathbf{Q}} - \langle I_I\rangle_{\mathbf{Q}} \tag{3.1}$$

where the subscript $\mathbf{Q}$ will subsequently be dropped for simplicity. $\langle I_S\rangle$ is given by

$$\begin{aligned}\langle I_S\rangle &= \log\det\left(\mathbf{1} + \sqrt{\rho_S}\, t^{s1}\, \mathbf{QT}\right) - t^{s1} r^{s1} n_T \\ &+ \log\det\left(\mathbf{1} + \sqrt{\rho_S}\, r^{s1}\, \mathbf{R} + \sqrt{\rho_i}\, r^{s2}\right) \\ &+ n_I \log\left(1 + \sqrt{\rho_i}\, t^{s2}\right) - t^{s2} r^{s2} n_I \end{aligned} \tag{3.2}$$

where the variables r^{s1}, r^{s2}, t^{s1}, and t^{s2} are given by the following system of four equations in four unknowns

$$r^{s1} = \frac{\sqrt{\rho_S}}{n_T} \sum_{a=1}^{n_T} \frac{[\mathbf{TQ}]_a}{1+\sqrt{\rho_S}\, t^{s1}\, [\mathbf{TQ}]_a} \tag{3.3}$$

$$t^{s1} = \frac{\sqrt{\rho_S}}{n_T} \sum_{\alpha=1}^{n_R} \frac{R_\alpha}{1+\sqrt{\rho_S}\, R_\alpha r^{s1} + \sqrt{\rho_i} r^{s2}} \tag{3.4}$$

$$r^{s2} = \frac{\sqrt{\rho_i}}{1+\sqrt{\rho_i}\, t^{s2}} \tag{3.5}$$

$$t^{s2} = \frac{\sqrt{\rho_i}}{n_I} \sum_{\alpha=1}^{n_R} \frac{1}{1+\sqrt{\rho_S}\, R_\alpha r^{s1} + \sqrt{\rho_i} r^{s2}} \tag{3.6}$$

Note that Eqs. 3.3, 3.4 3.6 can be written solely in terms of the eigenvalues of $[\mathbf{T}^s\mathbf{Q}]$ and $\mathbf{R}$, thus simplifying their solution. Similarly $\langle I_I \rangle$ is given by

$$\langle I_I \rangle = n_I \log\left(1+\sqrt{\rho_i}\, t^i\right) + n_R \log\left(1+\sqrt{\rho_i}\, r^i\right) - n_I t^i r^i \tag{3.7}$$

where the variables t^i and r^i are given by

$$\begin{aligned} r^i &= \frac{\sqrt{\rho_i}}{1+\sqrt{\rho_i}\, t^i} \\ t^i &= \frac{n_R\sqrt{\rho_i}}{n_I\left(1+\sqrt{\rho_i} r^i\right)} \end{aligned} \tag{3.8}$$

Solving for r^i and t^i and inserting into Eq. 3.7 we get

$$\langle I_I \rangle = (n_T - n_R)\log u + n_R \log(\rho_S + u) - n_T\left(1-u^{-1}\right) \tag{3.9}$$

with

$$u = \frac{1}{2}\left[1+\rho_S\left(\frac{n_R}{n_T}-1\right) + \sqrt{\left(1+\rho_S\left(\frac{n_R}{n_T}-1\right)\right)^2 + 4\rho_S}\right] \tag{3.10}$$

As expected from the form of I_I in Eq. 2.2, this result is the open-loop capacity in the large-antenna limit for the $n_R \times n_I$ random i.i.d. matrix G^i[**3, 11**].

3.2. Closed Loop Capacity: Optimizing the Signal Covariance Q. As discussed in the beginning of this section, when the transmitter has partial channel information in the form of correlations $\mathbf{T}$, the signal covariance $\mathbf{Q}$ can be optimized to maximize the average mutual information $\max_{\mathbf{Q}}\langle I \rangle_{\mathbf{Q}}$. This method was introduced for MIMO systems without interference in [**8**] and is also discussed in [**10**].

Below we briefly describe the method of optimizing over $\mathbf{Q}$ in the large array limit. We start by observing that before finding the optimal t and r values using Eqs. 3.3-3.6, $\langle I \rangle$ depends on $\mathbf{Q}$ only through the first term in Eq. 3.2. Expressing the determinant of this term in the eigen-basis of $\mathbf{T}$, it can be written as

$$\det\left(\mathbf{1}+\sqrt{\rho_S} t^{s1}\tilde{\mathbf{Q}}\tilde{\mathbf{T}}\right) \tag{3.11}$$

where $\tilde{\mathbf{T}}$ is a diagonal matrix with the eigenvalues of $\mathbf{T}$, T_a for $a = 1, \ldots, n_T$ on the diagonals and $\tilde{\mathbf{Q}}$ the original matrix $\mathbf{Q}$ expressed in the eigen-basis of $\mathbf{T}$. Noting that for any non-negative definite matrix A, $\det(A) \leq \prod_a A_{aa}$, where A_{aa} are the diagonal elements, [**2**] we get

$$\det\left(\mathbf{1} + \sqrt{\rho_S} t^{s1} \tilde{\mathbf{Q}} \tilde{\mathbf{T}}\right) \leq \prod_{a=1}^{n_T} \left(1 + \sqrt{\rho_S} t^{s1} \tilde{Q}_{aa} T_a\right) \tag{3.12}$$

with equality when $\tilde{\mathbf{Q}}$ is diagonal. Thus the $\mathbf{Q}$ maximizing $\langle I \rangle_{\mathbf{Q}}$ is simultaneously diagonalizable with $\mathbf{T}$. This result has been shown [**14**] to hold in general in the absence of interference for arbitrary antenna numbers (but can be generalized to hold in the presence interference). Here we have shown this to be true in the presence of interference, at least within the large antenna number approximation.

Once the optimal eigen-basis of $\mathbf{Q}$ has been determined to be the same as $\mathbf{T}$ one needs to find its optimal eigenvalues q_a for $a = 1, \ldots, n_T$. $\langle I \rangle_{\mathbf{Q}}$ has to be optimized subject to the power constraint $\mathrm{Tr}\mathbf{Q} = n_T$. This constraint is enforced by adding a Lagrange multiplier to $\langle I \rangle_{\mathbf{Q}}$, i.e.

$$\langle I \rangle_{\mathbf{Q}} \to \langle I \rangle_{\mathbf{Q}} - \Lambda \left(\sum_{a=1}^{n_T} q_a - n_T \right) \tag{3.13}$$

The optimal eigenvalues of $\mathbf{Q}$ are then given by

$$q_a = \left[\frac{1}{\Lambda} - \frac{1}{\sqrt{\rho_S}\, T_a t^{s1}} \right]_+ \tag{3.14}$$

where $[x]_+ = x\Theta(x)$ and T_i is the corresponding eigenvalue of $\mathbf{T}$, with $\Theta(x)$ the Heaviside Θ-function. Λ is determined by imposing the power constraint

$$\mathrm{Tr}\left\{\mathbf{Q}\right\} = \sum_{a=1}^{n_T} q_a = n_T \tag{3.15}$$

Essentially, the optimization of $\mathbf{Q}$ amounts to waterfilling over the modes of $\mathbf{T}$ rather than the instantaneous channel $\mathbf{G}^{S\dagger}\mathbf{G}^{S}$ itself. Thus, the statistically waterfilled, closed-loop capacity of the link is found by solving for five unknowns $(t^{s1}, r^{s1}, t^{s2}, r^{s2}, \Lambda)$ from Eqs. 3.3-3.6 and 3.15 and using them to evaluate Eq. 3.2.

4. Applications

We now apply the ergodic capacity equations of the previous section to a number of representative situations with aim to describe the basic characteristics and dependencies of the MIMO capacity for correlated channels and interferers. For concreteness, we assume that the antennas form a uniform linear antenna array and for the cases of correlated antennas we use the following simple model for the correlation coefficient dependence on the antenna separation and angle spread [**13, 15**]

$$c_{kl}(d_\lambda, \delta) = \int_{-\pi}^{\pi} \frac{d\phi}{\sqrt{2\pi\delta^2}} \exp\left[i2\pi(k-l)d_\lambda \sin(\phi)\right] \exp\left[-\frac{\phi^2}{2\delta^2}\right] \tag{4.1}$$

where $d_\lambda = d_{min}/\lambda$ is the nearest neighbor antenna spacing in wavelengths, δ is the angle spread in radians measured from the vertical to the array. $k, l = 1, \dots n$ are the indices of the antennas.

4.1. Closed loop vs. open loop capacity for spatially correlated transmission. We first analyze the benefits of statistical waterfilling, i.e. knowing $\mathbf{T}$ at the transmitter. For simplicity we neglect interferers and set $\rho_{\tilde{i}} = 0$. In addition we assume that $\mathbf{R} = \mathbf{1}_{n_R}$. This situation corresponds to a receiving mobile array deep inside the clutter and a transmitting base-station array with correlations due to finite angle spread.

4.1.1. *Ergodic Capacity.* In this case Eq. 3.1 becomes

$$\langle I \rangle = \sum_{a=1}^{n_T} \log\left(1 + \sqrt{\rho_S} t^{s1} q_a T_a\right) + n_R \log\left(1 + \sqrt{\rho_S} r^{s1}\right) - n_T t^{s1} r^{s1} \tag{4.2}$$

To get the closed loop capacity we need to optimize over r^{s1}, t^{s1} and q_a. After standard manipulation we need to solve for r^{s1} below numerically

$$\frac{n_R r^{s1} \sqrt{\rho_S}}{1 + \sqrt{\rho_S}} = \sum_{i=a}^{n_T} \left[1 - \frac{r^{s1}}{\sqrt{\rho_S} T_a}\right]_+ \tag{4.3}$$

and then

$$t^{s1} = \frac{n_R \sqrt{\rho_S}}{n_T \left(1 + \sqrt{\rho_S} r^{s1}\right)} \tag{4.4}$$

$$q_a = \left[\frac{1}{t^{s1} r^{s1}} - \frac{1}{t^{s1} \sqrt{\rho_S} T_a}\right]_+ \tag{4.5}$$

and then substitute the results into Eq. 4.2. Interestingly, applying these equations to find the optimality of beamforming (i.e. the stability of the solution with one non-zero q_a) produces extremely accurate results when compared with the exact criterion for the optimality of beamforming [**12**].

For the open loop case, $q_a \equiv 1$, so that one needs to solve only for t^{s1} and r^{s1} using Eq. 4.4 and

$$\frac{t^{s1}}{\sqrt{\rho_S}} = \frac{n_R}{n_T} - 1 + \frac{1}{n_T} \sum_{a=1}^{n_T} \frac{1}{1 + \sqrt{\rho_S} t^{s1} T_a} \tag{4.6}$$

We now wish to quantify the throughput gain due to channel feedback to the transmitter in the form of $\mathbf{T}$. In Fig. 1 the dependence of the open loop and closed loop capacity is plotted as a function of antenna spacing for three transmitting and receiving array configurations. It is interesting to point out that for the low SNR and angle-spread used, the closed loop capacity is substantially higher than the open loop capacity and in fact it is higher than the open loop capacity of uncorrelated channels. Furthermore, we note that the closed loop capacity is not convex and, especially for small arrays, has apparent discontinuities in its slope. This effect is due to the fact that at certain points (marked with solid dots in the figure) it is optimal for additional transmission modes q_a to become non-zero (see Eq. 4.5). This non-convex and even non-monotonic behavior of the closed loop capacity as a function of correlations has been shown to be present in exact solutions of multi-antenna arrays with few antennas in [**12**]. Finally the analytic solutions presented here are in very good agreement with simulated ergodic capacities (depicted with open circles, diamonds and asterisks in the figure).

4.1.2. *Higher Moments.* Once t^{s1}, r^{s1} and q_a have been determined from Eqs. 4.3, 4.4 and 4.6, the higher moments can be evaluated in a straightforward manner. Details on the derivation of the equations below are presented elsewhere.[**11**] Defining the following matrix

$$\mathbf{M_2} = \begin{pmatrix} -M_{r,2} & -1 \\ -1 & -M_{t,2} \end{pmatrix} \tag{4.7}$$

where, for general $p \geq 2$

$$M_{t,p} = \left(\frac{\rho_S}{n_T}\right)^{p/2} \sum_{a=1}^{n_T} \left[\frac{q_a T_a}{1 + \sqrt{\rho_S}\, q_a T_a t^{s1}}\right]^p \tag{4.8}$$

and

$$M_{r,p} = \left(\frac{\rho_S}{n_T}\right)^{p/2} \sum_{\alpha=1}^{n_R} \left[\frac{R_\alpha}{1 + \sqrt{\rho_S}\, R_\alpha r^{s1}}\right]^p \tag{4.9}$$

the variance $Var(I)$ can be expressed simply as:

$$\begin{aligned} Var(I) &= -\log|\det \mathbf{M_2}| \\ &= -\log|1 - M_{t,2}M_{r,2}| \end{aligned} \tag{4.10}$$

Similarly, the skewness can be expressed as:

$$\begin{aligned} Sk(I) &= \frac{4}{\det \mathbf{M_2}^3}\left(M_{t,3}^2 M_{r,2}^3 + M_{r,3}^2 M_{t,2}^3 - 2M_{t,3}M_{r,3}\right) \\ &- \frac{3}{\det \mathbf{M_2}^2}\left(M_{t,2}^2 M_{r,4} + M_{r,2}^2 M_{t,4} + 2M_{t,3}M_{r,3}\right) \end{aligned} \tag{4.11}$$

Generally, while the average mutual information is $O(n)$, where n is the order of the number of antennas in each array, the variance is $O(1)$ and the skewness is $O(1/n)$. In fact, all higher moments in this asymptotic expansion are much smaller ($O(1/n^2)$ or smaller). This suggests that for large antenna numbers the capacity distribution can be described accurately by its mean and its variance. This means that the capacity distribution approaches a Gaussian. Of course, in principle for few antennas this is not expected to be true, since all moments will be of the same order. However we find that even in this case the variance together with the average describe the distribution very accurately. We see this by comparing numerically the Gaussian distribution with average $\langle I\rangle_{\mathbf{Q}}$ and variance $Var(I)$ calculated using Eqs. 4.2 and 4.10 with the simulated distribution resulting from the generation of 10^6 random matrix realizations.

We consider the case of a correlated transmitting array and an uncorrelated receiver array, both with 3 elements. The correlations of the transmitter are the same as in section 4.1 and Fig. 1 for antenna spacing $d_\lambda = 1$. We compare the derived distribution with mean capacity and its variance calculated analytically as above, with the distribution resulting from the generation of 10^6 random matrix realizations with the proper statistics. Fig. 2 shows the results, which indicate an impressive agreement. We also compare the closed loop capacity with covariance feedback with the full closed loop capacity, in which case the instantaneous channel matrix is known at the transmitter. We see that the mean throughput of the two closed loop schemes are within 5% of each other even for low signal to noise ratio.

4.2. Capacity distribution in the presence of interference known at the receiver. In this section we briefly discuss the case of a MIMO system in the presence of interferers. For simplicity we assume that all channel coefficients are i.i.d.

4.2.1. *Ergodic Capacity.* In this case the ergodic capacity of Eq. 3.1 simplifies to $\langle I\rangle = \langle I_S\rangle - \langle I_I\rangle$, where $\langle I_I\rangle$ is given by Eq. 3.9 and $\langle I_S\rangle$ becomes:

$$\begin{aligned}\langle I_S\rangle &= n_T\left(\log\frac{\sqrt{\rho_S}}{r^{s1}} + \frac{r^{s1}}{\sqrt{\rho_S}} - 1\right) + n_I\left(\log\frac{\sqrt{\rho_i}}{r^{s2}} + \frac{r^{s2}}{\sqrt{\rho_i}} - 1\right)\\ &+ n_R\log\left[1+\sqrt{\rho_i}r^{s2}+\sqrt{\rho_S}r^{s1}\right] \qquad (4.12)\end{aligned}$$

with r^{s1}, r^{s2} given by

$$\frac{n_R}{1+\sqrt{\rho_i}r^{s2}+\sqrt{\rho_S}r^{s1}} = \frac{n_I}{\sqrt{\rho_i}}\left(\frac{1}{r^{s2}} - \frac{1}{\sqrt{\rho_i}}\right) = \frac{n_T}{\sqrt{\rho_S}}\left(\frac{1}{r^{s1}} - \frac{1}{\sqrt{\rho_S}}\right) \qquad (4.13)$$

The above equations can be reduced to a simple cubic equation.

4.2.2. *Higher Moments.* The variance, which can be written in a straight forward manner in terms of r^{s1}, r^{s2} and u of Eq. 3.10 as:

$$\begin{aligned}Var(I) = &- \log\left|1-\frac{n_T}{n_R}\left(1-\frac{r^{s1}}{\sqrt{\rho_S}}\right)^2 - \frac{n_I}{n_R}\left(1-\frac{r^{s2}}{\sqrt{\rho_i}}\right)^2\right|\\ &- \log\left|1-\frac{n_R}{n_I}\left(\frac{\rho_i}{u+\rho_i}\right)^2\right| + 2\log\left|1+\frac{\rho_i}{u+\rho_i}\left(1-\frac{r^{s2}}{\sqrt{\rho_i}}\right)\right| \qquad (4.14)\end{aligned}$$

To compare the above equations with simulations we set for simplicity all arrays equal in size $n_T = n_R = n_I = n$ and consider $n = 2, 3$. Fig. 3 displays the cumulative distribution of the I for these two systems, for signal to interference and noise ratio $SINR = \rho_S/(1+\rho_i) = 10$ and interference to noise ratio $\rho_i = 1$. Also displayed are the simulated distributions for the same cases. The agreement between analytic and simulated distributions is decent for two antennas and impressive for three (less than 1% difference). This analysis suggests that for a few number of antennas the mutual information average and its variance are sufficient to describe the full distribution.

5. Conclusion

In conclusion, we have presented an analytic approach to calculate the statistics of the mutual information of MIMO systems in the presence of spatially correlated channels, and with interference with known channel at the receiver. In addition, we have used this method to find the optimal signal covariance $\mathbf{Q}$ and thus analytically calculate the closed loop ergodic capacity covariance feedback is available at the transmitter. The ergodic capacity was calculated to demonstrate the applicability of this approach. These methods, although formally valid for large antenna numbers, apply with very high accuracy to arrays with only few numbers of antennas. To demonstrate this we have simulated the distributions of small antenna arrays with spatial correlations and in the presence of interferers and have compared them with the corresponding distributions that we get by calculating the mean mutual information and its variance.

This analytic approach provides the framework and a simple tool to accurately analyze the statistics of throughput of even small arrays in the presence of arbitrary channel correlations, as well as interferers with known channel at the receiver.[**11**]

References

[1] G. J. Foschini and M. J. Gans, "On limits of wireless communications in a fading environment when using multiple antennas," *Wireless Personal Communications*, vol. 6, pp. 311–335, 1988.

[2] I. E. Telatar, "Capacity of multi-antenna gaussian channels," *European Transactions on Telecommunications and Related Technoilogies*, vol. 10, no. 6, pp. 585–596, November 1999.

[3] S. Verdú and S. Shamai, "Spectral efficiency of CDMA with random spreading," *IEEE Trans. Inform. Theory*, vol. 45, no. 2, pp. 622–640, March 1999.

[4] P. B. Rapajic and D. Popescu, "Information capacity of a random signature multiple-input multiple-output chanel," *IEEE Trans. Commun.*, vol. 48, no. 8, p. 1245, August 2000.

[5] A. Lozano and A. M. Tulino, "Capacity of multiple-transmit multiple-receive antenna architectures," *IEEE Trans. Inform. Theory*, 2002.

[6] D. N. Tse and S. V. Hanly, "Linear multiuser receivers: Effective interference, effective bandwidth and user capacity," *IEEE Trans. Inform. Theory*, vol. 45, no. 2, p. 641, March 1999.

[7] C. N. Chuah, D. Tse, J. Kahn, and R. A. Valenzuela, "Capacity scaling in MIMO wireless systems under correlated fading," *IEEE Trans. Inform. Theory*, vol. 48, no. 3, p. 637, March 2002.

[8] A. L. Moustakas, H. U. Baranger, L. Balents, A. M. Sengupta, and S. H. Simon, "Communication through a diffusive medium: Coherence and capacity," *Science*, vol. 287, pp. 287–290, January 2000.

[9] A. M. Sengupta and P. P. Mitra, "Distributions of singular values for some random matrices," *Phys. Rev. E*, vol. 60, no. 3, pp. 3389–3392, 1999.

[10] ——, "Capacity of multivariate channels with multiplicative noise: I. random matrix techniques and large-n expansions for full transfer matrices," *LANL arXiv:physics/0010081*, October 2000.

[11] A. L. Moustakas, S. H. Simon, and A. M. Sengupta, "MIMO capacity through correlated channels in the presence of correlated interferers and noise: A (not so) large N analysis," *accepted for publication in IEEE Trans. of Inform. Theory, also available at http://mars.bell-labs.com/*, 2003.

[12] S. H. Simon and A. L. Moustakas, "Optimizing MIMO systems with channel covariance feedback," *IEEE J. Select. Areas Commun.*, vol. 21, no. 3, April 2003.

[13] D. Chizhik, F. Rashid-Farrokhi, J. Ling, and A. Lozano, "Effect of antenna separation on the capacity of BLAST in correlated channels," *IEEE Commun. Lett.*, vol. 4, no. 11, pp. 337–339, November 2000.

[14] S. A. Jafar and A. Goldsmith, "Channel capacity and beamforming for multiple transmit and receive antennas with covariance feedback," *Preprint*, 2001.

[15] R. M. Buehrer, "Generalized equations for spatial correlation for low to moderate angle-spread," *Proc. of 10th annual Virginia Tech Symposium on Wireless Personal Communications*, pp. 70–77, July 2000.

Lucent Technologies, Bell Labs, 67 Whippany Rd, Whippany, NJ 07981
E-mail address: arislm@lucent.com

Lucent Technologies, Bell Labs, 600 Mountain Ave, Murray Hill, NJ 07974
E-mail address: shsimon@lucent.com

Lucent Technologies, Bell Labs, 600 Mountain Ave, Murray Hill, NJ 07974
Current address: Physics Dept, Rutgers Univ, 136 Frelinghuysen Rd, Piscataway, NJ 08854
E-mail address: anirvans@physics.rutgers.edu

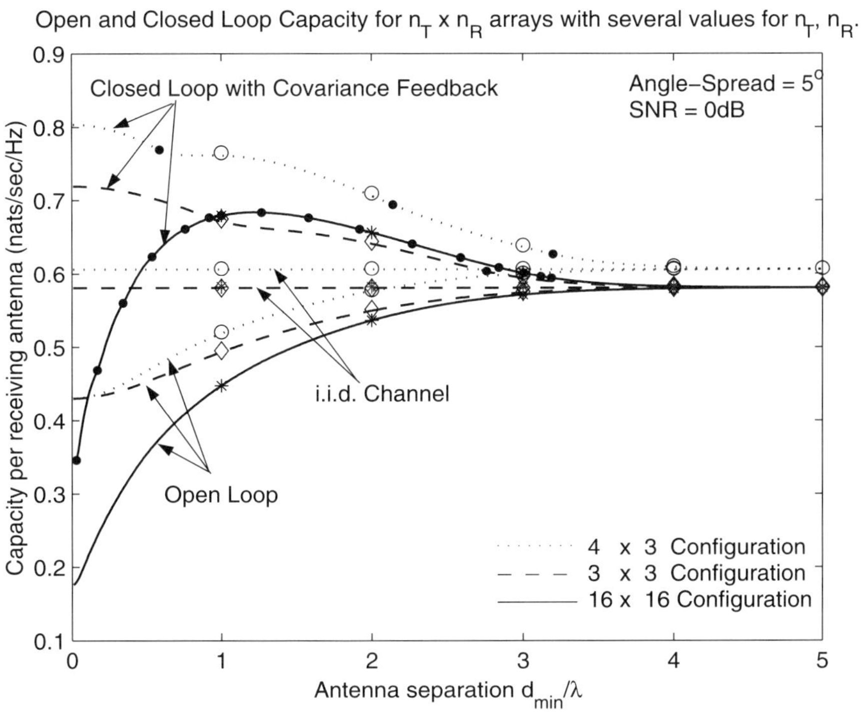

FIGURE 1. Plot of capacity per receiving antenna for 3 different transmitting array-receiving array configurations. These three configurations are a 16-element transmitting array to a 16-element receiving array, a 3-element transmitting array to a 3-element receiving array and a 4-element transmitting array to a 3-element receiving array. In all cases the receiving array has no antenna correlations ($\mathbf{R} = \mathbf{1}$). The transmitting array is a uniformly spaced array with neighboring antennas spaced by d_{min}. The spatial correlations between antennas are calculated from Eq. 4.1 with angle-spread $\delta = 5^o$. The three lower curves indicated as "Open Loop" depict the open loop capacity per receiving antenna for the three configurations. At small antenna separation the 4×3 and 3×3 capacities approach each other. The top three curves indicated as "Closed Loop ..." represent the closed loop capacities with covariance feedback: the transmitter adjusts to the *typical* channel. The dots represent the values of antenna separation at which an additional non-zero transmitting mode (q_a in Eq. 4.5) becomes optimal. Especially for the smaller arrays these points coincide to apparent kinks in the capacity. The horizontal straight lines are the capacities per antenna in the absence of any antenna correlations ($\mathbf{T} = \mathbf{R} = \mathbf{1}$). Note the slight increase in the capacity per receiving antenna when the transmitter array is increased by one antenna. The open circles, diamonds and asterisks represent the values of the simulated mean capacities of the 4×3, 3×3 and 16×16 configurations correspondingly, evaluated for a representative number of antenna spacings using 10^6 channel realizations. We see that the agreement with analytical results is very good.

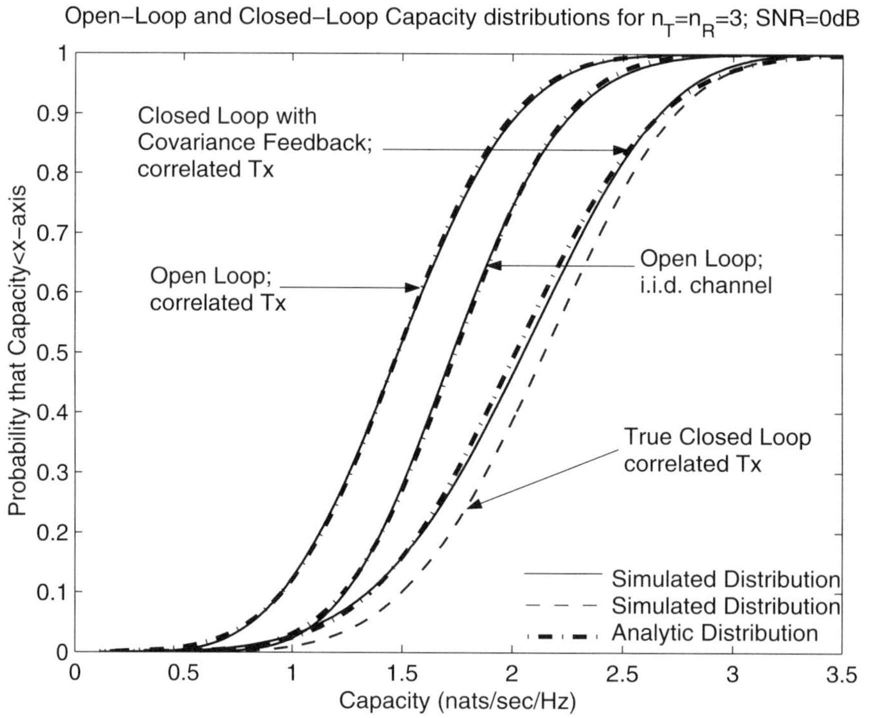

FIGURE 2. Cumulative distribution (CDF) of mutual information for 3 antenna transmitting and receiving arrays in the absence of interferers ($\rho_i = 0$). Each closely spaced pair of curves corresponds to a simulated mutual information distribution (solid curves) and a normal distribution (dash-dotted curves) with mean and variance calculated from Eqs. 4.2 and 4.10. The agreement between analytic estimate and simulated distributions is impressive. This is especially so given that $n_T = n_R = 3$ while the analytic method described in this paper is formally valid for large antenna numbers. The three pairs of curves plotted correspond to the three corresponding curves plotted in Fig. 1 with $n_T = n_R = 3$, and $SNR = 0dB$, $d_\lambda = 1$ and $\delta = 5^o$. The open loop curve assumes no feedback to the transmitter, while the closed loop curve takes into account optimal covariance feedback. As an example, the analytically calculated skewness from Eq. 4.11 for the closed loop, open loop and i.i.d. cases for $SNR = 10dB$ are $Sk = -0.15$, $Sk = -0.12$ and $Sk = 0.05$, respectively, which are all small. For comparison, the simulated skewness for the i.i.d. case is $Sk_{sim} = 0.074 \pm 0.007$. The dashed curve represents the full closed loop capacity distribution, where the full correlated channel matrix is fed back to the transmitter. The mean throughput of the two closed loop schemes are within 5% of each other.

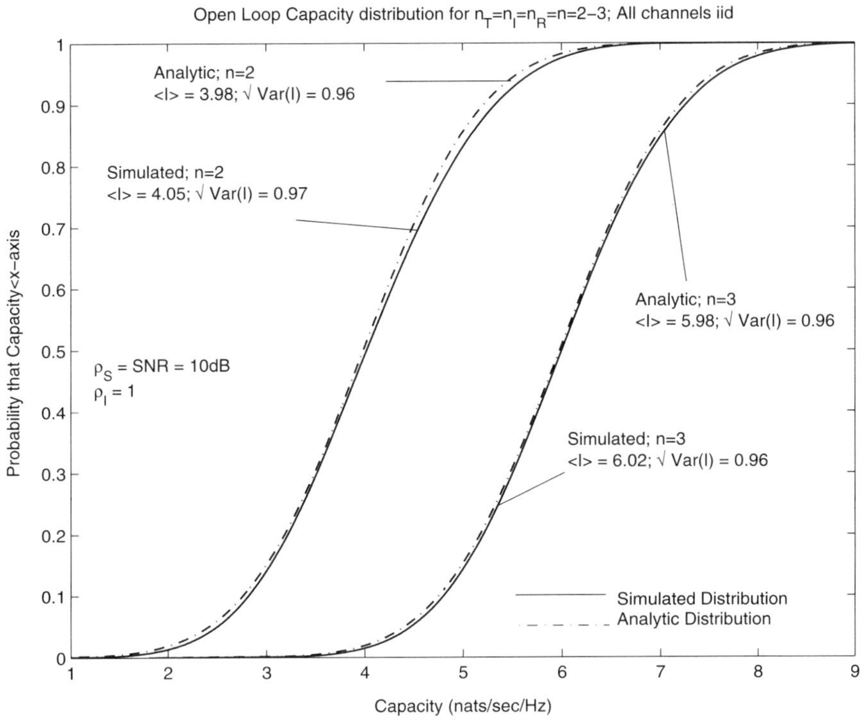

FIGURE 3. Cumulative distribution (CDF) of the mutual information for 2- and 3-antenna arrays for all transmitter, receiver and interferer with i.i.d. signal and interfering channels ($\mathbf{R} = \mathbf{T} = \mathbf{1}$). For both $n = 2$ and $n = 3$ we plot the CDF of the simulated mutual information (solid lines) and a Gaussian distribution with average and variance analytically calculated from Eq. 3.9, 4.12 and Eq. 4.14. We also display the average $\langle I \rangle$ and the square root of the variance Var(I) of the simulated curves. The error bar in the simulated values of the variance is approximately 2%. Despite the relatively large variance the analytical results acquired from a large antenna number approximation agree quite well with simulations.

DIMACS Series in Discrete Mathematics
and Theoretical Computer Science
Volume **62**, 2003

Optimality of Beamforming in Multiple Transmitter Multiple Receiver Communication Systems with Partial Channel Knowledge

Steven H. Simon and Aris L. Moustakas

ABSTRACT. In this paper we consider a narrowband point-to-point MIMO communication system with n_{T} transmit antennas and n_{R} receive antennas, and we assume that the receiver has perfect knowledge of the channel. We consider two canonical cases of partial channel feedback: (a) where the mean channel is fed back to the transmitter and the channel covariance is assumed to be white with known magnitude, and (b) the covariance (assumed factorizable) of the channel is fed back to transmitter, and we assume the mean channel is zero. In both cases we are able to derive an exact analytic condition (both necessary and sufficient) that determines when beamforming is optimal for maximizing the average throughput.

1. Introduction

The large capacities available to communication systems employing multiple antennas has been the subject of much research in recent years. Interesting questions arise in cases where the receiver has perfect knowledge of the channel (as might be achieved by pilots in practice) but the transmitter has only partial information about the channel (as would have to be gained from some sort of feedback). It has been seen in many cases that exploiting such partial channel information at the transmitter can increase link capacities substantially. In this paper we will focus on two canonical cases of channel feedback : (a) mean feedback, where the mean of the channel is known to the transmitter and the covariance of the channel is assumed to be white with known magnitude, and (b) covariance feedback, where the full (factorizable) covariance of the channel is known and the average channel is assumed to be zero. These two cases appear to be realistic and have been considered before in a number of publications[**1, 2, 3**]. In this paper, we focus on the question of whether beamforming transmission strategies are optimal for maximizing the ergodic (average) channel capacity given the partial channel knowledge. (Some very limited results have been previously obtained on the harder question

2000 *Mathematics Subject Classification.* Primary 94A15, 94A17; Secondary 94A05.
Key words and phrases. Transmit Diversity; Scheduling; Multiple Antennas, Side Information.

of whether beamforming transmission is optimal for maximizing some particular outage capacity[**4, 5**]).

The question of beamforming optimality has been addressed in a number of publications. Much of this work has been concerned with the simple case where there are $n_{\rm T}$ transmit antennas and a single receive antenna[**6, 7, 8, 4**]. More recently certain limited results have been obtained for the more general case with multiple transmit and multiple receive antennas. Jafar et al[**3**] considered the case of covariance feedback in the special case where the channel is assumed to have i.i.d. rows and correlated columns, and managed to derive upper and lower bounds on the optimality condition for beamforming. Later, the current authors[**5**] derived the exact optimality condition in this special case. In the present paper we shall obtain a result that will apply to a general channel covariance which can be written as a direct product of arbitrarily correlated columns (transmitting antenna elements) and correlated rows (receiving antenna elements). In addition, we will also obtain an exact beamforming optimality condition for the case of mean channel feedback where the covariance of the channel is assumed to be white.

2. Definitions

In the narrowband MIMO problem - $n_{\rm T}$ transmitters and $n_{\rm R}$ receivers - the channel is defined by

$$y_\alpha = \sum_{i=1}^{n} G^*_{i\alpha} x_i + \eta_\alpha \tag{1}$$

where y_α is α^{th} component of the vector of $n_{\rm R}$ complex received signals, x_i is the ith component of the $n_{\rm R}$ dimensional complex vector of transmitted signal, $G^*_{i\alpha}$ is the complex channel from the ith transmitter to the αth receiver, and η_α is the complex noise at the αth receiver. We will try to keep the notation that italic roman indices $i, j, \ldots$ represent the transmitter antennas and take values $1 \ldots n_{\rm T}$ whereas Greek indices $\alpha, \beta, \ldots$ represent the receiver end and go from $1, \ldots, n_{\rm R}$. In this paper, boldface quantities denote matrices or vectors (hopefully usage will make clear which one is which) so we can also write $\mathbf{y} = \mathbf{G}^\dagger \mathbf{x} + \boldsymbol{\eta}$ (with the $\dagger$ indicating transpose-conjugate). We will assume that the noise is an i.i.d. Gaussian random vector, $E\{\eta^*_\alpha \eta_\beta\} = \nu \delta_{\alpha\beta}$ where $E\{\cdot\}$ denotes the expectation value, ν is the noise power per receiver, and δ is the Kronecker delta function. (Generalizing the calculations below to a nontrivial noise correlation matrix (in place of the δ) is trivial in the covariance feedback problem, but is nontrivial for mean feedback). For simplicity, we assume in this paper that the noise power per receiver is of unit amplitude $\nu = 1$ (all other quantities need to be measured in units of the noise power).

The transmitted signal covariance matrix is given by $E\{x^*_i x_j\} = Q_{ij}$ with Q_{ij} a non-negative definite Hermitian matrix, with eigenvalues q_i, for $i = 1, \ldots, n_{\rm T}$. The constraint of fixed total transmitted power P is expressed in terms of $\mathbf{Q}$ as $\mathrm{Tr}\{\mathbf{Q}\} = \sum_{i=1}^{n_{\rm T}} q_i = P$.

For a given instantiation of $\mathbf{G}$, if the receiver knows the channel (which in practice is achieved by transmitting pilot signals), the mutual information $I[(\mathbf{y}, \mathbf{G}); \mathbf{x}]$ is given by

$$I = \log\det(\mathbf{1} + \mathbf{G}^\dagger \mathbf{Q} \mathbf{G}) = \mathrm{Tr}\log(\mathbf{1} + \mathbf{G}^\dagger \mathbf{Q} \mathbf{G}) \tag{2}$$

where $\mathbf{1}$ is the identity matrix. Throughout this paper we will measure information in nats/sec/Hz where one nat is equal to e bits ($e = 2.718\ldots$). That is, the log in this equation is defined to be a natural log.

The quantity we are concerned with is the so-called ergodic or mean average of this information over instantiations of the channel within some ensemble of channels which we write as $\langle I \rangle$. We will write brackets $\langle\rangle$ to denote an average over instantiations in a given ensemble of instantiations. We would like to maximize $\langle I \rangle$ over all possible transmission covariances $\mathbf{Q}$. More specifically, in this paper, we would like to know when $\mathbf{Q}$ being of rank one (ie, a beamforming solution) is optimal, and what that particular beamforming solution is.

3. Concavity and Optimality Condition

An essential part of the following calculations rely on the concavity of the mutual information (Eq. 2) with respect to the transmission covariance. To show this, we vary the transmission covariance slightly by writing

$$\mathbf{Q} = \mathbf{Q}^0 + \epsilon \mathbf{Q}^1 \tag{3}$$

where both $\mathbf{Q}$ and $\mathbf{Q}_0$ are nonnegative definite Hermitian matrices with $\mathrm{Tr}\mathbf{Q} = \mathrm{Tr}\mathbf{Q}^0 = P$, and $\mathbf{Q}^1$ has trace zero. We can then write Eq. 2 as

$$\begin{aligned} I &= \mathrm{Tr}\log(\mathbf{1} + \mathbf{G}^\dagger[\mathbf{Q}^0 + \epsilon\mathbf{Q}^1]\mathbf{G}) \\ &= \mathrm{Tr}\log\left([\mathbf{1} + \mathbf{G}^\dagger\mathbf{Q}^0\mathbf{G}][\mathbf{1} + (\mathbf{1} + \mathbf{G}^\dagger\mathbf{Q}^0\mathbf{G})^{-1}\epsilon\mathbf{G}^\dagger\mathbf{Q}^1\mathbf{G})]\right) \\ &= I_0 + \mathrm{Tr}\log\left([\mathbf{1} + (\mathbf{1} + \mathbf{G}^\dagger\mathbf{Q}^0\mathbf{G})^{-1}\epsilon\mathbf{G}^\dagger\mathbf{Q}^1\mathbf{G})]\right) \end{aligned}$$

where $I_0 = \mathrm{Tr}\log(\mathbf{1} + \mathbf{G}^\dagger\mathbf{Q}^0\mathbf{G})$ is the value of I at $\epsilon = 0$. We now expand to second order in ϵ to obtain

$$\begin{aligned} I &= I_0 + \epsilon\mathrm{Tr}\left[(\mathbf{1} + \mathbf{G}^\dagger\mathbf{Q}^0\mathbf{G})^{-1}\mathbf{G}^\dagger\mathbf{Q}^1\mathbf{G}\right] \\ &- \frac{\epsilon^2}{2}\mathrm{Tr}\left[(\mathbf{1} + \mathbf{G}^\dagger\mathbf{Q}^0\mathbf{G})^{-1}\mathbf{G}^\dagger\mathbf{Q}^1\mathbf{G}(\mathbf{1} + \mathbf{G}^\dagger\mathbf{Q}^0\mathbf{G})^{-1}\mathbf{G}^\dagger\mathbf{Q}^1\mathbf{G}\right] + \ldots . \end{aligned}$$

We see that the trace in the second order term is of the form $\mathrm{Tr}[\mathbf{MNMN}]$ where both $\mathbf{M}$ and $\mathbf{N}$ are Hermitian matrices. Using the cyclic property of the trace, this term can be rewritten as $\mathrm{Tr}[(\sqrt{\mathbf{M}}\,\mathbf{N}\sqrt{\mathbf{M}}\,)^2]$ which is nonnegative definite. Therefore, we have the mutual information being a concave function of $\mathbf{Q}$ at every allowable value of $\mathbf{Q}$. Thus, the average of the mutual information over any ensemble is also everywhere concave.

Since the mutual information is everywhere concave, the optimality condition for maximizing the average of the mutual information can be written as follows: A nonnegative definite Hermitian matrix $\mathbf{Q}^0$ is optimal if and only if

$$\frac{d\langle I\rangle}{d\epsilon} = \left\langle \mathrm{Tr}\left[(\mathbf{1} + \mathbf{G}^\dagger\mathbf{Q}^0\mathbf{G})^{-1}\mathbf{G}^\dagger\mathbf{Q}^1\mathbf{G}\right]\right\rangle \leq 0 \tag{4}$$

for all zero trace Hermitian matrices $\mathbf{Q}^1$ such that $Q^0 + \epsilon Q_1$ remains nonnegative definite for infinitesimal ϵ. Here, we are simply saying that any allowable small change of $\mathbf{Q}^0$ at fixed power decreases the mutual information (allowable means that the new $\mathbf{Q}$ is still Hermitian and nonnegative definite). The concavity property tells us that optimality with respect to small changes in $\mathbf{Q}$ implies global optimality. Note that the $\langle\rangle$ brackets in Eq. 4 represent average over the ensemble of channels. The derived optimality condition will, of course, depend on the definition of the ensemble.

If $\mathbf{Q}^0$ corresponds to beamforming (ie, has rank one) we can use the identities of Appendix A to reduce the optimality condition to

$$\frac{d\langle I\rangle}{d\epsilon} = \left\langle \text{Tr}\left[\mathbf{G}^\dagger\mathbf{Q}^1\mathbf{G}\right] - \frac{\text{Tr}\left[\mathbf{G}^\dagger\mathbf{Q}^0\mathbf{G}\mathbf{G}^\dagger\mathbf{Q}^1\mathbf{G}\right]}{1+\text{Tr}[\mathbf{G}^\dagger\mathbf{Q}^0\mathbf{G}]}\right\rangle \leq 0. \tag{5}$$

The calculational difficulty here is that the channel matrix appears in the denominator. The first step in attacking this problem is to move the denominator into the exponent to give the following equivalent beamforming optimality condition

$$\left\langle \text{Tr}\left[\mathbf{G}^\dagger\mathbf{Q}^1\mathbf{G}\right] - \int_0^\infty dx\; e^{-x(1+\text{Tr}[\mathbf{G}^\dagger\mathbf{Q}^0\mathbf{G}])}\;\text{Tr}\left[\mathbf{G}^\dagger\mathbf{Q}^0\mathbf{G}\mathbf{G}^\dagger\mathbf{Q}^1\mathbf{G}\right]\right\rangle \leq 0 \tag{6}$$

which will be our calculational starting point below.

4. Covariance Feedback

If the covariance of the channel is fed back to the transmitter, we then assume a channel model where $\mathbf{G}$ has zero mean and is normally distributed with that given covariance. In this paper we assume a factorizable form of the covariance $\mathbf{T}\otimes\mathbf{R}$ with $\mathbf{T}$ being an n_{T} dimensional nonnegative definite Hermitian matrix and $\mathbf{R}$ being an n_{R} dimensional nonnegative definite Hermitian matrix (so that $\mathbf{T}\otimes\mathbf{R}$ is an $n_{\text{T}}n_{\text{R}}$ dimensional matrix, see Eq. 8 below). This factorizable form has been used by many authors[**1, 2**] and has been found to be a good representation of MIMO channels in highly scattering environments[**2**]. Further, it can be shown that this factorizable form is the first term in a controlled approximation for diffusive environments[**1**]. The matrices $\mathbf{T}$ and $\mathbf{R}$ become nontrivial when there are non-negligible antenna correlations at the transmitter or receiver respectively. We will sometimes use the notation that T_i and R_α are the eigenvalues of the matrices $\mathbf{T}$ and $\mathbf{R}$.

Thus we have $\mathbf{G}$ being $\mathcal{N}(\mathbf{0},\mathbf{T}\otimes\mathbf{R})$ such that

$$\langle\mathbf{G}\rangle = 0 \tag{7}$$

and

$$\left\langle G_{i\alpha}G^*_{j\beta}\right\rangle = R_{\alpha\beta}T_{ij} \quad . \tag{8}$$

Equivalently, we could write have written $\mathbf{G} = \sqrt{\mathbf{T}}\;\mathbf{Z}^{(iid)}\sqrt{\mathbf{R}}$ where $\mathbf{Z}$ is an i.i.d. Gaussian random matrix with unit covariance.

The ensemble average of any function F of $\mathbf{G}$, which we write as $\langle F(\mathbf{G})\rangle$, can be written more explicitly as

$$\langle F(\mathbf{G})\rangle = \int d\mu(\mathbf{G})e^{-\text{Tr}[\mathbf{G}^\dagger\mathbf{T}^{-1}\mathbf{G}\mathbf{R}^{-1}]}\,F(\mathbf{G}) \tag{9}$$

which is shorthand for

$$\frac{1}{\det[\mathbf{R}\otimes\mathbf{T}]}\prod_{i=1}^{n_{\text{T}}}\prod_{\alpha=1}^{n_{\text{R}}}\left[\frac{1}{\pi}\int d\text{Re}G_{i\alpha}\int d\text{Im}G_{i\alpha}\right]e^{-\text{Tr}[\mathbf{G}^\dagger\mathbf{T}^{-1}\mathbf{G}\mathbf{R}^{-1}]}\,F(\mathbf{G}) \tag{10}$$

We also note that the integration measure is correctly defined so that $\langle 1\rangle = 1$, and so that Eq. 8 is satisfied.

For this situation of covariance feedback, we will use to our advantage the fact that the optimal transmission covariance $\mathbf{Q}$ must be diagonal in the same basis as $\mathbf{T}$ (ie, $\mathbf{Q}$ and $\mathbf{T}$ have the same eigenvectors). This fact has been proven previously in references [**3, 5**]. Although these particular proofs considered the case of $\mathbf{R} = \mathbf{1}$, it is easy to see that they apply equally well to the case of nontrivial $\mathbf{R}$. Furthermore, it is proven in these works that an optimal $\mathbf{Q}$ must have values along the diagonal that are ordered in the same manner as the eigenvalues of $\mathbf{T}$. In other words the eigenvector corresponding to the largest eigenvalue of $\mathbf{T}$ is also the eigenvector corresponding to the largest eigenvalue of $\mathbf{Q}$. Similarly, the eigenvector corresponding to the second largest eigenvalue of $\mathbf{T}$ is also the eigenvector corresponding to the second largest eigenvalue of $\mathbf{Q}$; and so forth. Thus, if a rank one (beamforming) $\mathbf{Q}^0$ is optimal, then it must be of the form

$$\mathbf{Q}^0 = P\,\mathbf{t}_1 \otimes \mathbf{t}_1^\dagger \tag{11}$$

where $\mathbf{t}_1$ is the normalized n_{T}-dimensional eigenvector of the matrix $\mathbf{T}$ corresponding to the largest eigenvalue T_1 of this matrix. (In more explicit notation $Q^0_{ij} = P[t_1]_i[t_1]^*_j$).

Since the optimal $\mathbf{Q}$ must be diagonal in the $\mathbf{T}$ basis, in the beamforming condition Eq. 6, one need only consider $\mathbf{Q}^1$'s that are also diagonal in this basis. Furthermore, since we must have $\mathbf{Q}^0 + \epsilon\mathbf{Q}^1$ nonnegative definite, we need only consider $\mathbf{Q}^1$'s of the form (recall $\mathbf{Q}^1$ must be traceless)

$$\mathbf{Q}^1 = \sum_{i=2}^{n_{\mathrm{T}}} q_i \left\{ \mathbf{t}_i \otimes \mathbf{t}_i^\dagger - \mathbf{t}_1 \otimes \mathbf{t}_1^\dagger \right\} \tag{12}$$

where $\mathbf{t}^i$ are the eigenvectors of $\mathbf{T}$ and all $q_i > 0$.

Working in a basis where both $\mathbf{R}$ and $\mathbf{T}$ are diagonal (and $\mathbf{Q}^0$ and $\mathbf{Q}^1$ therefore), and using the above forms (Eq. 11 and 12) for $\mathbf{Q}^0$ and $\mathbf{Q}^1$, the optimality condition Eq. 6) is rewritten as

$$\begin{aligned} &\left\langle \mathrm{Tr}\left[\mathbf{G}^\dagger \mathbf{Q}^1 \mathbf{G}\right]\right\rangle = \left\langle \sum_{\alpha=1}^{n_{\mathrm{R}}} \sum_{j=2}^{n_{\mathrm{T}}} q_j (|G_{j\alpha}|^2 - |G_{1\alpha}|^2) \right\rangle \leq \\ &\left\langle \int_0^\infty dx\; e^{-x - Px\sum_{\alpha=1}^{n_{\mathrm{R}}} |G_{1\alpha}|^2} \sum_{\alpha,\beta=1}^{n_{\mathrm{R}}} P G^*_{1\alpha} G_{1\beta} \sum_{j=2}^{n_{\mathrm{T}}} q_j \left[G^*_{j\beta} G_{j\alpha} - G^*_{1\beta} G_{1\alpha}\right] \right\rangle . \end{aligned} \tag{13}$$

(Here we have used the fact that the covariance of $\mathbf{G}$ is diagonal in both j and α). Taking some of the averages, can be reduced to

$$\begin{aligned} &\sum_{\alpha=1}^{n_{\mathrm{R}}} \sum_{j=2}^{n_{\mathrm{T}}} q_j (T_j R_\alpha - T_1 R_\alpha) \leq \\ &\left\langle P \int_0^\infty e^{-x} \sum_{j=2}^{n_{\mathrm{T}}} q_j \left(\sum_{\alpha=1}^{n_{\mathrm{R}}} |G_{1\alpha}|^2 T_j R_\alpha - \sum_{\alpha,\beta=1}^{n_{\mathrm{R}}} \frac{d^2}{d(Px)^2} \right) e^{-Px\sum_{\alpha=1}^{n_{\mathrm{R}}} |G_{1\alpha}|^2} \right\rangle . \end{aligned} \tag{14}$$

Writing out the expectation on the right hand side explicitly as an integral we obtain

$$\begin{aligned} \mathrm{Tr}[\mathbf{R}] \sum_{j=2}^{n_{\mathrm{T}}} q_j (T_j - T_1) &\leq P \int d\mu(\mathbf{G}) e^{-\mathrm{Tr}[\mathbf{R}^{-1}\mathbf{G}^\dagger \mathbf{T}^{-1}\mathbf{G}]} \\ &\int_0^\infty dx\; e^{-x} \sum_{j=2}^{n_{\mathrm{T}}} q_j \left(\sum_{\alpha=1}^{n_{\mathrm{R}}} |G_{1\alpha}|^2 T_j R_\alpha - \frac{d^2}{d(Px)^2} \right) e^{-Px\sum_{\alpha=1}^{n_{\mathrm{R}}} |G_{1\alpha}|^2} \end{aligned} \tag{15}$$

Writing in the exponent

$$\mathrm{Tr}[\mathbf{R}^{-1}\mathbf{G}^{\dagger}\mathbf{T}^{-1}\mathbf{G}] = \sum_{\alpha=1}^{n_{\mathrm{R}}}\sum_{j=1}^{n_{\mathrm{T}}} |G_{j\alpha}|^2 T_j^{-1} R_\alpha^{-1} \tag{16}$$

the integrals over $\mathbf{G}$ can be performed explictly to yield

$$\begin{aligned}\mathrm{Tr}[\mathbf{R}] \sum_{i=2}^{n_{\mathrm{T}}} q_i (T_i - T_1) \leq \\ P\int_0^\infty dx\; e^{-x} \sum_{j=2}^{n_{\mathrm{T}}} q_j \left\{ \sum_{\alpha=1}^{n_{\mathrm{R}}} \frac{T_j R_\alpha}{T_1^{-1}R_\alpha^{-1} + Px} - \frac{d^2}{d(Px)^2}\right\} \prod_{\gamma=1}^{n_{\mathrm{R}}} \left[\frac{T_1^{-1}R_\gamma^{-1}}{T_1^{-1}R_\gamma^{-1} + Px}\right].\end{aligned} \tag{17}$$

There is only a single appearance of T_j on the right hand side for $j \neq 1$. The magnitude of this term can be easily bounded to be (see Appendix B)

$$P\int_0^\infty dx\; e^{-x} \prod_{\gamma=1}^{n_{\mathrm{R}}} \left[\frac{T_1^{-1}R_\gamma^{-1}}{T_1^{-1}R_\gamma^{-1} + Px}\right] \sum_{\alpha=1}^{n_{\mathrm{R}}} \frac{R_\alpha}{T_1^{-1}R_\alpha^{-1} + Px} < \mathrm{Tr}\mathbf{R} \tag{18}$$

which then allows us to symbolically write Eq. 17 as

$$\sum_{i=2}^{n_{\mathrm{T}}} q_i [A T_i - C] \leq 0 \tag{19}$$

with $A > 0$ due to the inequality Eq. 18 (where all dependence on q_i and T_i for $i \geq 2$ are explicit here). Since $T_2 \geq T_3 \geq T_4 \ldots$, using this form, one can see that if the the above optimality condition (Eq. 17) is true for the case of $q_2 = 1$ and all other q_i's equal to zero, then it is true for any selection of q_i's (recall that all $q_i > 0$) . It is then sufficient to only consider this one case of $q_2 = 1$. This is simply the statement that beamforming optimality is completely determined by testing whether transferring a small amount of weight from the strongest eigendirection (the $\mathbf{t}_1$ direction) to the next strongest direction (the $\mathbf{t}_2$ direction or q_2 direction) increases capacity or not (This principle has been deduced previously[**3, 5**]). Thus setting $q_2 = 1$ and all other $q_i = 0$, Eq. 17 is further simplified by integrating by parts twice (paying careful attention to the boundary terms) to give

$$T_2 \mathrm{Tr}[\mathbf{R}] - \frac{1}{P} \leq P\int_0^\infty dx\; e^{-x} \left\{ \sum_{\alpha=1}^{n_{\mathrm{R}}} \frac{T_2 R_\alpha}{T_1^{-1}R_\alpha^{-1} + Px} - \frac{1}{P^2}\right\} \prod_{\gamma=1}^{n_{\mathrm{R}}} \left[\frac{T_1^{-1}R_\gamma^{-1}}{T_1^{-1}R_\gamma^{-1} + Px}\right]. \tag{20}$$

For the simple case where $\mathbf{R} = \mathbf{1}$, a case that can be solved[**5**] with other methods (and previously studied by [**3**]), one can evaluate these integrals explicitly using the fact that for $A/P > 0$ we have [**9**]

$$P\int_0^\infty dx \frac{e^{-x}}{(A + Px)^m} = e^{A/P} P^{(1-m)} \Gamma[-m+1, A/P] \tag{21}$$

with Γ the incomplete gamma function. With some algebra, using the identity[**9**] $\Gamma[m+1, x] = m\Gamma[m, x] + x^m e^{-x}$ the result can be simplified to the trivial form

$$\frac{e^{-1/(T_1 P)} T_2 P (T_1 P)^{n_{\mathrm{R}}}}{1 + T_2 P} \leq \Gamma\left[-n_{\mathrm{R}}, 1/(T_1 P)\right]. \tag{22}$$

which agrees with previous derived results[**5**] for the special case of $\mathbf{R} = \mathbf{1}$. The current derivation, however, is simpler and far more general.

In the more general case where $\mathbf{R} \neq \mathbf{1}$, the integrals in Eq. 20 can still be performed analytically using Eq. 21 after decomposing products into partial fractions. We write

$$\prod_{\gamma=1}^{n_\mathrm{R}} \left[\frac{T_1^{-1} R_\gamma^{-1}}{T_1^{-1} R_\gamma^{-1} + Px} \right] = \frac{1}{\det[\mathbf{R}]} \sum_{\gamma=1}^{n_\mathrm{R}} \frac{T_1^{-1}}{T_1^{-1} R_\gamma^{-1} + Px} \prod_{\beta=1, \beta \neq \gamma}^{n_\mathrm{R}} \left[\frac{1}{R_\beta^{-1} - R_\gamma^{-1}} \right] \tag{23}$$

which enables us to write Eq. 20 as

$$\begin{aligned} T_2 \mathrm{Tr}[\mathbf{R}] - \frac{1}{P} \quad \leq \quad & \frac{PT_1^{-1}}{\det[\mathbf{R}]} \int_0^\infty dx \; e^{-x} \sum_{\gamma=1}^{n_\mathrm{R}} \left(\prod_{\beta=1, \beta\neq\gamma}^{n_\mathrm{R}} \left[\frac{1}{R_\beta^{-1} - R_\gamma^{-1}} \right] \right) \\ & \left[\frac{T_2 R_\gamma}{(T_1^{-1} R_\gamma^{-1} + Px)^2} - \frac{1}{P^2 (T_1^{-1} R_\gamma^{-1} + Px)} \right. \\ & \left. + \sum_{\alpha=1, \alpha\neq\gamma}^{n_\mathrm{R}} \frac{T_2 R_\alpha}{(T_1^{-1} R_\gamma^{-1} + Px)(T_1^{-1} R_\alpha^{-1} + Px)} \right]. \end{aligned} \tag{24}$$

Writing the last term as

$$\begin{aligned} & \frac{1}{(T_1^{-1} R_\gamma^{-1} + Px)(T_1^{-1} R_\alpha^{-1} + Px)} = \\ & \frac{T_1}{R_\alpha^{-1} - R_\gamma^{-1}} \left[\frac{1}{(T_1^{-1} R_\gamma^{-1} + Px)} - \frac{1}{(T_1^{-1} R_\alpha^{-1} + Px)} \right] \end{aligned}$$

and using Eq. 21 we then simplify Eq. 24 to

(25)

$$\begin{aligned} PT_2 \mathrm{Tr}[\mathbf{R}] - 1 \leq & \frac{1}{PT_1 \det[\mathbf{R}]} \sum_{\gamma=1}^{n_\mathrm{R}} \left(\prod_{\beta=1, \beta\neq\gamma}^{n_\mathrm{R}} \left[\frac{1}{R_\beta^{-1} - R_\gamma^{-1}} \right] \right) \\ & \left[e^{1/(T_1 R_\gamma P)} \left(PR_\gamma T_2 \, \Gamma[-1, 1/(T_1 R_\gamma P)] - \Gamma[0, 1/(T_1 R_\gamma P)] \right) \right. \\ & \left. + \sum_{\alpha \neq \gamma} \frac{P^2 T_1 T_2 R_\alpha}{R_\alpha^{-1} - R_\gamma^{-1}} \left(e^{1/(T_1 R_\gamma P)} \, \Gamma[0, 1/(T_1 R_\gamma P)] - e^{1/(T_1 R_\alpha P)} \, \Gamma[0, 1/(T_1 R_\alpha P)] \right) \right] \end{aligned}$$

which is our final analytic result for the general optimality condition for beamforming in a MIMO system with covariance feedback (where T_1 and T_2 are the largest and second largest eigenvalues of the transmission correlation matrix $\mathbf{T}$, and R_γ are the eigenvalues of the receiver correlation matrix $\mathbf{R}$). It is interesting to note that the beamforming condition at the transmitter depends not only on the eigenvalues of the correlation matrix at the transmitter T, but in general it also depends on the correlations of the receiving array.

5. Mean Feedback

If the mean of the channel is fed back to the transmitter, we then assume a channel model that has nonzero mean $\mathbf{H}$ and white covariance around this mean with some amplitude σ^2 which we assume is also known by the transmitter. Thus we have $\mathbf{G}$ being $\mathcal{N}(\mathbf{H}, \sigma^2 \mathbf{1} \otimes \mathbf{1})$ such that

$$\langle \mathbf{G} \rangle = \mathbf{H} \tag{26}$$

and

$$\langle (G_{i\alpha} - H_{i\alpha})(G_{j\beta} - H_{j\beta})^* \rangle = \sigma^2 \delta_{ij} \delta_{\alpha\beta} \tag{27}$$

Equivalently, we could write

$$\mathbf{G} = \mathbf{H} + \sigma \mathbf{Z}^{(iid)} \tag{28}$$

where $\mathbf{Z}$ is an i.i.d. Gaussian random matrix with unit covariance.

Again, the ensemble average of any function F of $\mathbf{G}$ we write as $\langle F(\mathbf{G}) \rangle$ can be written more explicitly as

$$\langle F(\mathbf{G}) \rangle = \int d\mu(\mathbf{Z}) e^{-\mathrm{Tr}[\mathbf{Z}^\dagger \mathbf{Z}]} F(\mathbf{H} + \sigma \mathbf{Z}) \tag{29}$$

which here is shorthand for

$$\prod_{i=1}^{n_\mathrm{T}} \prod_{\alpha=1}^{n_\mathrm{R}} \left[\frac{1}{\pi} \int d\mathrm{Re} Z_{i\alpha} \int d\mathrm{Im} Z_{i\alpha} \right] e^{-\mathrm{Tr}[\mathbf{Z}^\dagger \mathbf{Z}]} F(\mathbf{H} + \sigma \mathbf{Z}) \tag{30}$$

We note again that the integration measure is correctly defined so that $\langle 1 \rangle = 1$ and Eq. 27 is satisfied.

We now rewrite the beamforming optimality condition Eq. 6 as

$$\begin{aligned} \left\langle \mathrm{Tr}\left[(\mathbf{H} + \sigma \mathbf{Z})^\dagger \mathbf{Q}^1 (\mathbf{H} + \sigma \mathbf{Z}) \right] \right\rangle \le \Bigg\langle \int_0^\infty & dx \; e^{-x(1+\mathrm{Tr}[(\mathbf{H}+\sigma\mathbf{Z})^\dagger \mathbf{Q}^0 (\mathbf{H}+\sigma\mathbf{Z})]} \\ & \mathrm{Tr}\left[(\mathbf{H} + \sigma \mathbf{Z})^\dagger \mathbf{Q}^0 (\mathbf{H} + \sigma \mathbf{Z}) (\mathbf{H} + \sigma \mathbf{Z})^\dagger \mathbf{Q}^1 (\mathbf{H} + \sigma \mathbf{Z}) \right] \Bigg\rangle \end{aligned}$$

or

$$\begin{aligned} \mathrm{Tr}\left[\mathbf{H}^\dagger \mathbf{Q}^1 \mathbf{H} \right] \;\; \le \;\; & \int_0^\infty dx \int d\mu(\mathbf{Z}) e^{-\mathrm{Tr}[\mathbf{Z}^\dagger \mathbf{Z}] - x(1+\mathrm{Tr}[(\mathbf{H}+\sigma\mathbf{Z})^\dagger \mathbf{Q}^0 (\mathbf{H}+\sigma\mathbf{Z})]} \\ & \mathrm{Tr}\left[(\mathbf{H} + \sigma \mathbf{Z})^\dagger \mathbf{Q}^0 (\mathbf{H} + \sigma \mathbf{Z}) (\mathbf{H} + \sigma \mathbf{Z})^\dagger \mathbf{Q}^1 (\mathbf{H} + \sigma \mathbf{Z}) \right] \end{aligned}$$

where on the left hand side we have used the fact that $\mathrm{Tr}\mathbf{Q}^1 = 0$. With some algebraic manipulation we massage this into

$$\begin{aligned} \mathrm{Tr}\left[\mathbf{H}^\dagger \mathbf{Q}^1 \mathbf{H} \right] \;\; \le \;\; & \int_0^\infty dx \, e^{-x(1+\mathrm{Tr}[\mathbf{H}^\dagger \mathbf{Q}^0 \mathbf{M} \mathbf{H}])} \\ & \int d\mu(\mathbf{Z}) e^{-\mathrm{Tr}[\{\mathbf{Z}^\dagger + \mathbf{H}^\dagger x \sigma \mathbf{Q}^0 \mathbf{M}\} \mathbf{M}^{-1} \{\mathbf{Z} + \mathbf{M} x \sigma \mathbf{Q}^0 \mathbf{H}\}]} \\ & \mathrm{Tr}\left[(\mathbf{H} + \sigma \mathbf{Z})^\dagger \mathbf{Q}^0 (\mathbf{H} + \sigma \mathbf{Z}) (\mathbf{H} + \sigma \mathbf{Z})^\dagger \mathbf{Q}^1 (\mathbf{H} + \sigma \mathbf{Z}) \right] \end{aligned}$$

where we have defined the Hermitian matrix (see appendix A)

$$\mathbf{M} = (\mathbf{1} + x\sigma^2 \mathbf{Q}^0)^{-1} = \mathbf{1} - \frac{x\sigma^2 \mathbf{Q}^0}{1 + x\sigma^2 \mathrm{Tr}[\mathbf{Q}^0].} \tag{31}$$

Making the change of variables

$$\tilde{\mathbf{Z}} = \mathbf{Z} + \mathbf{M} x \sigma \mathbf{Q}^0 \mathbf{H} \tag{32}$$

this then becomes

$$
\text{Tr}\left[\mathbf{H}^\dagger\mathbf{Q}^1\mathbf{H}\right] \leq \int_0^\infty dx\, e^{-x(1+\text{Tr}[\mathbf{H}^\dagger\mathbf{Q}^0\mathbf{M}\mathbf{H}])}\int d\mu(\tilde{\mathbf{Z}})e^{-\text{Tr}[\tilde{\mathbf{Z}}^\dagger\mathbf{M}^{-1}\tilde{\mathbf{Z}}]}
$$
$$
\text{Tr}\left[(\mathbf{H}^\dagger\mathbf{M}+\sigma\tilde{\mathbf{Z}}^\dagger)\mathbf{Q}^0(\mathbf{M}\mathbf{H}+\sigma\tilde{\mathbf{Z}})(\mathbf{H}^\dagger\mathbf{M}+\sigma\tilde{\mathbf{Z}}^\dagger)\mathbf{Q}^1(\mathbf{M}\mathbf{H}+\sigma\tilde{\mathbf{Z}})\right]. \tag{33}
$$

This can now be integrated using Wick's theorem[**10, 11**] to yield the expression

$$
\text{Tr}\left[\mathbf{H}^\dagger\mathbf{Q}^1\mathbf{H}\right] \leq \int_0^\infty dx\, \frac{e^{-x(1+\text{Tr}[\mathbf{H}^\dagger\mathbf{Q}^0\mathbf{M}\mathbf{H}])}}{\det[\mathbf{M}^{-1}]}\Big\{ \text{Tr}\left[\mathbf{H}^\dagger\mathbf{M}\mathbf{Q}^0\mathbf{M}\mathbf{H}\mathbf{H}^\dagger\mathbf{M}\mathbf{Q}^1\mathbf{M}\mathbf{H}\right]
$$
$$
+\ \sigma^2\text{Tr}\left[\mathbf{Q}^0\mathbf{M}\right]\text{Tr}\left[\mathbf{H}^\dagger\mathbf{M}\mathbf{Q}^1\mathbf{M}\mathbf{H}\right]+\sigma^2\text{Tr}\left[\mathbf{Q}^1\mathbf{M}\right]\text{Tr}\left[\mathbf{H}^\dagger\mathbf{M}\mathbf{Q}^0\mathbf{M}\mathbf{H}\right]
$$
$$
+\ \sigma^2 n_{\text{R}}\text{Tr}\left[\mathbf{M}\mathbf{Q}^0\mathbf{M}\mathbf{H}\mathbf{H}^\dagger\mathbf{M}\mathbf{Q}^1\right]+\sigma^2 n_{\text{R}}\text{Tr}\left[\mathbf{M}\mathbf{Q}^1\mathbf{M}\mathbf{H}\mathbf{H}^\dagger\mathbf{M}\mathbf{Q}^0\right]
$$
$$
+\ \sigma^4 n_{\text{R}}\text{Tr}\left[\mathbf{Q}^0\mathbf{M}\right]\text{Tr}\left[\mathbf{Q}^1\mathbf{M}\right]+\sigma^4 {n_{\text{R}}}^2\text{Tr}\left[\mathbf{Q}^0\mathbf{M}\mathbf{Q}^1\mathbf{M}\right]\Big\}.
$$

We now define H_i to be the eigenvalues of $\mathbf{H}\mathbf{H}^\dagger$ and $\mathbf{h}_i$ to be the corresponding eigenvectors. We also assume the eigenvalues are ordered $H_1 > H_2 > \ldots$. Analogous to above covariance feedback case, the beamforming solution for this mean feedback case should be of the form

$$
\mathbf{Q}^0 = P\mathbf{h}_1 \otimes \mathbf{h}_1^\dagger \quad . \tag{34}
$$

We now write $\mathbf{Q}^1$ in the general form

$$
\mathbf{Q}^1 = \sum_{i=2}^{n_{\text{T}}} q_i\left\{\mathbf{h}_i\otimes\mathbf{h}_i^\dagger - \mathbf{h}_1\otimes\mathbf{h}_1^\dagger\right\} + \sum_{i,j=1;i\neq j}^{n_{\text{T}}} q_{ij}\mathbf{h}_i\otimes\mathbf{h}_j^\dagger \tag{35}
$$

with $q_{ij} = q_{ji}^*$. (Note that we have not taken $\mathbf{Q}^1$ to be diagonal yet for generality. Nonetheless, the restriction $q_i > 0$ still applies to assure that $\mathbf{Q}^0 + \epsilon\mathbf{Q}^1$ remains nonnegative definite).

Using this form, the above expression can be simplified (using the results of Appendix A) to yield

$$
\sum_{i=2}^{n_{\text{T}}} q_i(H_i - H_1) \leq P\int_0^\infty dx \frac{e^{-x[1+H_1P/(1+x\sigma^2P)]}}{(1+x\sigma^2P)^2}
$$
$$
\sum_{i=2}^{n_{\text{T}}} q_i\left\{\sigma^2 H_i - \frac{H_1^2}{(1+x\sigma^2P)^3} + \frac{\sigma^2H_1[x\sigma^2P-(1+2n_{\text{R}})]}{(1+x\sigma^2P)^2} + \frac{\sigma^4n_{\text{R}}[x\sigma^2P-n_{\text{R}}]}{1+x\sigma^2P}\right\}. \tag{36}
$$

Note that the off-diagonal terms of $\mathbf{Q}^1$ don't enter and therefore need not be considered (as one might have guessed).

Analogous to the argument given near Eq. 18, we use the inequality

$$
\sigma^2P\int_0^\infty dx\frac{e^{-x[1+H_1P/(1+x\sigma^2P)]}}{(1+x\sigma^2P)^2} \leq \int_0^\infty dx\frac{e^{-x}\sigma^2P}{(1+x\sigma^2P)^2} \leq 1 \tag{37}
$$

(where the last inequality is demonstrated in Eq. 44) to again allow us to symbolically write Eq. 36 as

$$\sum_{i=2}^{n_{\rm T}} q_i(AH_i - C) \leq 0 \tag{38}$$

with $A > 0$. Once again, we see that this inequality holds for arbitrary $\mathbf{Q}^1$ if and only if it holds for the case of $q_2 = 1$ and all other $q_i = 0$. As with the above case, we can then reduce our beamforming optimality condition to

$$\begin{aligned}(H_2 - H_1) \leq P \int_0^\infty dx \frac{e^{-x[1+H_1P/(1+x\sigma^2P)]}}{(1+x\sigma^2P)^2} \\ \left\{\sigma^2 H_2 - \frac{H_1^2}{(1+x\sigma^2P)^3} + \frac{\sigma^2 H_1[x\sigma^2P-(1+2n_{\rm R})]}{(1+x\sigma^2P)^2} + \frac{\sigma^4 n_{\rm R}[x\sigma^2P - n_{\rm R}]}{1+x\sigma^2P}\right\}\end{aligned} \tag{39}$$

which is our final analytic result for optimality of beamforming in the mean feedback case. (Here H_1 and H_2 are the largest and second largest eigenvalues of the mean channel squared $\mathbf{HH}^\dagger$, and σ is the fluctuation of the channel around this average).

6. Summary

We have shown how one can determine whether or not beamforming is optimal for MIMO systems when partial information about the channel is known at the transmitter. We have given analytic results for the cases of channel covariance feedback and mean channel feedback. Both of these cases seem realizable for practical systems. Further discussion and interpretation of these results will be presented in a forthcoming paper.

Appendix A. Identities for Rank One Matrices

A rank one Hermitian matrix $\mathbf{M}$ can be written as $M_{ij} = v_i v_j^*$. We then have the identity

$$\mathbf{M}^2 = \mathbf{M}\,\mathrm{Tr}\mathbf{M} \tag{40}$$

which is easily seen since $\sum_j M_{ij}M_{jk} = v_i v_k^* \sum_j |v_j|^2 = M_{ik}\mathrm{Tr}\mathbf{M}$. Using this result, we can establish that

$$(\mathbf{1}+\mathbf{M})^{-1} = \mathbf{1} - \frac{\mathbf{M}}{1+\mathrm{Tr}\mathbf{M}} \tag{41}$$

which can easily be checked since

$$(\mathbf{1}+\mathbf{M})\left[\mathbf{1} - \frac{\mathbf{M}}{1+\mathrm{Tr}\mathbf{M}}\right] = \mathbf{1}+\mathbf{M} - \frac{\mathbf{M}+\mathbf{M}^2}{1+\mathrm{Tr}\mathbf{M}} = \mathbf{1} \tag{42}$$

Finally, we note another useful identity that

$$\det[\mathbf{1}+\mathbf{M}] = 1 + \mathrm{Tr}\mathbf{M} \tag{43}$$

which is most easily proven by simply working in a basis where $\mathbf{M}$ is diagonal.

Appendix B. A Bound

$$\begin{aligned} & P\int_0^\infty dx\, e^{-x}\prod_{\gamma=1}^{n_R}\left[\frac{T_1^{-1}R_\gamma^{-1}}{T_1^{-1}R_\gamma^{-1}+Px}\right]\frac{R_\alpha}{T_1^{-1}R_\alpha^{-1}+Px} \\ < \; & P\int_0^\infty dx\, e^{-x}\frac{T_1^{-1}}{(T_1^{-1}R_\alpha^{-1}+Px)^2} \\ = \; & R_\alpha\int_0^\infty dx\, e^{-x}\frac{PR_\alpha T_1}{(1+PxR_\alpha T_1)^2} < R_\alpha \end{aligned}$$

The last inequality follows since by integration by parts

$$\begin{aligned} \int_0^\infty dx\frac{e^{-x}A}{(1+Ax)^2} &= 1-\int_0^\infty dx\frac{e^{-x}}{1+Ax} \\ &= \int_0^\infty dx\frac{e^{-x}Ax}{1+Ax} < \int_0^\infty dx\, e^{-x} = 1 \end{aligned} \tag{44}$$

with the inequality holding for $A > 0$.

References

[1] A. L. Moustakas *et al.*, "Communication through a diffusive medium: Coherence and capacity," *Science*, vol. 287, pp. 287–290, January 2000.

[2] F. Ling, D. Chizhik, F. Rashid-Fanokhi, and A. Lozano, "Antenna separation and capacity of blast in correlated channels," *IEEE-APS Conference on Antennas and Propagation for Wireless Communication*, pp. 183–185, 2000.

[3] S. Vishwanath S. A. Jafar and A. Goldsmith, "Channel capacity and beamforming for multiple transmit and receive antennas with covariance feedback," *ICC2001, Helsinki Finland*, 2001.

[4] S. H. Simon and A. L. Moustakas, "Optimizing MISO systems with partial channel knowledge," *submitted to IEEE Trans. of Inform Theory*, submitted. *Also available from http://mars.bell-labs.com.*

[5] S. H. Simon and A. L. Moustakas, "Optimizing MIMO antenna systems with channel covariance feedback," *IEEE Journal on Selected Areas in Communications*, to be published. *Also available from http://mars.bell-labs.com.*

[6] A. Narula *et al.*, "Efficient use of side information in multiple-antenna data transmission over fading channels," *IEEE Journal on Selected Areas in Communications*, vol. 16, no. 8, pp. 1423–1436, October 1998.

[7] E. Visotsky and U. Madhow, "Space-time transmit precoding with imperfect feedback," *IEEE Trans. of Inform Theory*, vol. 47, no. 6, pp. 2632–2639, Sept 2001.

[8] A. J. Syed and A. J. Goldsmith, "On optimality of beamforming for multiple antenna systems with imperfect feedback," *International Symposium on Information Theory (ISIT) 2001.*

[9] I. S. Gradshteyn and I. M. Ryzhik, *Table of Integrals, Series, and Products*, Academic Press, London, 5 edition, 1994.

[10] L. H. Ryder, *Quantum Field Theory*, Cambridge University Press, 1985.

[11] A. L. Moustakas, A. Sengupta, and S. H. Simon, "MIMO Capacity Distribution in the Presence of Correlated Signal, Interference and Noise : A (not so) Large N analyis" *submitted to IEEE Trans. of Inform Theory. Also Available at http://mars.bell-labs.com, May 2002.*

Lucent Technologies, Murray Hill NJ
E-mail address: shsimon@lucent.com

Lucent Technologies, Whippany NJ
E-mail address: arislm@lucent.com

Part II

Capacity of Broadcast Channels and Dirty-Paper Coding

DIMACS Series in Discrete Mathematics
and Theoretical Computer Science
Volume **62**, 2003

Duality, Dirty Paper Coding, and Capacity for Multiuser Wireless Channels

Nihar Jindal, Sriram Vishwanath, Syed Jafar, and Andrea Goldsmith

ABSTRACT. We introduce techniques that shed new light on the capacity analysis of multiuser channels, and apply these techniques to broadcast and multiple access channels. We first determine a duality between broadcast and multiple access channels that can be used to obtain the capacity region and optimal transmission strategy for one channel based on the capacity-achieving transmission strategy and region for the dual channel. This duality result is applicable to additive Gaussian noise and fading channels for several different notions of fading channel capacity, including ergodic capacity, outage capacity, and minimum rate capacity. Duality provides a powerful connection between the broadcast channel and multiple access channel capacity regions that solves many open capacity problems, and also greatly simplifies the calculation of known regions. We next consider the dirty paper coding strategy of Costa for broadcast channels. Dirty paper coding exploits the known interference between users in a broadcast channel to presubtract out this interference. We apply this coding strategy to broadcast channels with multiple antennas at the transmitter and receiver (the MIMO channel). We show that dirty paper coding provides significant rate gains over the high-data-rate (HDR) strategy used in current cellular systems. We next turn our attention to the MIMO broadcast channel capacity region. Since this channel is in general non-degraded, its capacity region remains an unsolved problem. We first establish a duality between the achievable region of the MIMO broadcast channel using dirty paper coding and the capacity region of the dual MIMO multiple-access channel, which is easy to compute. We then show that dirty paper coding achieves sumrate capacity of the MIMO broadcast channel. The proof exploits duality, dirty paper coding, and clever upper bounding techniques.

Key words and phrases. Broadcast channels, multiple access channels, capacity region, duality, fading channels, MIMO, dirty paper coding, Sato upper bound, sumrate capacity.

Work supported by ONR grants N00014-99-1-0578 and N00014-02-1-0003. The work of S. Vishwanath was supported by a Stanford Graduate Fellowship.

1. Introduction

The main focus of this paper is to establish a connection between the capacity regions of broadcast channels (BCs) and multiple access channels (MACs), then exploit this connection to solve new capacity problems. Despite extensive study of the BC and MAC independently, no relationship between them has previously been discovered. We show that the Gaussian MAC and BC are essentially duals of each other. As a result, the capacity regions of the BC and the MAC with the same channel gains and the same noise power at every receiver can be derived from one another. In particular, we show that the capacity region of a fading or constant Gaussian BC is equal to the union of the capacity regions of the dual Gaussian MAC where the union is taken over power constraints on the individual users of the MAC that sum up to the BC power constraint. This result allows us to directly compute the capacity region of a Gaussian BC from the capacity region of its dual Gaussian MAC. We also show that the capacity region of the Gaussian MAC equals the intersection of capacity regions of the dual Gaussian BC where the intersection is taken over appropriately scaled channel gains and the total BC power equals the sum of powers of the dual MAC users. Moreover, we obtain an explicit relationship between the optimum power allocation scheme and the decoding order used to achieve points along the boundary of the dual MAC and BC capacity regions. The relationships between the capacity regions and optimal transmission strategies for dual MAC and BCs holds for several different notions of fading channel capacity, including ergodic capacity, outage capacity, and minimum rate capacity.

The duality relationships between the capacity regions of the Gaussian MAC and BC are quite powerful and interesting. First they indicate a direct connection between known capacity results for dual MAC and broadcast channels. More importantly, there are several BCs and MACs where the capacity region and optimal transmission strategy for one channel is either unknown or very hard to compute, whereas the regions are known and easy to compute for the dual channel. In these cases we can use duality to obtain new capacity results. In particular, we will show that the minimum rate capacity region and optimal transmission strategy for a fading MAC can be obtained from the dual BC capacity region, yet no direct method exists to compute this region.

The capacity of a broadcast channel with multiple antennas at the transmitter and receiver (the MIMO BC) is an open problem due to the lack of a general theory on the capacity of non-degraded broadcast channels. Pioneering work in this area by Caire and Shamai [1] developed an achievable set of rates for a broadcast channel with two transmit antennas and one receive antenna at each node based on the dirty paper coding result of Costa [2]. This coding strategy allows a channel with interference known at the transmitter to achieve the same data rate as if the interference did not exist. The original coding strategy was described as writing on dirty paper, where the structure introduced by the "dirt" is exploited in the code design. Computing the corresponding set of achievable rates for the MIMO BC is extremely complex, especially for a larger number of antennas at either the transmitter or the receivers.

We establish a duality between the achievable region of the MIMO BC obtained using dirty paper coding and the capacity region of the dual MIMO MAC for any number of transmit and receive antennas. This result greatly simplifies calculation of the achievable region for the MIMO BC obtained with dirty paper coding, since we can use recently developed iterative waterfilling techniques to obtain the capacity region for the dual MIMO MAC capacity region [4] and then apply duality [3]. We also show that dirty paper coding achieves the sumrate point of the MIMO BC channel capacity region (maximum achievable sum of all users' rates). This proof applies the Sato upper bound [5] to the broadcast channel by allowing joint signal detection at all receivers and assuming worst-case noise, a technique first proposed in [1] that we generalize to multiple transmit and receive antennas.

The remainder of this paper is organized as follows. Section 2 introduces the duality between broadcast and MAC channels and shows how this can be used to compute their dual capacity regions. Section 3 describes the different notions of capacity associated with multiuser fading channels, including the ergodic, outage, and minimum rate capacities. Sections 4 and 5 derive the ergodic and minimum rate capacity regions, respectively, for fading broadcast and MAC channels along with their duality relationship. Section 6 defines the dual MIMO MAC and BCs. Sections 7 outlines the achievable rate region for a broadcast MIMO fading channel with dirty paper coding, and illustrates the gains of dirty paper coding relative to the high-data-rate (HDR) scheme currently used in cellular systems. Section 8 establishes a duality between the dirty paper region for the MIMO BC and the capacity region of the dual MIMO MAC. Section 9 derives an upper bound on the MIMO BC capacity region based on joint detection and worst-case noise, and shows that this bound is achievable at the sumrate point using dirty paper coding. Thus, the sumrate capacity of the MIMO BC is obtained for any number of users, transmit antennas, and receive antennas. Our conclusions are discussed in Section 8.

2. Duality between Broadcast and Multiple Access Channels

In this section we establish a duality between broadcast and multiple access channels and show how this duality can be used to compute the capacity of one channel from the capacity of its dual. A discrete-time BC, where one transmitter sends information to M receivers, is described mathematically by

$$Y_j[i] = \sqrt{h_j[i]}X[i] + n_j[i], \tag{2.1}$$

where $X[i]$ is the transmitted signal, $Y_j[i]$ is the signal received by the jth user, $n_j[i]$ is the receiver noise sample at time i of the jth user, and $h_j[i]$ is the time-varying channel power gain at time i of the jth user. A discrete-time MAC, where many transmitters send information to one receiver, is described mathematically by

$$Y[i] = \sum_{j=1}^{M} \sqrt{h_j[i]}X_j[i] + n[i], \tag{2.2}$$

where $X_j[i]$ is the signal transmitted at time i by the jth transmitter, $h_j[i]$ is the channel power gain between the jth transmitter and the receiver, $Y[i]$ is the received signal at time i, and $n[i]$ is the receiver noise sample at time i. We assume in our system models that the noise power of all the receivers in the BC and the single receiver in the MAC are equal to σ^2. Also, the term $h_j[i]$ is the channel power gain of receiver j in the BC (downlink) and $h_j[i]$ is also the channel power gain of transmitter j in the MAC (uplink). We call this BC the **dual** of the MAC, and vice versa. In our broadcast and MAC models the channel gains can be constant or changing with i (fading). We assume in both cases that all channel gains $h_j[i]$ are known to the transmitter(s) and receiver(s) at time i.

The dual channels have several key differences as well as some key similarities. Specifically, the BC has a single power constraint associated with the transmitter, whereas the MAC has a different individual power constraint for each user. In addition, the interference signal on the BC has the same channel gain as the desired signal, whereas in the MAC these signals are received with different powers (the near-far effect). Despite these differences, a superposition coding strategy is optimal for both channels, and the optimal decoders for each channel exploit successive decoding and interference cancellation. The duality relationship between the two channels is based on exploiting their similar encoding and decoding strategies while bridging their differences by summing the individual MAC power constraints to obtain the BC power constraint and scaling the BC gains to achieve the near-far effect of the MAC.

We now state the explicit duality result, proved in [9], that the capacity region of a Gaussian BC can be characterized in terms of the capacity region of the dual MAC. For simplicity the theorem is stated for a two-user channel, but it applies to any number of users.

THEOREM 2.1. *The capacity region of a constant Gaussian BC with power P is equal to the union of capacity regions of the dual MAC with power P_1 and P_2 such that $P_1 + P_2 = P$:*

$$\mathcal{C}_{BC}(P; h_1, h_2) = \bigcup_{0 \leq P_1 \leq P} \mathcal{C}_{MAC}(P_1, P - P_1; h_1, h_2). \tag{2.3}$$

In [9] it is also shown that the boundary of each MAC capacity region touches the boundary of the BC capacity region at a *different* point. More generally, each point along the BC capacity region boundary is intersected by a different MAC capacity region boundary. The BC and MAC power schemes corresponding to the intersection at rate point (R_1, R_2) are related by the following duality power transformation [9]:

$$\begin{aligned} R_1^{BC} &= \log\left(1 + \frac{h_1 P_1^{BC}}{\sigma^2}\right) = \log\left(1 + \frac{h_1 P_1^{MAC}}{h_2 P_2^{MAC} + \sigma^2}\right) = R_1^{MAC} \\ R_2^{BC} &= \log\left(1 + \frac{h_2 P_2^{BC}}{h_2 P_1^{BC} + \sigma^2}\right) = \log\left(1 + \frac{h_2 P_2^{MAC}}{\sigma^2}\right) = R_2^{MAC}. \end{aligned} \tag{2.4}$$

Thus, given the power pair (P_1, P_2) that achieves rates (R_1, R_2) for one channel, the corresponding power allocation to achieve these rates on the dual channel must satisfy (2.4). Moreover, the decoding order to achieve a given rate (R_1, R_2) on the MAC is the opposite of the optimal encoding order to achieve this rate on the dual BC: in the BC the user with the best channel gain is decoded last, whereas in the MAC the user with the best channel gain is decoded first. More details on these explicit power and decoding order transformations can be found in [9]. The duality relationship between the Gaussian BC and MAC defined by Theorem 2.1 is illustrated in Figure 1.

We now describe how the capacity region of the MAC can be characterized in terms of the capacity region of the dual BC. This theorem is also proved in [9].

THEOREM 2.2. *The capacity region of a constant Gaussian MAC is equal to the intersection of the capacity regions of the scaled dual BC over all scalings:*

$$\mathcal{C}_{MAC}(P_1, P_2; h_1, h_2) = \bigcap_{\alpha > 0} \mathcal{C}_{BC}\left(\frac{P_1}{\alpha} + P_2; \alpha h_1, h_2\right). \tag{2.5}$$

Moreover, the optimal power allocation and decoding order for the MAC can be obtained by reversing the dual BC encoding order and applying the power transformation in (2.4). Thus, the optimal transmission strategy to achieve any point on the MAC capacity region can be obtained from the dual BC channel. Note that since MAC region is a pentagon, the BC channels characterized by $\alpha = 0$, $\alpha = h_2/h_1$ and $\alpha = \infty$ are sufficient to form the pentagon. If $\alpha = h_2/h_1$, the channel gains of both users are the same and the BC capacity region is bounded by a straight line segment because the capacity region can be achieved by time-sharing. It can be shown [9] that as $\alpha \to 0$, $R_1 \to \log(1 + \frac{h_1 P_1}{\sigma^2})$ and

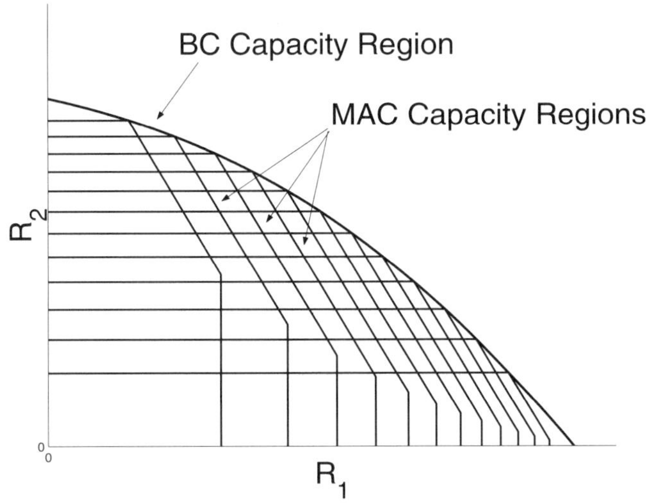

FIGURE 1. Gaussian BC Capacity Region as Union of Dual MAC Regions

$R_2 \to \infty$. Similarly, as $\alpha \to \infty$, $R_1 \to \infty$ and $R_2 \to \log(1 + \frac{h_2 P_2}{\sigma^2})$. These two regions bound the vertical and horizontal line segments, respectively, of the MAC capacity region. All scaled BC capacity regions except the $\alpha = h_2/h_1$ channel intersect the MAC at exactly one of the two corner points of the MAC region. The $\alpha = h_2/h_1$ channel intersects the MAC region along its time-sharing line. These boundaries along with the relationship defined by Theorem 2.2 is illustrated in Figure 2.

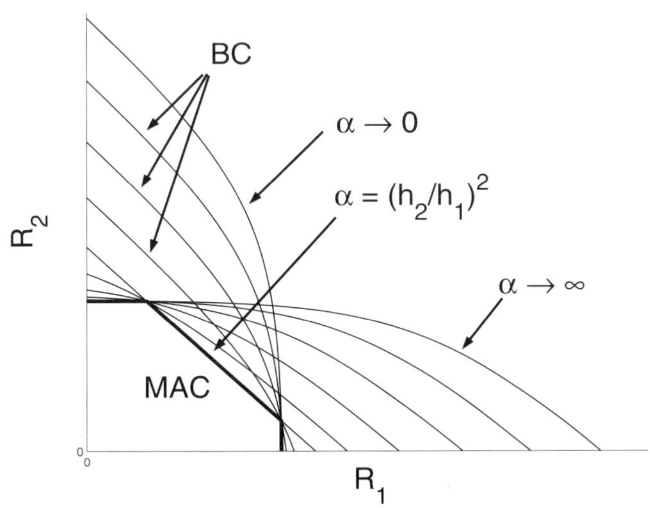

FIGURE 2. Gaussian MAC Capacity Region as Intersection of Dual BC Regions

3. Capacity of Fading Multiuser Channels

Fading channels exhibit random amplitude variations over time. Capacity for such channels depends on what is known about the channel fading at both transmitter and receiver, and whether or not capacity can be averaged over many channel states or must be achieved in each state. We focus on the case where both transmitter and receiver track the channel state perfectly and instantaneously. We also assume a slowly fading channel relative to codeword length, so the channel is constant for the transmission of a single codeword. For this channel there are four different definitions of capacity region for multiuser channels:

- **Ergodic capacity region:** The maximum rate region averaged over all channel states.
- **Outage capacity region:** The maximum rate region that can be achieved in all non-outage fading states subject to some outage probability. Under an outage probability of zero this is called the zero-outage capacity region.
- **Minimum rate capacity region:** The maximum rate region averaged over all fading states subject to a minimum rate for each user required in every fading state.
- **Minimum rate capacity region with outage:** The maximum rate region averaged over all fading states subject to some minimum rate for each user required in all non-outage fading states, with some nonzero outage probability.

The ergodic capacity region for fading broadcast and multiple access channels has been derived in [10] and [11] respectively. The optimal power allocation to achieve this capacity corresponds to a multi-level water-filling over both time (i.e. fading states) and users. As expected, users are allocated the most power when their channels are strong, and little, if any, power when their channels are weak. This results in a channel-dependent delay, which may not work for delay-constrained applications like voice and video.

The outage capacity region for fading broadcast and multiple access channels has been derived in [12] and [13, 14], respectively. For outage capacity each user maintains a constant rate some percentage of time with no data transmitted the rest of the time. The optimal power allocation over nonoutage must effectively invert the fading to eliminate channel variations such that a constant rate can be maintained. Clearly the ergodic capacity region exceeds the outage capacity region, since outage capacity has an additional constraint of constant rate transmission in all fading states. Since the weak channel states will effectively drive the maximum constant rate that can be maintained, outage capacity fails to fully exploit good channel states. However, outage capacity has the advantage that with fixed rate transmission, there is no channel-dependent delay.

Minimum rate capacity combines the concepts of ergodic and outage capacity to maximize the rate averaged over all channel states while maintaining some minimum rate in every fading state. Minimum rate capacity for fading broadcast channels has been derived in [15]. The optimal power allocation for the broadcast channel is a two-step process, where first the minimum power required to achieve the minimum rates in all fading states is allocated, and then the excess power to maximize average rate in excess of the minimum rate is allocated. The optimal allocation of the excess power is a multi-level water-filling based on effective noise that incorporates the minimum rate constraints. We will use these results in Section 5 to obtain the minimum rate capacity, optimal power allocation, and optimal decoding order of the MAC using duality.

The minimum rates associated with the minimum rate capacity region must be within the zero-outage capacity region of the channel. Thus, the boundary of the minimum rate capacity region lies between the ergodic and zero-outage boundaries. The difference between the ergodic and zero-outage capacity regions is a good indicator of the degradation in capacity due to minimum rate constraints. If the zero-outage capacity region is much smaller than the ergodic capacity region, the minimum rate capacity region is generally significantly smaller than the ergodic capacity region. Alternatively, if the zero-outage capacity region is close to the ergodic capacity region, then clearly the minimum rate capacity region will be close as well. More details on these relationships for the fading BC can be found in [15].

4. Fading Channel Capacity and Duality

For fading channels the ergodic capacity region is obtained by averaging over the constant capacity regions. Thus, the duality relationships obtained for BC and MAC Gaussian channels in the previous section extend to the ergodic capacity regions for fading BC and MAC channels. This extension is captured in the following theorems, proved in [9].

THEOREM 4.1. *The capacity region of a fading Gaussian BC with power constraint* $\overline{P}$ *is equal to the union of ergodic capacity regions of the dual MAC with power constraints* $\overline{P}_1$ *and* $\overline{P}_2$ *such that* $\overline{P}_1 + \overline{P}_2 = \overline{P}$:

$$\mathcal{C}_{BC}(\overline{P}; H_1, H_2) = \bigcup_{0 \leq \overline{P}_1 \leq \overline{P}} \mathcal{C}_{MAC}(\overline{P}_1, \overline{P} - \overline{P}_1; H_1, H_2). \tag{4.1}$$

The ergodic BC capacity region obtained by a union of ergodic capacity regions of the dual MAC is illustrated in Figure 3.

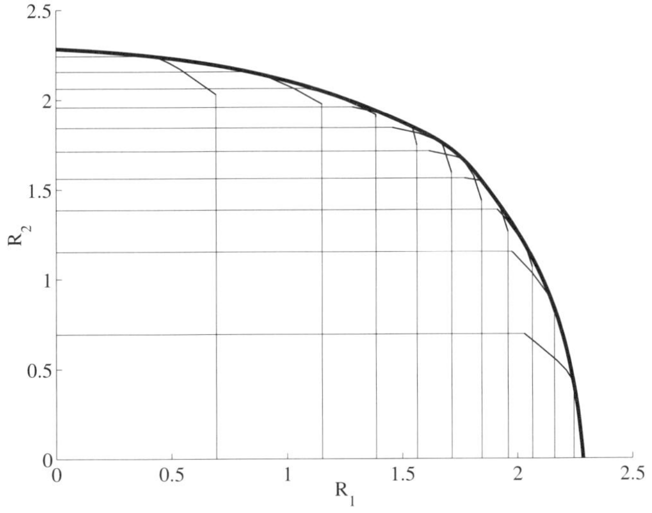

FIGURE 3. Ergodic BC Capacity Region as Union of Dual MAC Regions

The next theorem shows that the ergodic capacity region of the fading MAC can be obtained as the intersection of the regions for the dual BC.

THEOREM 4.2. *The ergodic capacity region of a fading MAC is equal to the intersection of the ergodic capacity regions of the scaled dual BC over all scalings:*

$$\mathcal{C}_{MAC}(\overline{P}_1, \overline{P}_2; H_1, H_2) = \bigcap_{\alpha > 0} \mathcal{C}_{BC}(\frac{\overline{P}_1}{\alpha} + \overline{P}_2; \alpha H_1, H_2). \tag{4.2}$$

The ergodic capacity region of the fading MAC obtained by intersecting the ergodic capacity regions of the dual BC is shown in Figure 4.

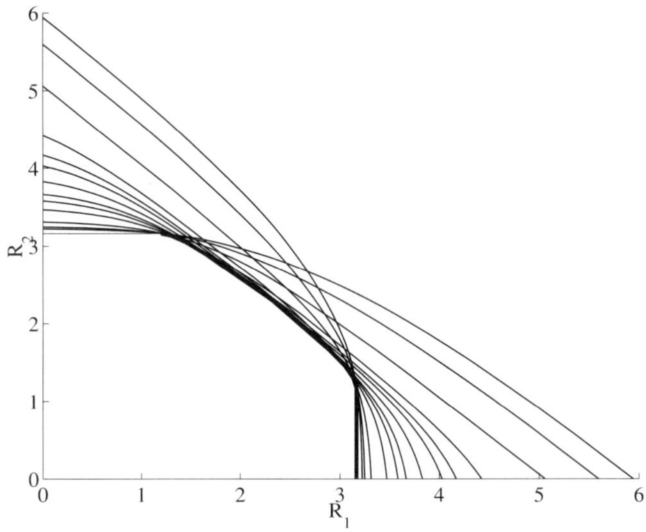

FIGURE 4. Ergodic MAC Capacity Region as Intersection of Dual BC Regions

Note that in addition to the capacity region duality captured in the above two theorems, explicit power and decoding order transformations between the regions are also obtained in [9]. Thus, given a point on the capacity region and optimal transmission strategy (power allocation and decoding order) for one channel, the corresponding optimal transmission strategy for achieving that same rate point on the dual channel can be explicitly obtained.

The duality of the Gaussian MAC and BC is a general result that also holds for all definitions of fading channel capacity given in the previous section. The outage and minimum rate definitions of capacity differ from the ergodic capacity definition since they have restrictions on the *instantaneous* transmission rates (i.e. the state-by-state rates). However, as shown in [9], the duality power transformation for these ergodic regions preserves state-by-state rates (i.e. the same instantaneous rates are achieved in the dual channels). Thus, Theorems 4.1 and 4.2 hold for minimum rate and outage capacity as well as ergodic capacity. In the next two sections we use duality to obtain new capacity results from known results for the dual channels.

5. Minimum Rate Capacity

In this section we show that the minimum rate capacity region of the MAC and BC are duals of each other, so that the capacity region and capacity-achieving transmission strategy for one channel can be found from the other dual channel. We then use this result to find the minimum rate capacity region of the MAC using the known minimum rate capacity region of the BC obtained in [15]. The minimum rate capacity of the MAC and BC, denoted respectively by $\mathcal{C}_{MAC}^{min}(\overline{P}_1, \overline{P}_2, \mathbf{R}^*; H_1, H_2)$ and $\mathcal{C}_{BC}^{min}(\overline{P}_1 + \overline{P}_2, \mathbf{R}^*; H_1, H_2)$, is defined as the maximum ergodic capacity that can be obtained while ensuring that a set of minimum rates $\mathbf{R}^* = (R_1^*, \ldots, R_M^*)$ is maintained for all users in *all* fading states. Clearly the minimum rates themselves must be in the zero-outage capacity of the channel for the rates to be achievable in all fading states. Every feasible $\mathbf{R}^*$ defines a different minimum rate capacity region (i.e. the minimum rate capacity region is a function of $\mathbf{R}^*$ in addition to the channel and power constraints).

We now state two theorems, proved in [9], indicating the duality between the minimum rate capacity region of the fading MAC and BC.

THEOREM 5.1. *The minimum rate capacity region of a fading Gaussian BC with perfect channel information at the transmitter and receivers is given by:*

$$\mathcal{C}_{BC}^{min}(\overline{P}, \mathbf{R}^*; H_1, H_2) = \bigcup_{0 \le \overline{P}_1 \le \overline{P}} \mathcal{C}_{MAC}^{min}(\overline{P}_1, \overline{P} - \overline{P}_1, \mathbf{R}^*; H_1, H_2), \tag{5.1}$$

for all $\mathbf{R}^* \in \mathcal{C}_{BC}^0(\overline{P}; H_1, H_2)$.

THEOREM 5.2. *The minimum rate capacity region of a fading Gaussian MAC with perfect channel information at the transmitters and receiver is given by:*

$$\mathcal{C}_{MAC}^{min}(\overline{P}_1, \overline{P}_2, \mathbf{R}^*; H_1, H_2) = \bigcap_{\alpha \ge 0} \mathcal{C}_{BC}^{min}(\frac{\overline{P}_1}{\alpha} + \overline{P}_2, \mathbf{R}^*; \alpha H_1, H_2) \tag{5.2}$$

for all $\mathbf{R}^* \in \mathcal{C}_{MAC}^0(\overline{P}_1, \overline{P}_2; H_1, H_2)$.

While these theorems characterize the relationship between the dual capacity regions, we also wish to obtain an explicit transformation between the optimal transmission strategies. The power and decoding order transformations to achieve any point on the minimum rate capacity region of the fading MAC from the corresponding region of the dual BC that intersects that point is obtained in [15]. Thus we can explicitly characterize the MAC minimum rate capacity region and its corresponding capacity-achieving transmission strategy without having to directly solve for it. Specifically, these results are obtained by applying the power and decoding transformations to the known optimal transmission strategies of the dual BC obtained in [15].

6. Multi-Antenna Multiple Access and Broadcast Channels

Adding multiple antennas to the transmitter and receiver has recently been shown to greatly increase the capacity of wireless channels [16, 17]. As a result, the capacity and capacity region for single and multiuser multiple input multiple output (MIMO) systems is of great interest. While the capacity region of the MIMO MAC has been fully characterized [7,17,18], the capacity of the MIMO BC is difficult to obtain due to a lack of general theory on the capacity of non-degraded broadcast channels. However, in the next few sections we show that duality combined with a novel encoding technique called dirty paper coding can shed much light on the achievable rates of MIMO BCs and in fact determine their sumrate capacity point.

Consider a K-user MIMO Gaussian BC in which receiver k has r_k receive antennas and the transmitter has t transmit antennas. The $r_k \times t$ matrix $\mathbf{H_k}$ defines the channel gains from the transmitter to the r_k antennas of receiver k. Each receiver has additive white Gaussian noise with unit variance. The dual MIMO MAC channel is arrived at by converting the receivers in the BC into transmitters in the MAC and converting the

t-antenna transmitter into a t-antenna receiver with additive noise of unit variance. The channel gains of the dual MAC are the same as that of the broadcast channel. Specifically, if $\mathbf{H_k}$ is the $t \times r_k$ matrix defining the channel gains from t-antenna transmitter to the kth receiver with r_k antennas in the BC, then $\mathbf{H_k}^\dagger$ is the $r_k \times t$ matrix defining the channel gains from transmitter k with r_k antennas to the t-antenna receiver in the dual MAC.

The MIMO BC is a nondegraded broadcast channel due to the multiple receive antennas. Thus, receivers are not necessarily "better" or "worse" than one another. The capacity region of a nondegraded broadcast channel is an open problem in general. However, the sumrate capacity of the MIMO BC for two transmit antennas ($t = 2$) and two users with one receive antenna each ($r_1 = r_2 = 1$) was obtained in pioneering work by Caire and Shamai [1]. In that work a set of achievable rates (the achievable region) was obtained by using the dirty paper coding technique of Costa [2] (also known as coding for non-causally known interference). It was also shown that the dirty paper coding technique at the sumrate point equals the Sato upper bound with joint receiver decoding and worst-case noise. Thus, dirty paper coding achieves the sumrate capacity of the MIMO BC for the two-user case with $t = 2$ and $r_1 = r_2 = 1$. However, computing the dirty paper coding region is extremely complex and this approach does not appear to work for the more general class of channels (any number of users with any number of transmit and receive antennas). However, we will see shortly that exploiting duality can greatly simplify this calculation. First, however, we must describe dirty paper coding in more detail.

7. Dirty Paper Coding

The basic premise of dirty paper coding is that if interference to a given user is known in advance, the encoding strategy can exploit the structure of the interference such that the capacity is the same as if there was no interference at all. The encoding strategy cleverly distributes the codewords based on the interference, and the decoder must know how to read these codewords. Dirty paper coding is a natural technique to use on the broadcast channel since the interference between all users is known. An achievable region for the MIMO BC based on dirty paper coding was first proposed in [1]. In [19], the region was extended to the more general multiple-user, multiple-antenna case using the following extension of the "dirty paper result" [2] to the vector case:

LEMMA 7.1. *[Yu, Cioffi] Consider a channel with* $\mathbf{y}_k = \mathbf{H}_k\mathbf{x}_k + \mathbf{s}_k + \mathbf{n}_k$, *where* $\mathbf{y}_k$ *is the received vector,* $\mathbf{x}_k$ *the transmitted vector,* $\mathbf{s}_k$ *the vector Gaussian interference, and* $\mathbf{n}_k$ *the vector white Gaussian noise. If* $\mathbf{s}_k$ *and* $\mathbf{n}_k$ *are independent and non-causal knowledge of* $\mathbf{s}_k$ *is available at the transmitter but not at the receiver, then the capacity of the channel is the same as if* $\mathbf{s}_k$ *is not present.*

The transmission strategy under dirty paper coding first picks a codeword for receiver 1. The transmitter then chooses a codeword for receiver 2 with full (non-causal) knowledge of the codeword intended for receiver 1. Therefore receiver 2 does not see the codeword intended for receiver 1 as interference. Similarly, the codeword for receiver 3 is chosen such that receiver 3 does not see the signals intended for receivers 1 and 2 as interference. This process continues for all K receivers. Since the ordering of the users clearly matters in such a procedure, the following is an achievable set of rates

$$R_{\pi(i)} = \frac{1}{2}\log\frac{|\mathbf{I} + \mathbf{H}_{\pi(i)}(\sum_{j\geq i}\mathbf{\Sigma}_{\pi(j)})\mathbf{H}_{\pi(i)}^\dagger|}{|\mathbf{I} + \mathbf{H}_{\pi(i)}(\sum_{j>i}\mathbf{\Sigma}_{\pi(j)})\mathbf{H}_{\pi(i)}^\dagger|}, \quad i = 1, \ldots, K,$$

where $\pi(i)$ is a permutation that depends on the encoding order. The dirty-paper region $\mathcal{C}_{\text{dirtypaper}}(P, \mathbf{H})$ is defined as the union of all such rates vectors over all covariance matrices $\mathbf{\Sigma}_1, \ldots, \mathbf{\Sigma}_K$ such that $\text{Tr}(\mathbf{\Sigma}_1 + \ldots \mathbf{\Sigma}_K) = \text{Tr}(\mathbf{\Sigma}_x) \leq P$ and over all permutations $(\pi(1), \ldots, \pi(K))$. The transmitted signal is $\mathbf{x} = \mathbf{x_1} + \ldots + \mathbf{x_K}$ and the input covariance matrices are of the form $\mathbf{\Sigma}_i = \mathbb{E}[\mathbf{x_i}\mathbf{x_i}^\dagger]$.

Many cellular systems use a technique called high data rate (HDR) on the downlink for high rate data transmission. HDR is an adaptive technique whereby the user with the best channel gain is allocated the total bandwidth and power. When the channel changes, the bandwidth and power are reallocated, and fairness is achieved over time if all users have the same channel statistics. Dirty paper coding is a more efficient technique since all users are active at the same time, and the power allocation and encoding/decoding order is optimized relative to the channel. We have recently computed the gains of dirty paper coding relative to HDR. Note that computing the dirty paper coding region directly is intractable, however the duality method described below allows this computation to be made based on the dual MIMO MAC, whose capacity is easy to compute. The capacity gains of dirty paper coding relative to HDR are illustrated in Figure 5, where we see up to a factor of seven capacity increase.

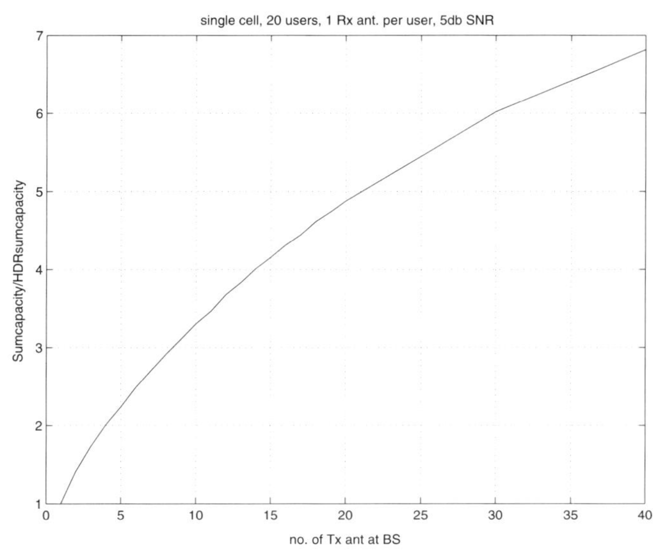

FIGURE 5. Capacity Gain of Dirty Paper Coding over HDR

While Figure 5 indicates that dirty paper coding exhibits large gains over other techniques, we would like to see if it actually achieves Shannon capacity. In order to address this question, we will first establish a duality between the dirty paper achievable region for the MIMO BC and the MIMO MAC capacity region.

8. MIMO MAC Capacity Region

The capacity region of a general MIMO MAC was obtained in [7, 17, 18]. We now describe this capacity region for the dual MIMO MAC of our BC model. For any set of powers $(P_1, \ldots, P_K)$, the capacity of the MIMO MAC is

$$\mathcal{C}_{\text{MAC}}(P_1, \ldots, P_K; \mathbf{H}^\dagger) \triangleq \bigcup_{\{\text{Tr}(\mathbf{P}_1) \leq P_i \ \forall i\}} \Big\{ (R_1, \ldots, R_K) : \tag{8.1}$$

$$\sum_{i \in S} R_i \leq \frac{1}{2} \log |\mathbf{I} + \sum_{i \in S} \mathbf{H}_i^\dagger P_i \mathbf{H}_i| \ \forall S \subseteq \{1, \ldots, M\} \Big\} \tag{8.2}$$

For $P > 0$, we denote by $\mathcal{C}_{\text{union}}(P, \mathbf{H}^\dagger)$

$$\mathcal{C}_{\text{union}}(P, \mathbf{H}^\dagger) = \bigcup_{\sum_{i=0}^{K} P_i \leq P} \mathcal{C}_{\text{MAC}}(P_1, \ldots, P_K; \mathbf{H}^\dagger). \tag{8.3}$$

It can be easily shown that this region is the capacity region of a MAC when the transmitters have a sum power constraint instead of individual power constraints but are not allowed to cooperate.

Our next theorem, proved in [20], indicates that the capacity region of the MIMO MAC with a total power constraint of P for the K transmitters is the same as the dirty paper region of the dual MIMO BC with power constraint P. In other words, any rate vector that is achievable in the dual MAC with power constraints $(P_1, \ldots, P_K)$ is in the dirty paper region of the BC with power constraint $\sum_{i=1}^{K} P_i$. Conversely, any rate vector that is in the dirty paper region of the BC is also in the dual MIMO MAC region with the same total power constraint.

THEOREM 8.1. *The dirty paper region of a MIMO BC channel with power constraint P is equal to the the capacity region of the dual MIMO MAC with sum power constraint P.*

$$\mathcal{C}_{\text{dirtypaper}}(P, \mathbf{H}) = \mathcal{C}_{\text{union}}(P, \mathbf{H}^\dagger).$$

In addition to the capacity region relationships, we also wish to obtain an explicit transformation between the transmission strategies of dirty paper coding for the MIMO BC and the corresponding capacity region of the MIMO MAC. These explicit transformations will be briefly summarized here, details can be found in [20]. Suppose we are given a rate point in the MIMO MAC, and the corresponding input covariance matrix and decoding order that achieves this rate point. We need to find the corresponding input covariance matrix and encoding order to achieve the same rate point on the MIMO BC using dirty paper coding. It is quite difficult to find the input covariance matrix that yields the same rates and also satisfies the power constraint. It turns out that the correct transformation defines an effective channel and then flips it to obtain the covariance matrix. The encoding order for the MIMO BC under dirty paper coding is the reverse of the decoding order for the MIMO MAC. Explicit formulas for these transformations can be found in [20]. Interestingly, this transformation also shows that beamforming (transmit covariance matrix of rank one) is optimal for dirty paper coding on the broadcast channel with multiple transmit antennas but only one receive antenna at each user. That is because the dual MIMO MAC channel has one transmit antenna and thus a rank one covariance matrix, and the transformation defined in [20] preserves this rank.

Similarly, suppose a rate point achievable using dirty paper coding on the MIMO BC and its corresponding input covariance matrix and encoding order is known. The covariance matrix for the dual MIMO MAC is also obtained by a "flipping of the effective channel" and decoding based on the reverse order of the BC dirty paper encoding. In general the MIMO MAC capacity region is easy to compute, since it is a convex region. In fact, an iterative waterpouring formula was recently obtained that computes this region very efficiently [4]. However, the dirty paper coding achievable region is quite hard to compute since it is nonconvex [1]. Thus, while the dirty paper coding region for the MIMO BC cannot typically be computed directly for a general number of users and antennas, this region is easily computed by exploiting duality and the transformations between the MIMO MAC and this dirty paper coding region [3].

9. Sato Upper Bound and Sumrate Capacity

In the previous sections we showed that the dirty paper region of the BC and the union of the dual MAC capacity regions are equal. Now we show that dirty paper coding is the sumrate capacity achieving strategy for the MIMO BC. We do this based on Sato's upper bound for the capacity region of general BCs [5]. This bound utilizes the capacity of the *cooperative system* whereby the different receivers cooperate to decode the transmitted signal. Since the cooperative system is the same as the BC, but with receiver coordination, the capacity of the cooperative system ($\mathcal{C}_{\text{coop}}(P, \mathbf{H})$) is an upper bound on the BC capacity region ($\mathcal{C}_{\text{BC}}(P, \mathbf{H})$). We now show that the bound can be tightened by introducing noise correlation.

Since the capacity region of a general BC depends only on the marginal transition probabilities (i.e. $p(y_i|x)$) and not on the entire joint distribution $p(y_1, \ldots, y_K|x)$, we can introduce correlation between the noise vectors at different receivers of the BC without affecting the BC capacity region. This correlation does, however, affect the capacity of the cooperative system, which is still an upper bound on the sumrate of the BC. By searching over all feasible positive definite noise covariance matrices, we get the following bound for sum rate capacity, based on the capacity of the cooperative system with the worst case noise:

$$\mathcal{C}_{\text{BC}}^{\text{sumrate}}(P, \mathbf{H}) \leq \inf_{\Sigma_z} \max_{\Sigma_x} \frac{1}{2} \log |I + \Sigma_z^{-1/2} H \Sigma_x H^T \Sigma_z^{-1/2}|.$$

An explicit formula for the resulting worst case noise is found in [20]. By using the fact that the capacity region of the dual MIMO MAC equals the capacity region of the dirty paper region of the MIMO BC (and therefore the maximum sumrate of the MAC and the dirty paper region are equal), we are able to show using Lagrangian duality that this bound is tight for the MIMO BC. The proof that dirty paper coding achieves the Sato upper bound, and therefore equals the sumrate capacity of the MIMO BC, is depicted pictorially in Figure 6. Note that a special case of this proof for a single antenna at each receiver was obtained independently in [21]. This proof is somewhat simpler since beamforming is optimal in this case, and therefore the covariance matrix transformations described in Section 8 are not needed.

Figure 7 shows the dirty paper region, Sato upper bound, and sumrate point where the two meet. We conjecture that dirty paper coding achieves the full capacity region of the MIMO BC and not just the sum rate point. We have obtained a proof of this result under the assumption that Gaussian inputs are optimal for the MIMO BC, but we have yet to prove this assumption. More details about this conjecture, our proof, and the one missing link can be found in [22]

10. Conclusions

Duality is a power technique to relate capacity regions for broadcast and MAC channels. Duality holds for many notions of capacity, including ergodic, outage, and minimum

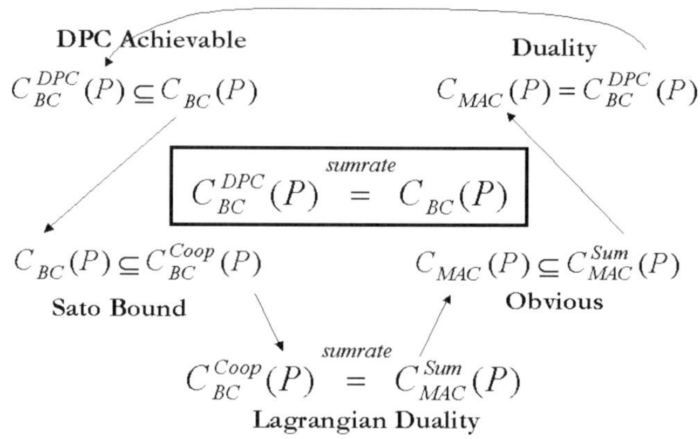

FIGURE 6. MIMO BC Sumrate Capacity Proof

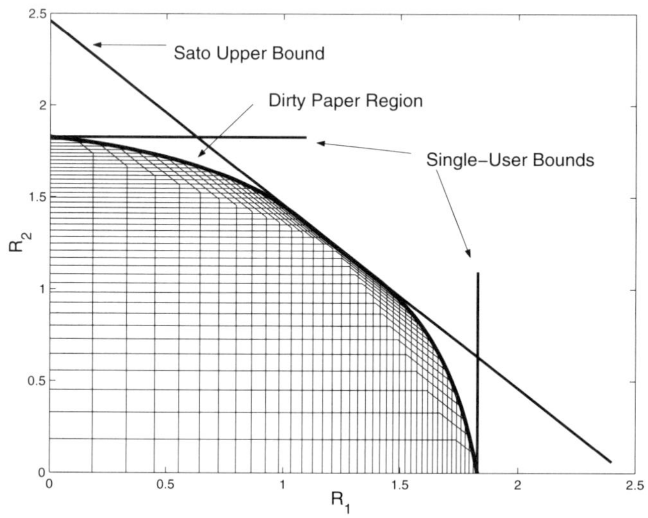

FIGURE 7. Dirty Paper Region and Sato Upper Bound for MIMO BC

rate capacity. Duality can be used to show the connection between capacity regions and optimal transmission strategies for BC and MAC channels. More importantly, it can be used to obtain new capacity results previously thought intractable for the MAC or BC when results are known and/or easily computable for the dual channel. This concept is used to obtain the minimum rate capacity region of a fading MAC channel. We then turn our attention to the MIMO BC. We introduce the notion of dirty paper coding for this channel, which subtracts out the effect of known interference and thus achieves much higher rates than other techniques such as HDR. Unfortunately, the rates achievable using dirty paper coding are hard to compute. However, we show that a duality exists between the achievable rates of the MIMO BC under dirty paper coding and the capacity region of the dual MIMO MAC under a sum power constraint. Since this MAC region is easy to compute, we can obtain the rates of the MIMO BC under dirty paper coding quite easily using duality. Moreover, using duality and Sato's upper bound with joint detection and worst case noise, we show that dirty paper coding achieves the sumrate point of the MIMO BC capacity region. We conjecture that dirty paper coding achieves the entire region, but this proof requires Gaussian inputs to be optimal for this channel, which we have not yet shown.

References

[1] G. Caire and S. Shamai, "On Achievable Rates in a Multi-antenna Broadcast Downl ink", 38th Annual Allerton Conference on Commun., Control and Computing, Monticello, I L, Oct. 4 - 6, 2000.

[2] M. Costa, "Writing on Dirty Paper", IEEE. Trans. Inf. Theory, v. 29, pp. 439-4 41, May 1983.

[3] N. Jindal, S. Jafar, S. Vishwanath, and A. Goldsmith, "Sum power iterative water-filling for vector broadcast channels," *Proc. Asilomar Conf. on Signls., Syst., Compt.*, Oct. 2002.

[4] W. Yu, W. Rhee, S. Boyd, J. Cioffi, "Iterative Water-filling for Vector Multipl e Access Channels", pp. 322, Proc. IEEE Int. Symp. Inf. Theory, (ISIT), Washington DC, June 24-29, 2001.

[5] H. Sato, "An Outer Bound on the Capacity Region of Broadcast Channel", *IEEE. Trans. Inf. Theory*, v. 24, pp. 374-377, May 1978.

[6] G. Caire and S. Shamai, "On the Achievable Throughput of a Multi-antenna Gaussi an Broadcast Channel", Submitted to *IEEE Trans. on Inform. Theory*, July 2001.

[7] S. Verdú, "Multiple-access channels with memory with and without frame synchronism," *IEEE Trans. Info. Theory,* pp. 605-619, May 1989.

[8] S. Verdú, "The capacity region of the symbol-asynchronous Gaussian multiple-access channel," *IEEE Trans. Info. Theory,* pp. 733-751, July 1989.

[9] N. Jindal, S. Vishwanath, and A. J. Goldsmith, "On the duality of multiple-access and broadcast channels," *Proc. IEEE Allerton Conf. Commun. Cont. Comput.*, Oct. 2001. Also submitted to *IEEE Trans. Inform. Theory.*

[10] L. Li and A. Goldsmith, "Capacity and optimal resource allocation for fading broadcast channels: Part I: Ergodic capacity", IEEE Trans. Inform. Theory, vol. 47, pp 1083-1102, March 2001.

[11] D.N. Tse and S. Hanly, "Multi-Access Fading Channels: Part I: Polymatroid Struct ure, Optimal Resource Allocation and Throughput Capacities", IEEE Trans. Inform. Theory, v. 44, pp 2796-2815, Nov. 1998.

[12] L. Li and A. Goldsmith, "Capacity and optimal resource allocation for fading broadcast channels: Part II: Outage capacity", IEEE Trans. Inform. Theory, vol. 47, pp 1103-1127, March 2001.

[13] S.V. Hanly and D. N. Tse, "Multiaccess fading channels- Part II:Delay-limited capacities", IEEE Trans. Inform. Theory, vol. 44, pp. 2816-2831, Nov. 1998.

[14] L. Li and A. Goldsmith, "Outage capacities and optimal power allocation for fading multiple-access channels", in revision for IEEE Trans. Inform. Theory. See also *Proc. IEEE Intl. Symp. Inform. Theory,* pg. 305, June 2000.

[15] N. Jindal and A. Goldsmith, "Capacity and optimal power allocation for fading broadcast channels with minimum rates", *Proc. IEEE Globecom,* pp. 1292-1296, Nov. 2001. Also in revision for *IEEE Trans. Inform. Theory.*

[16] G. J. Foschini. Layered space-time architecture for wireless communication in fading environments when using multi-element antennas. *Bell Labs Techn. J.*, 41–59, Autumn 1996.

[17] E. Telatar, "Capacity of Multi-antenna Gaussian Channels," European Trans. on Telecomm. ETT, 10(6):585-596, November 1999.

[18] W. Yu, W. Rhee, S. Boyd, J. Cioffi, "Iterative Water-filling for Vector Multipl e Access Channels", pp. 322, Proc. IEEE Int. Symp. Inf. Theory, (ISIT), Washington DC, June 24-29, 2001.

[19] W. Yu and J. Cioffi, "Trellis Precoding for the Broadcast Channel", Proc. IEEE Globecom, pp. 1344-1338, Nov., 2001.

[20] S. Vishwanath, N. Jindal, and A. J. Goldsmith, "On the capacity of multiple input multiple output broadcast channels," To appear: *Proc. IEEE Intl. Conf. Commun.*, April 2002. Also submitted to *IEEE Trans. Inform. Theory.*

[21] P. Viswanath and D. Tse, "Sum Capacity of the Multi-Antenna Broadcast Channel", *Proc. IEEE Int. Symp. Inf. Theory (ISIT)*, pp. 497, Lausanne, Switzerland, Jun. 30- Jul. 5, 2002. Also submitted to *IEEE Trans. on Inform. Theory.*

[22] S. Vishwanath, G. Kramer, S. Shamai, S. A. Jafar, and A. J. Goldsmith, "Capacity bounds for Gaussian vector broadcast channels," *Proc. DIMACS Workshop Signal. Proc. Wireless Trans.*, Oct. 7-9, 2002.

DEPARTMENT OF ELECTRICAL ENGINEERING, MAIL CODE 9515, STANFORD UNIVERSITY, STANFORD, CA 94305-9515

E-mail address: {njindal,sriram,syed,andrea}@wsl.stanford.edu

DIMACS Series in Discrete Mathematics
and Theoretical Computer Science
Volume **62**, 2003

On the Capacity of the Multiple Antenna Broadcast Channel

David Tse and Pramod Viswanath

ABSTRACT. The capacity region of the multiple antenna (transmit and receive) broadcast channel is considered. We propose an outer bound to the capacity region by converting this nondegraded broadcast channel into a degraded one with users privy to the signals of users ordered below them. We extend our proof techniques in the characterization of the sum capacity of the multiple antenna broadcast channel to evaluate this outer bound with Gaussian inputs. Our main result is the observation that if Gaussian inputs are optimal to the constructed degraded channel, then the capacity region of the multiple antenna broadcast channel is characterized.

1. Introduction

We consider a memoryless broadcast channel to model the downlink of a wireless system with N antennas at the base station and K users with M_i receive antennas with user i. Focusing on one particular time instant, denote the received vector of symbols at receiver i by $\mathbf{y}_i$ and $\mathbf{y}_{\rm dl} \stackrel{\rm def}{=} (\mathbf{y}_1, \ldots, \mathbf{y}_K)^t$. Observe that $\mathbf{y}$ is a column vector of length $\sum_{i=1}^K M_i$. They are related by:

$$\mathbf{y}_{\rm dl} = \begin{bmatrix} \mathbf{y}_1 \\ \vdots \\ \mathbf{y}_K \end{bmatrix} = \begin{bmatrix} H_1^\dagger \\ \vdots \\ H_K^\dagger \end{bmatrix} \mathbf{x}_{\rm dl} + \begin{bmatrix} \mathbf{z}_1 \\ \vdots \\ \mathbf{z}_K \end{bmatrix}. \tag{1.1}$$

Here H_i is a fixed $N \times M_i$ matrix with $H_i^\dagger(j,k)$ entry denoting the (flat fading) channel gain from the jth transmit antenna to the kth receive antenna of user i, assumed to be known at both the transmitter and receivers. $\mathbf{x}_{\rm dl}$ is the vector input to the antenna array with an average total power constraint of P. The additive noise $\mathbf{z}_j$ at user j is zero mean, covariance I, complex circular symmetric Gaussian. Each user j has access only to the received symbol $\mathbf{y}_j$. We observe that the specification of the covariance of the noise $\mathbf{z}_j$ to be identity is not a restriction; colored additive noise can be viewed as white after an appropriate whitening transform on the corresponding channel ($H_j^\dagger$) to user j.

For $N > 1$, this is a vector Gaussian broadcast channel, and unlike the scalar Gaussian broadcast channel ($N = 1$ and $M_i = 1$ for each user i), it is in general

---

1991 *Mathematics Subject Classification.* Primary 94A05, 94A14; Secondary 94B60.
Key words and phrases. Broadcast Channel, Multiple Antennas, Downlink, Capacity.

not degraded and the capacity region is unknown. Recent works [**3, 21, 18, 22**] obtained the *sum capacity* of this channel. This paper focuses on the entire capacity region. We propose a natural outer bound by converting the multiple antenna broadcast channel into a *degraded* one by making users privy to the signals of users ordered below them. The capacity region of degraded broadcast channels is known (Theorem 14.6.2 in [**4**]), and is parameterized by certain auxiliary random variables, one for each of the users. A simple extension of the proof techniques in [**21**] allows us to evaluate the achievable region of this degraded multiple antenna broadcast channel with jointly Gaussian parameters (auxiliary random variables and the input). We show that this achievable region is contained within the *achievable* region of the multiple antenna Gaussian broadcast channel. We conclude that the characterization of the capacity region is complete if jointly Gaussian inputs are optimal for degraded multiple antenna Gaussian broadcast channels (the single antenna case was solved in [**1**] using the celebrated *entropy-power inequality* for the first time in information theory).

In Sections 2 and 3 we develop a conservation law associated with a simple change of variable between two *reciprocal* Gaussian linear estimation problems and characterize the achievable region of the multiple antenna broadcast channel in the context of a class of Costa precoding strategies. Section 4 introduces the degraded outer bound. We evaluate the achievable region of the degraded outer bound with jointly Gaussian inputs as the capacity region of a reciprocal multiple access channel with *degraded transmitters* in Section 5. Finally, we extend the convex duality result of Section 3.4 in [**21**] to conclude that the capacity region of the degraded transmitters multiple access channel is equal to the region achievable with Costa precoding in the multiple antenna broadcast channel.

Related results are reported in [**19**].

Notations: we use lower case letters to denote scalars, upper case letters to denote matrices, and boldface to denote vectors. $\mathcal{C}N(\mu, Q)$ denotes a complex circular symmetric Gaussian distribution with mean μ and covariance matrix Q.

2. Two Gaussian Estimation Problems: Change of Variable and a Conservation Law

In this section, we consider two related Gaussian mean square estimation problems. Focusing on a change of variable between the variances of the Gaussian random variables to be estimated in the two problems, we show that it satisfies a conservation law. This is an extension of our observation in [**21**] where we had observed an equivalence between the performance of a class of receive and transmit strategies when the role of transmitters and receivers are reversed for the Gaussian broadcast channel of (1.1). This equivalence has been observed in seemingly different contexts in the literature.

(1) In the context of the capacity of a point-to-point multiple transmit, multiple receive antenna channel, [**15**] shows that the capacity is unchanged when the role of the transmitters and receivers is interchanged. The author calls this a *reciprocity* result.
(2) In the context of the degraded Gaussian broadcast channel, [**9**] shows that the capacity region is contained in the capacity region of the corresponding multiple access channel with the transmit power constraint of

the broadcast channel translated to the sum of powers in the multiple access channel. The authors name their result a *duality* connection.

(3) The recent work of [**18**] where the initial duality observation of [**9**] is extended to the multiple antenna case.

(4) In the context of the downlink of a multiple antenna system employing simple linear beamforming strategies followed by single user receivers by the users, [**17**] and [**11**] show that the optimal choice of transmit and receive beamforming vectors is closely related to a *virtual uplink problem.*

Our change of variable followed by the conservation law allows us to succinctly generalize the observations above and gives a simple characterization of the largest achievable region with Gaussian inputs in Marton's strategy, [**10**]. An independent and similar derivation of the duality in the context of *linear* beamforming strategies with $M_i = 1$ for all i is presented in [**14**].

2.1. Two Linear Gaussian Estimation Problems. We present two closely related estimation problems involving Gaussian random variables. We use the minimum mean square error in the estimation as the performance measure. Next we introduce a change of variable in the powers of the random variables being estimated to preserve the same performance in both the problems. Finally we show that this change of variable satisfies a simple conservation law.

2.1.1. *Primal Estimation Problem.* We focus on the vector Gaussian channel:

$$\mathbf{y}_{\text{dl}} = H^\dagger \mathbf{x}_{\text{dl}} + \mathbf{z}, \tag{2.1}$$

with $H^\dagger$ being a fixed matrix of dimension k by n. The additive noise $\mathbf{z}$ is $\mathcal{C}N(0, Q)$ where Q is a covariance matrix of dimension k. Here $\mathbf{x}_{\text{dl}}$ and $\mathbf{y}_{\text{dl}}$ are notational abuses since they are also used in (1.1). However, this abuse is on purpose and in the context of this subsection, the careful reader can view these variables as different from those defined in (1.1). We consider vector inputs $\mathbf{x}_{\text{dl}}$ (of length n) of the form:

$$\mathbf{x}_{\text{dl}} = \sum_{i=1}^{k} x_i \mathbf{v}_i,$$

where $x_1, \ldots, x_k$ are zero mean Gaussian random variables with variance $q_1, \ldots, q_k$ respectively. Here $\mathbf{v}_1, \ldots, \mathbf{v}_k$ are fixed vectors (of length n) and normalized so that their l_2 norm is unity. We are interested in the problem of linear mean square estimation of the x_i's from $\mathbf{y}_{\text{dl}}$ and focus on the following class of separate, linear, estimation strategies, $1 \leq i \leq k$:

$$\hat{x}_i = \mathbf{u}_i^\dagger \mathbf{y}_{\text{dl}},$$

where $\mathbf{u}_i$ is a fixed vector (of length k). With this linear filter bank, the signal to interference ratio (SIR) of the random variable x_i is defined, and computed as, (see Section 3.2 in [**8**], a comprehensive tutorial article),

$$\text{SIR}_i \stackrel{\text{def}}{=} \frac{q_i \mid \mathbf{u}_i^\dagger H^\dagger \mathbf{v}_i \mid^2}{\mid \mathbf{u}_i^\dagger Q \mathbf{u}_i \mid^2 + \sum_{j \neq i} q_j \mid \mathbf{u}_i^\dagger H^\dagger \mathbf{v}_j \mid^2}. \tag{2.2}$$

Denoting $\mathbf{b} \stackrel{\text{def}}{=} (b_1, \ldots, b_k)$ where

$$b_i \stackrel{\text{def}}{=} \frac{\text{SIR}_i}{(1 + \text{SIR}_i) \mid \mathbf{u}_i^\dagger H^\dagger \mathbf{v}_i \mid^2},$$

we can rewrite (2.2) in matrix notation as:

$$(I - \operatorname{diag}\{b_1, \ldots, b_K\} B)\, \mathbf{q} = D\mathbf{b}\,. \tag{2.3}$$

Here we have written the diagonal matrix D (of size k) with diagonal elements,

$$D_{ii} = \mathbf{u}_i^\dagger Q \mathbf{u}_i.$$

We have also written B as a $k \times k$ matrix with (i,j)th component equal to $\mathbf{u}_j^\dagger H^\dagger \mathbf{v}_i$. As in Section 2.5 of [**8**], a positive solution to $\mathbf{q}$ that satisfies (2.3) exists if and only if the Perron-Frobenius eigenvalue of $\operatorname{diag}\{b_1, \ldots, b_k\} B$ is less than 1. This characterizes the *performance region* (as the vector $\mathbf{b}$) of the bank of linear transmit and receive strategies $\mathbf{v}_i, \mathbf{u}_i,\ i = 1 \ldots k$ and the corresponding component-wise minimum transmit power required as

$$\mathbf{q}_{\min} \stackrel{\text{def}}{=} (I - \operatorname{diag}\{b_1, \ldots, b_k\} B)^{-1} D\mathbf{b}\,. \tag{2.4}$$

2.1.2. *Reciprocal Estimation Problem.* We now turn to another linear Gaussian estimation problem, one that is *reciprocal* to (2.1).

$$\mathbf{y}_{\text{ul}} = H\mathbf{x}_{\text{ul}} + \mathbf{w}, \tag{2.5}$$

Here $\mathbf{w} \sim \mathcal{C}N(0, I)$ of length n and the random vector $\mathbf{x}_{\text{ul}}$ (of length k) is of the form:

$$\mathbf{x}_{\text{ul}} = \sum_{i=1}^{k} \tilde{x}_i \mathbf{u}_i,$$

where $\mathbf{u}_i$'s are fixed vectors (of length k) and $\tilde{x}_1, \ldots, \tilde{x}_k$ are zero mean Gaussian random variables with variances $p_1, \ldots, p_k$ respectively. We consider the problem of estimating these random variables from $\mathbf{y}_{\text{ul}}$ through the bank of linear transforms:

$$\mathbf{v}_i^\dagger \mathbf{y}_{\text{ul}}, \quad 1 \le i \le k.$$

Observe that the dimensions of the received and transmit vectors of this estimation problem are reversed when compared to those of (2.1). Further, the role of transmit and receiver filters ($\mathbf{u}_i, \mathbf{v}_i$'s) is also reversed.

The SIR of $\tilde{x}_i$ with this transmission-reception pair can be written as

$$\text{SIR}_i \stackrel{\text{def}}{=} \frac{p_i \mid \mathbf{v}_i^\dagger H \mathbf{u}_i \mid^2}{1 + \sum_{j \neq i} p_j \mid \mathbf{v}_i^\dagger H \mathbf{u}_j \mid^2}\,. \tag{2.6}$$

Denoting $\mathbf{a} \stackrel{\text{def}}{=} (a_1, \ldots, a_k)^t$ where

$$a_i \stackrel{\text{def}}{=} \frac{\text{SIR}_i}{(1 + \text{SIR}_i) \mid \mathbf{v}_i^\dagger H \mathbf{u}_i \mid^2},$$

we can rewrite (2.6) in matrix notation as:

$$(I - \operatorname{diag}\{a_1, \ldots, a_k\} A)\, \mathbf{p} = \mathbf{a}\,. \tag{2.7}$$

We observe that the $k \times k$ matrix A has (i,j)th component equal to $\mid \mathbf{v}_j^\dagger H \mathbf{u}_i \mid^2$. and hence that $B = A^t$. The achievable performance region with this choice of transmission-reception strategy is the set of $\mathbf{a}$ such that the Perron-Frobenius eigenvector of $\operatorname{diag}\{a_1, \ldots, a_k\} B^t$ is less than one. Further, the component-wise minimal transmit power vector required to achieve this performance is

$$\mathbf{p}_{\min} \stackrel{\text{def}}{=} \left(I - \operatorname{diag}\{a_1, \ldots, a_k\} B^t\right)^{-1} \mathbf{a}\,. \tag{2.8}$$

2.1.3. *A Conservation Law.* We first observe that the achievable performance region in both the cases is the *same*. This is seen by noting that the Perron Frobenius eigenvalues of $\operatorname{diag}\{a_1,\dots,a_k\}A$ and $\operatorname{diag}\{a_1,\dots,a_k\}A^t$ are equal. Thus for equal performance, with $a_1,\dots,a_k$ equal to $b_1,\dots,b_k$, we have the following change of variable from the minimum variances (powers) of the random variables required in the first problem to the ones in the reciprocal, i.e.,

$$\mathbf{q}_{\min} \longmapsto \mathbf{p}_{\min}.$$

We now claim our key result: this change of variable satisfies the conservation law:

$$\sum_{i=1}^{k} \mathbf{q}_{\min}(i) = \sum_{i=1}^{k} \mathbf{p}_{\min}(i)\mathbf{u}_i^\dagger Q \mathbf{u}_i. \tag{2.9}$$

In particular, when the noise $\mathbf{z}$ in the primal estimation problem is white (so that the covariance matrix $Q = I$) and normalizing the transmit vectors $\mathbf{u}_1,\dots,\mathbf{u}_k$ to have unit norm, (2.9) says that the *sum* of the minimum transmit powers required in both the estimation problems to maintain a performance level (kept the same for both the problems) are equal.

To see (2.9), denote $\mathbf{e} = (1,\dots,1)$ and observe that

$$\begin{aligned}
\mathbf{e}^t\mathbf{q}_{\min} &= \mathbf{e}^t\left(I - \operatorname{diag}\{b_1,\dots,b_K\}B\right)^{-1} D\mathbf{b}\ , \text{ from (2.8)}\ ,\\
&= \sum_{j=1}^{\infty} \mathbf{e}^t\left(\Pi_{i=1}^{j}\operatorname{diag}\{b_1,\dots,b_k\}B\right) D\mathbf{b}\ , \text{ since } \mathbf{b} \text{ is achievable}\ ,\\
&= \sum_{j=1}^{\infty} \mathbf{b}^t B \operatorname{diag}\{b_1,\dots,b_k\} B \cdots \operatorname{diag}\{b_1,\dots,b_k\} B\, D\mathbf{b}\ ,\\
&= \sum_{j=1}^{\infty} \mathbf{b}^t D B^t \operatorname{diag}\{b_1,\dots,b_k\} \cdots B^t \operatorname{diag}\{b_1,\dots,b_k\} B^t\, \mathbf{b}\ ,\\
&= \sum_{j=1}^{\infty} \mathbf{e}^t D\left(\Pi_{i=1}^{j}\operatorname{diag}\{b_1,\dots,b_k\}B^t\right)\mathbf{b}\ ,\\
&= \mathbf{e}^t D\left(I - \operatorname{diag}\{b_1,\dots,b_K\}B^t\right)^{-1}\mathbf{b}\ ,\\
&= \mathbf{e}^t D\mathbf{p}_{\min},
\end{aligned} \tag{2.10}$$

where the last step follows from (2.4) since $\mathbf{b} = \mathbf{a}$ and $B^t = A$. Though our transmission-reception strategy was parameterized by the bank of linear filters $\mathbf{u}_i, \mathbf{v}_i$'s, we observe that the performance depended only on them through the elements of the matrices B (in the original problem) and A (in the reciprocal problem). Further, the conservation law is solely dependent on the fact that $B = A^t$.

3. Successive Cancellation and Costa Precoding

In the previous section, we studied certain estimation problems on the channels (2.1) and (2.5). In this section, we study certain strategies of *reliable communication* on these two channels. We leverage the observation that the conservation law in (2.9) was due to the fact that $B^t = A$ by expanding our set of transmission-reception strategies that retain this property. We consider two strategies: successive cancellation for the reciprocal channel in (2.5) and Costa precoding for the primal channel in (2.1).

3.1. Successive Cancellation. Consider the reciprocal channel of (2.5)

$$\mathbf{y}_{\mathrm{ul}} = H\mathbf{x}_{\mathrm{ul}} + \mathbf{w},$$

with k independent data streams being communicated through this channel. We retain the bank of linear transmit and receive filters ($\mathbf{u}_i, \mathbf{v}_i$'s) but now demodulate symbols of the data streams using successive cancellation (this is a multiuser receiver in contrast to single user receiver structures of the previous section). Suppose we cancel in the order $1, \dots, k$, then the SIR of data stream i with this receiver structure is

$$\text{(3.1)} \qquad \mathrm{SIR}_i = \frac{p_i \mid \mathbf{u}_i^\dagger H \mathbf{v}_i \mid^2}{1 + \sum_{j>i} p_j \mid \mathbf{u}_j^\dagger H \mathbf{v}_i \mid^2},$$

with the signals from data streams $1, \dots, j-1$ decoded and perfectly canceled. In the matrix notation of (2.7), we have

$$\text{(3.2)} \qquad (I - \mathrm{diag}\,\{a_1, \dots, a_k\}\,\mathcal{U}(A))\,\mathbf{p} = \mathbf{a}\,.$$

Here we have introduced the notation $\mathcal{U}(A)$ to indicate the upper triangular part of A, i.e., the (i,j)th entry of $\mathcal{U}(A)$ is A_{ij} if $j \geq i$ and 0 otherwise. This scheme allows joint reliable communication of the data streams at rates $R_1, \dots, R_k$ with

$$R_i = \log\left(1 + \mathrm{SIR}_i\right).$$

3.2. Costa Precoding. Successive cancellation is a *receiver-centric* strategy. A *transmitter-centric* strategy is to precode the data streams at the transmitter so that performance is improved over the linear single user strategies of the previous section. Here we consider a precoding strategy built on a result of Costa [**5**]. Consider the scalar point-to-point channel:

$$y = x + s + z$$

where s, z are independent Gaussian noise with s known to the transmitter but not to the receiver and z known to no one. Costa showed that the capacity of this channel is the same as that of the AWGN channel $y = x + z$, i.e. having the side information on s at the transmitter is as powerful as knowing S both at the transmitter *and* the receiver.

We consider the primary channel in (2.1)

$$\mathbf{y}_{\mathrm{dl}} = H^\dagger \mathbf{x}_{\mathrm{dl}} + \mathbf{z},$$

and consider the input composed of k independent data streams as before:

$$\mathbf{x}_{\mathrm{dl}} = \sum_{i=1}^{k} x_i \mathbf{v}_i.$$

We also retain the bank of linear filters $\mathbf{u}_j$'s as before and the received signal at the output of the jth filter is

$$\mathbf{u}_j^\dagger \mathbf{y}_{\mathrm{dl}} = x_j \mathbf{u}_j^\dagger H^\dagger \mathbf{v}_j + \sum_{l<j} x_l \mathbf{u}_j^\dagger H^\dagger \mathbf{v}_l + \sum_{l>j} x_l \mathbf{u}_j^\dagger H^\dagger \mathbf{v}_l + \mathbf{u}_j^\dagger \mathbf{z}.$$

We use Gaussian i.i.d. inputs for $x_1, \dots x_k$, and perform Costa precoding for each data stream i, encoding the information in x_i treating the interference $\sum_{j>i} \sqrt{q_j} x_j \mathbf{u}_i^\dagger H^\dagger \mathbf{v}_j$ from data streams $i+1, \dots, k$ as side information S known at the transmitter and the interference from the other data streams as Gaussian noise. The SIR seen by data stream i is

$$\mathrm{SIR}_i = \frac{q_i \mid \mathbf{u}_i^\dagger H^\dagger \mathbf{v}_i \mid^2}{\mathbf{u}_i^\dagger Q \mathbf{u}_i^\dagger + \sum_{j=1}^{i-1} q_j \mid \mathbf{u}_i^\dagger H \mathbf{v}_j \mid^2}, \tag{3.3}$$

for each $i = 1, \ldots, k$. Note that data stream i now only sees interference from data streams $1, \ldots, i-1$, in contrast to the linear beamforming strategy where it sees interference from all other data streams. Hence, joint reliable communication of the data streams is possible at rates

$$R_i = \log\left(1 + \mathrm{SIR}_i\right),\ i = 1 \ldots k. \tag{3.4}$$

In the matrix notation of (2.3), we can write this as

$$\left(I - \mathrm{diag}\left\{b_1, \ldots, b_k\right\} \mathcal{L}(A^t)\right) \mathbf{q} = D\mathbf{b}\,. \tag{3.5}$$

As in the previous section, D is the diagonal matrix with diagonal elements $D_{ii} = \mathbf{u}_i^\dagger Q \mathbf{u}_i^\dagger$. Here we defined $\mathcal{L}(A)$ as a matrix whose (k, j)th component is equal to A_{kj} if $k \geq j$ and 0 otherwise. Observing that $\mathcal{L}(A^t) = \left(\mathcal{U}(A)\right)^t$, we see that the performance of the primary and reciprocal channels are equivalent (in terms of the SIRs achievable and the minimum transmit power required to achieve it) even with this extended set of transmission-reception strategies.

3.3. Broadcast and Multiple Access Channels. We can now see the application of the receiver-centric successive cancellation scheme to a multiple access channel where the transmitters do not cooperate and the Costa precoding scheme to a broadcast channel where the receivers do not cooperate. This allows us to derive a one-one correspondence in achievable rate tuples between broadcast and multiple access channels that are reciprocal to each other. We begin with the broadcast channel of (1.1).

3.3.1. *Multiple Antenna Broadcast Channel.* We would like to fit the primary channel of (2.1) to correspond to the broadcast channel of interest (see (1.1)). We proceed with the following steps.

- The first step towards this correspondence is to let $n = N$, the number of transmit antennas and $k = \sum_{i=1}^K M_i$ the total number of receive antennas, summed over all the K users.
- The k data streams are divided up with the ith user receiving data from M_i of the streams.
- To account for the fact that the users cannot cooperate with each other, we impose the following condition on the receive filters $\mathbf{u}_1, \ldots, \mathbf{u}_k$. The first M_1 filters which correspond to the data streams associated with user 1 are nonzero in only the first M_1 components. More generally, the M_i filters corresponding to the data streams associated with user i are nonzero in only M_i components; these components are $\sum_{j<i} M_j + 1, \ldots, \sum_{j\leq i} M_j$.
- The broadcast channel in (1.1) is specified only through the marginal distributions of $\mathbf{z}_1, \ldots, \mathbf{z}_K$ (each having identity covariance). The capacity region of any broadcast channel depends only on the marginal channels from the transmitter to the receivers (Theorem 14.6.1 of [**4**]). In this context, this means we can choose the joint distribution of $\mathbf{z}_1, \ldots, \mathbf{z}_K$ to be $\mathcal{C}N(0, Q)$ with the covariance matrix satisfying the individual covariance constraints of $\mathbf{z}_1, \ldots, \mathbf{z}_K$ to all be identity.

By this constraint on the receive filters, we ensure that there is no cooperation among the users in implementing the receiver filter bank. In the transmit side, we use Costa precoding, ordering the data streams from k through 1. We can now calculate the achievable rate tuple for this broadcast channel: user i rate is the sum of the rates of the M_i data streams associated with it and equal to

$$R_i = \sum_{l=1}^{M_i} \log\left(1 + \mathrm{SIR}_{\left(\sum_{j<i} M_j + l\right)}\right).$$

The expression for SIR_l is given in (3.3). What is the total power transmit from the base station to achieve this rate? It is equal to the sum of the powers in the k data streams and that is equal to

$$\mathbb{E}\left[\mathbf{x}_{\mathrm{dl}}^\dagger \mathbf{x}_{\mathrm{dl}}\right] = \sum_{i=1}^{k} q_i \mathbf{v}_i^\dagger \mathbf{v}_i = \sum_{i=1}^{k} q_i,$$

since $\mathbf{v}_i$'s have unit l_2 norm.

3.3.2. *Multiple Antenna Multiple Access Channel.* We now turn to the reciprocal channel of (2.5) retaining the choice of the filter banks $\mathbf{u}_i, \mathbf{v}_i$'s. As we did for the broadcast channel, we take $n = N$ and number of data streams $k = \sum_{i=1}^{K} M_i$ with user i messages contained over M_i of the data streams. Now, the role of the filters $\mathbf{u}_i$'s is at the transmitter and with the specification as above (with the first M_1 filters having nonzero values in only the first M_1 components, and so on), we see that the transmit strategy using the linear filter bank does not require the users to *cooperate.* To summarize, we consider this transmit strategy combined with successive cancellation at the receiver, ordering data streams 1 through k on the following multiple access channel:

$$\mathbf{y}_{\mathrm{ul}} = \sum_{i=1}^{K} H_i \mathbf{x}_i + \mathbf{w}. \tag{3.6}$$

Here $\mathbf{w}$ is $\mathcal{C}N(0, I)$ and the transmit vector of user i is

$$\mathbf{x}_i = \sum_{l=1}^{M_i} x_{il} \mathbf{u}_{\sum_{j<i} M_j + l}.$$

Here $x_{i1}, \ldots, x_{iM_i}$ are complex Gaussian with zero mean and unit variance. User i can now communicate at rate R_i given by

$$R_i = \sum_{l=1}^{M_i} \log\left(1 + \mathrm{SIR}_{\sum_{j<i} M_j + l}\right). \tag{3.7}$$

The value of SIR_l is given in (3.1).

3.4. The Inner Bound: An Application of the Conservation Law. Fix a set of $\mathbf{u}_i, \mathbf{v}_i$'s and successive cancellation in the multiple access channel and Costa precoding in the broadcast channel. We would like to achieve the same set of SIR values for each data stream. Then the rate tuple achieved in the reciprocal broadcast and multiple access channels are the same. We choose the the powers of the data streams ($q_1, \ldots, q_k$ in the broadcast channel and $p_1, \ldots, p_k$ in the multiple access channel) to be the component-wise minimal ones to meet the SIR targets with the fixed linear filters. The total transmit power in the broadcast channel is

$$P \stackrel{\text{def}}{=} \sum_{i=1}^{k} q_i.$$

The conservation law (2.9) allows us to derive the following constraint on the transmit signal in the multiple access channel

$$\begin{aligned} P &= \sum_{i=1}^{k} p_i \mathbf{u}_i^\dagger Q \mathbf{u}_i, \\ &= \sum_{i=1}^{k} p_i \mathbf{u}_i^\dagger \mathbf{u}_i, \quad (3.8) \\ &= \sum_{l=1}^{K} \mathbb{E}\left[\mathbf{x}_l^\dagger \mathbf{x}_l\right]. \quad (3.9) \end{aligned}$$

In (3.8) we used the property of the nonzero components of $\mathbf{u}_i$'s and the fact that the covariance matrix Q has to satisfy the individual covariance constraints of $\mathbf{z}_1, \ldots, \mathbf{z}_K$ to be all identity. The step of (3.9) follows by the definition of $\mathbf{x}_l$ to be the transmit signal from user l, which is equal to the sum of the signals of the M_l data streams that correspond to user l.

Hence a *cost* constraint of the type in (3.9) is met by the transmit strategy on the multiple access channel. Suppose we impose the cost constraint in (3.9) on the multiple access channel of (3.6). Linear transmit strategies of the users followed up with successive cancellation at the receiver are *optimal* for this channel. In particular, for a given set of transmit filters $\mathbf{u}_i$'s and powers p_i's, we can restrict the choice of receive filters $\mathbf{v}_i$ to be the linear MMSE (minimum mean square error) filter for data stream i treating signals from data streams $1, \ldots, i-1$ to be nonexistent and treating signals from data streams $k+1, \ldots, k$ as noise, for every i, i.e., the unnormalized vector $\mathbf{v}_k$ has the expression

$$\mathbf{v}_i = \left(I + \sum_{j=i+1}^{k} p_j \mathbf{u}_j H_l \mathbf{u}_j^\dagger\right)^{-1} H_l u_i \,, \quad (3.10)$$

(See Section 3 of [**20**]). Here l is the user to which the ith stream belongs, i.e., $\sum_{j=1}^{l-1} M_j < i \leq \sum_{j=1}^{l} M_j$. Then as we vary over all transmit strategies ($\mathbf{u}_i, p_i$'s), along with all reordering of the users in the successive cancellation scheme, the capacity region of the multiple access channel in (3.6) is achieved. This capacity region can be expressed as the closed convex hull of

$$\bigcup_{\Sigma_1,\ldots,\Sigma_K : \operatorname{tr}\left[\sum_{i=1}^{K}\Sigma_i\right] \leq P} \left\{(R_1, \ldots, R_K) : \sum_{i\in S} R_i \leq \log\det\left(I + \sum_{i \in S} H_i \Sigma_i H_i^\dagger\right), \right. \quad (3.11)$$
$$\left. \forall S \subseteq \{1, \ldots, K\}\right\}.$$

We can now conclude the following. Consider the broadcast channel in (1.1) with a power constraint on the total power transmit out of the base station equal to P. Consider the reciprocal multiple access channel in (3.6) with the cost constraint in (3.9). Then, there is a one-one correspondence (through the change of variable) between a Costa precoding scheme for the broadcast channel (meeting the power constraint) and a successive cancellation scheme for the multiple access channel (meeting the cost constraint) such that they achieve the same rate tuple. Since the successive cancellation schemes achieve the capacity region of the multiple access channel and the Costa precoding schemes form an achievable region of the broadcast channel, we have shown that the capacity region of the broadcast channel is inner bounded by that of the reciprocal multiple access channel (with the corresponding cost constraint).

Although we introduced the broadcast transmission scheme in terms of Costa precoding, it should be noted that the achievable rates are in fact the same as an achievable region for general broadcast region first proposed by Marton [**10**] (this strategy provides the largest known achievable region for a general broadcast channel).

THEOREM 3.1 (Marton). *Fix the joint distribution $p(u_1,\ldots,u_K,\mathbf{x})$ for some auxiliary random variables $u_1,\ldots,u_K$ (with no constraints on the cardinality of their alphabets) and $\mathbf{x}$ is a random variable on the input alphabets. The following is an achievable rate region:*
(3.12)

$$\left\{(R_1,\ldots,R_k):\sum_{s\in S}R_s\le\sum_{s\in S}(I(u_s;y_s)-H(u_s))+H(u_s,\ s\in S),\ \forall S\subseteq\{1,\ldots,k\}\right\}.$$

Let $x_1,\ldots,x_k$ be i.i.d. $\mathcal{CN}(0,1)$ random variables. If we let $\mathbf{x}=\sum_{i=1}^k\sqrt{q_i}x_i\mathbf{v}_i$ and

$$u_i\stackrel{\text{def}}{=}\sqrt{q_i}x_i\mid\mathbf{u}_i^\dagger H^\dagger\mathbf{v}_i\mid+\frac{q_i\mid\mathbf{u}_i^\dagger H^\dagger\mathbf{v}_i\mid^2}{\mathbf{u}_i^\dagger\mathbf{u}_i+\sum_{j=1}^{i-1}q_j\mid\mathbf{u}_i^\dagger H^\dagger\mathbf{v}_j\mid^2}\sum_{j=i+1}^k\sqrt{q_j}x_j\mathbf{u}_i^\dagger H^\dagger\mathbf{v}_j\,.$$

Then it can be seen by a direct calculation that the resulting rate point in the Marton's region for data stream i is exactly

$$\log\left(1+\mathrm{SIR}_i\right),$$

where SIR_i is given in (3.3). Now if we maintain the restrictions on the receiver filters $\mathbf{u}_i$'s as in Section 3.3.1, and identify the M_l of the data streams as corresponding to user l, then the Marton region with these inputs yields the same rate tuples as the Costa precoding strategy defined above. The conceptual connection between Marton's and Costa's result is that they are both based on a random binning encoding technique.

4. Degraded Receivers Outer Bound

The capacity region of nondegraded broadcast channels (such as (1.1)) is unknown. The best known outer bound to a general broadcast channel is also due to Marton (Theorem 5 in [**10**]) but is rather complicated to evaluate. In this section, we develop a natural outer bound to the multiple antenna broadcast channel by converting it into a *degraded* broadcast channel. We first order users according to the permutation π. We form the degraded version by letting user π_i be privy to the signals of all users $\pi_j,\ i<j\le K$. Let us denote the new received signal of user π_i by $\tilde{\mathbf{y}}_{\pi_i}=(\mathbf{y}_{\pi_i},\ldots\mathbf{y}_{\pi_K})$. In other words, the new user π_i has access to the

signals of all users $\pi_{i+1}, \ldots, \pi_K$:

$$\tilde{\mathbf{y}}_{\pi_i} = \begin{bmatrix} H_{\pi_i}^\dagger \\ \vdots \\ H_{\pi_K}^\dagger \end{bmatrix} \mathbf{x}_{\text{dl}} + \begin{bmatrix} \mathbf{z}_{\pi_i} \\ \vdots \\ \mathbf{z}_{\pi_K} \end{bmatrix}. \tag{4.1}$$

Here we have denoted the overall additive noise vector seen by the new user π_i by $\mathbf{z}^{(\pi_i)} = \left(\mathbf{z}_{\pi_i}, \mathbf{z}_{\pi_{i+1}}, \ldots, \mathbf{z}_{\pi_K}\right)$. This broadcast channel can be depicted through the Markov chain relation:

$$\mathbf{x}_{\text{dl}} \to \tilde{\mathbf{y}}_{\pi_1} \to \tilde{\mathbf{y}}_{\pi_2} \to \cdots \to \tilde{\mathbf{y}}_{\pi_K}. \tag{4.2}$$

While the broadcast channel of (1.1) was defined only through the marginal distributions of the noises $\mathbf{z}_1, \ldots, \mathbf{z}_K$ to all be $\mathcal{C}N(0, I)$, this degraded broadcast channel depends on the joint distribution of the noises. We take the noise vector seen by the user π_i in this degraded channel $\mathbf{z}^{(\pi,i)}$ to be $\mathcal{C}N(0, Q^{(\pi,i)})$ where the covariance matrix $Q^{(\pi,i)}$ of $\mathbf{z}^{(\pi,i)}$ has to satisfy the property that the constituent random vectors $\mathbf{z}_{\pi_i}, \ldots \mathbf{z}_{\pi_K}$ all have covariance less than or equal to (in the sense of the partial order on the cone of p.s.d. matrices) identity. This forms a restriction on the block diagonal matrices of $Q^{(\pi,i)}$ ($K - i + 1$ of them) to be less than or equal to the identity matrix. We denote this constraint by

$$Q^{(\pi,i)} \preceq^{(i)} I. \tag{4.3}$$

Denoting the capacity region of this degraded broadcast channel by

$$\mathcal{C}_\pi \left(Q^{(\pi,1)}, \ldots, Q^{(\pi,K)}\right)$$

we have an outer bound to the capacity region of the broadcast channel of (1.1):

$$\mathcal{C}_{\text{bc}} \subseteq \bigcap_{\pi} \bigcap_{Q^{(\pi,1)}, \ldots, Q^{(\pi,K)}} \mathcal{C}_\pi \left(Q^{(\pi,1)}, \ldots, Q^{(\pi,K)}\right). \tag{4.4}$$

Let us fix a permutation π, the joint distribution of the noises denoted by $Q^{(\pi,i)}, i = 1 \ldots K$ and consider the degraded broadcast channel in (4.1) and (4.2). An achievable region is given by the Costa coding strategies described in Section 3.2. Let us denote this region by $\mathcal{R}_{\text{costa}}^{\text{degraded}}$.

The capacity region of a general degraded broadcast channel is known and in this case is expressed as the closed convex hull of the arbitrary union over all joint distributions of $\mathbf{x}_{\text{dl}}$ and auxiliary random variables $\mathbf{r}_{\pi_1}, \ldots, \mathbf{r}_{\pi_K}$ (with a power constraint on the transmit signal $\mathbf{x}_{\text{dl}}$) of the region

$$\left\{(R_1, \ldots, R_K) : R_{\pi_i} \leq I\left(\mathbf{r}_{\pi_i}; \tilde{\mathbf{y}}_{\pi_i} \mid \mathbf{r}_{\pi_{i+1}}, \ldots, \mathbf{r}_{\pi_K}\right),\ i = 1 \ldots K\right\}. \tag{4.5}$$

A succinct characterization of this union region is quite complicated, but evaluating it for jointly Gaussian input distributions (i.e., $\mathbf{x}_{\text{dl}}, \mathbf{r}_1, \ldots, \mathbf{r}_K$ are jointly Gaussian) is quite easy. In particular, with the choice,

$$\mathbf{r}_i = \sum_{j=1}^{\sum_{l=i+1}^{K} M_l} x_{ij} \mathbf{v}_{ij},\ \forall i = 1 \ldots K, \tag{4.6}$$

$$\mathbf{x}_{\text{dl}} = \sum_{i=1}^{K} \sqrt{q_i} \mathbf{r}_i. \tag{4.7}$$

Here x_{ij}'s are i.i.d. zero mean, unit variance Gaussian random variables. With this choice of the parameters, the individual rates in (4.5) are easily evaluated and can be shown to equal $\mathcal{R}_{\text{costa}}^{\text{degraded}}$, the region achieved by Costa precoding.

For the scalar degraded Gaussian broadcast channel, [**1**] proves that the jointly Gaussian inputs are optimal. This proof also used, for the first time in information theory, the celebrated *entropy power inequality*. However, we dont know if jointly Gaussian inputs are still optimal for the multiple antenna degraded broadcast channel. We state this as a conjecture.

CONJECTURE 4.1. Jointly Gaussian inputs of the form (4.6) and (4.7) are optimal for the degraded multiple antenna broadcast channel in (4.2).

5. Degraded Transmitters Multiple Access Channel

We can now look at the channel reciprocal to the degraded broadcast of (4.1) with Costa coding strategies and use the conservation law to evaluate the achievable region denoted by $\mathcal{C}_\pi\left(Q^{(\pi,1)},\ldots,Q^{(\pi,K)}\right)$. Here we have fixed the permutation of users π and the corresponding noise covariances $Q^{(\pi,i)}$'s. The reciprocal channel can be viewed as a multiple access channel, as in Section 3.3.2. The degraded broadcast channel of (4.1) allowed the new user π_i to be privy to the signals of all the users $\pi_{i+1},\ldots,\pi_K$. In the reciprocal channel the role of transmit and receive strategies are reversed, and the user π_i has "access" to the antennas of users $\pi_{i+1},\ldots,\pi_K$. Precisely, the multiple access channel of interest is:

$$\mathbf{y}_{\text{ul}} = \sum_{i=1}^{K} H^{(\pi,i)}\mathbf{x}_{\pi,i} + \mathbf{w}. \tag{5.1}$$

Here $\mathbf{y}_{\text{ul}}$ is the received signal at the N receive antennas, as in (3.6). The difference from (3.6), however, is that the users have more transmit antennas. Specifically, user π_i here has $\sum_{j=i}^{K} M_{\pi_j}$ transmit antennas with the channel denoted by

$$H^{(\pi,i)} = \left[H_{\pi_i} H_{\pi_{i+1}} \cdots H_{\pi_K}\right],$$

and the transmit signal is denoted by $\mathbf{x}_{\pi,i}$ (a vector of length $\sum_{j=i}^{K} M_{\pi_j}$). Analogous to the degraded nature of the receivers in the broadcast channel of (4.1), here the *transmitters* are degraded with user π_i transmitting signals in the "same channel" as that of users $\pi_{i+1},\ldots,\pi_K$. This suggests us to name (5.1) as a *degraded transmitters* multiple access channel (MAC).

What is the cost constraint on the input in this degraded transmitters MAC? As we saw in the application of the conservation law in Section 3.4, the colored noise in the primary channel combined with the cost constraint translates into a cost constraint on the reciprocal. In this context, where we consider linear transmit and receive filers with successive cancellation in the degraded transmitters MAC and the same linear strategies with Costa precoding in the degraded receivers broadcast channel, the same performance is obtained by imposing the following cost constraint on the MAC of (5.1):

$$\sum_{i=1}^{K} \mathbb{E}\left[\mathbf{x}_{\pi,i}^{\dagger} Q^{(\pi,i)} \mathbf{x}_{\pi,i}\right] \leq P.$$

Analogous to (3.11), the capacity region of the (5.1) can be written as the closed convex hull of the union over all covariance matrices $\Sigma^{(\pi,1)}, \ldots, \Sigma^{(\pi,K)}$ of the regions (5.2)

$$\left\{(R_1,\ldots,R_K) : \sum_{i\in S} R_{\pi_i} \le \log\det\left(I + \sum_{i\in S} H^{(\pi,i)}\Sigma^{(\pi,i)}H^{(\pi,i)\dagger}\right),\ \forall S \subseteq \{1,\ldots,K\}\right\}.$$

The cost constraint on the inputs translates into

$$\operatorname{tr}\left[\sum_{i=1}^{K} Q^{(\pi,i)}\Sigma^{(\pi i)}\right] \le P.$$

Here $Q^{(\pi,i)}$ is a covariance matrix with the constraint that the block diagonal matrices (of sizes corresponding to those of the noise vectors) are all less than or equal to identity, i.e., $Q^{(\pi,i)} \preceq^{(i)} I$, using the notation from $Q^{(\pi,i)} \preceq^{(i)} I$, using the notation from (4.3). We now observe that this capacity region in (5.2) is larger than that of the multiple access channel in (3.11). This is easily seen by restricting the the user π_i with channel $H^{(\pi,i)}$ to signal only on the first M_{π_i} components, i.e., each π_i user though provided additional antennas (corresponding to all the degraded users $\pi_{i+1},\ldots,\pi_K$) signals only on its on M_{π_i} antennas. With this restriction, we have now reduced this degraded MAC to the multiple antenna MAC of (3.6).

5.1. Finding the Right Cost Matrix. Capacity regions of multiple access and broadcast channels are closed, convex and bounded and hence can be equivalently represented by the half spaces that contain them (Theorem 11.5 in [**12**]). We can hence focus on the linear functional

$$\sum_{i=1}^{K} a_i R_i, \tag{5.3}$$

parameterized by the vector $(a_1,\ldots,a_K)$. Referring to the degraded receivers bound (see (4.4)), we need to show that value of the linear functional in (5.3) optimized over all Costa coding strategies over both the broadcast channels (the original one and the degraded one which serves as the outer bound) are equal. To achieve this we need an appropriate ordering π of the users and a choice of the noise covariances $Q^{(\pi,i)}$'s. Using the conservation law, this is equivalent to showing that there exists an ordering π and cost matrix $Q^{(\pi,i)}$ such that the values of the linear functional in (5.3) maximized over the MAC of (3.6) and over the degraded transmitters MAC of (5.1) are equal.

Without loss of generality, let us suppose the ordering $a_1 \le a_2 \le \cdots \le a_K$. We pick the ordering of users π to be the same as the ordering of the linear functional, so that π is simply the identity permutation. Henceforth, we assume that the users are simply ordered from 1 through K and drop the notation of π in our variables. The capacity regions of Gaussian multiple access channels have special combinatorial properties. In particular, they belong to a class of polyhedrons known as *polymatroids* ([**16**]). Considering the value of the linear functional maximized over the capacity region of the MAC of (3.6), and using the polymatroidal property we can write this as:

$$\max \sum_{i=1}^{K} a_i R_i \text{ over rate tuples in the region in (3.11)}$$

$$= \max_{\sum_{i=1}^{K} \operatorname{tr}[\Sigma_i] \le P} \sum_{i=1}^{K} (a_i - a_{i-1}) \log\det \left(I + \sum_{j=i}^{K} H_j \Sigma_j H_j^\dagger \right) \tag{5.4}$$

Here we have written $a_0 = 0$. On the other hand, the value of the linear functional maximized over the capacity region of the degraded transmitters MAC of (5.1) is

$$\max \sum_{i=1}^{K} a_i R_i \text{ over rate tuples in the region in (5.2)}$$

$$= \max_{\sum_{i=1}^{K} \operatorname{tr}\left[\Sigma^{(i)} Q^{(i)}\right] \le P} \sum_{i=1}^{K} (a_i - a_{i-1}) \log\det \left(I + \sum_{j=i}^{K} H^{(j)} \Sigma^{(j)} H^{(j)\dagger} \right). \tag{5.5}$$

Our proof is complete if we can construct the cost functions $Q^{(i)}$'s with $Q^{(i)} \preceq^{(i)} I$, such that the expression in (5.5) equals that in (5.4). This is summarized in the following theorem:

THEOREM 5.1. *Value of the expression* (5.5) *minimized over* $Q^{(i)} \preceq^{(i)} I$, *is equal to the value of the expression* (5.4), $\forall i = 1 \dots K$.

We do this by a slight generalization of a corresponding result in Sections 3.3 and 3.4 of [**21**] where we characterized the sum capacity of this channel. We briefly mention the main step here and move the proof to the next section. The key step is the choice of the cost function $Q^{(i)}$ that provides user i no benefit by jointly encoding its signal across the "antennas" of the users $i+1, \dots, K$. Specifically, we choose the cost functions $Q^{(i)}$ such that the optimal user i input is *block diagonal*, i.e., there is no cooperative encoding of the user i signal over the antennas of users $i+1, \dots, K$. Since user i rate has a weight a_i no more than that of the weights to users $i+1, \dots, K$, there is no benefit to letting user i transmit any power on the antennas of users $i+1, \dots, K$. This means that each user signals only on its own antennas and this performance is equal to that of the MAC in (5.4). The proof provides a convexity theoretic justification on the lines of the calculation in Section 3.4 of [**21**].

5.2. Proofs.

5.2.1. *Proof of Theorem 5.1.* We start with the value of the linear functional maximized on the degraded MAC capacity region. Our first observation is that the map

$$\left(\Sigma^{(1)}, \dots, \Sigma^{(K)}\right) \mapsto \sum_{i=1}^{K} (a_i - a_{i-1}) \log\det \left(I + \sum_{j=i}^{K} H^{(j)} \Sigma^{(j)} H^{(j)\dagger} \right)$$

is jointly concave in $\left(\Sigma^{(1)}, \dots, \Sigma^{(K)}\right)$. This has been shown earlier in various forms (with some forms weaker than the statement above) in the literature. See [**15**] or Lemma II.4 of [**6**] or Prop. 4.2 of [**20**]. Further, for every fixed

$$Q^{(1)}, \dots, Q^{(i-1)}, Q^{(i+1)}, \dots, Q^{(K)},$$

the map

$$Q^{(i)} \mapsto \max_{\sum_{j=1}^{K} \mathrm{tr}\left[\Sigma^{(j)} Q^{(j)}\right] \le P} \sum_{i=1}^{K} (a_i - a_{i-1}) \log\det\left(I + \sum_{j=i}^{K} H^{(j)} \Sigma^{(j)} H^{(j)\dagger}\right)$$

is *convex* for every $i = 1 \ldots K$. This result has also been proved in the literature though not exactly in this form. One way to derive this is to combine Lemma II.3 of [**6**] with the observation that a point-wise supremum of convex functions yields a convex function. Thus the objective function is concave in the input distribution and convex in the cost functions. We use this saddle function property (Theorem 36.3 in [**12**]) to alternate the minimization (over the cost functions) and maximization (over the input covariance matrices). Starting with (5.5), we use the saddle function theorem to write the minimization over all $Q^{(1)}, \ldots, Q^{(K)}$ over all p.s.d. matrices with the constraints $Q^{(i)} \preceq^{(i)}$ as follows:

$$\min_{Q^{(i)}, i=2,\ldots K} \max_{\sum_{i=2}^{K} \mathrm{tr}\left[\Sigma^{(i)} Q^{(i)}\right] + p_1 \le P} \min_{Q^{(1)}} \max_{\mathrm{tr}\left[\Sigma^{(1)} Q^{(1)}\right] = p_1}$$

$$\sum_{i=1}^{K} (a_i - a_{i-1}) \log\det\left(I + \sum_{j=i}^{K} H^{(j)} \Sigma^{(j)} H^{(j)\dagger}\right). \tag{5.6}$$

The key step in our proof is contained in the following lemma. It is a slight generalization of a corresponding result derived in characterizing the sum capacity of this channel (in Section 3.3 of [**21**]).

LEMMA 5.2. *Fix $b_1, \ldots, b_k \ge 0$ and $1 \le i \le K$. Consider*

$$C_1 = \min_{Q^{(i)} \succeq 0, Q^{(i)} \preceq^{(i)} I} \max_{\Sigma^{(i)} \succeq 0, \mathrm{tr}\left[Q^{(i)} \Sigma^{(i)}\right] \le P} \sum_{j=1}^{k} b_j \log\det\left(S_j + H^{(i)} \Sigma^{(i)} H^{(i)\dagger}\right). \tag{5.7}$$

Here $S_1, \ldots, S_k$ are positive definite matrices of the (same) appropriate dimension. Then C_1 is equal to

$$C_2 = \max_{\Sigma_i, \ldots, \Sigma_K \succeq 0, \sum_{i=1}^{K} \mathrm{tr}[\Sigma_i] \le P} \sum_{j=1}^{k} b_j \log\det\left(S_j + \sum_{l=i}^{K} H_l \Sigma_l H_l^\dagger\right) \tag{5.8}$$

with appropriate dimensions of $\Sigma_j, i \le j \le K$. Further, $C_1 = C_2$ for every i.

In other words, the lemma says that the minimizing $Q^{(i)}$ in (5.7) makes the corresponding $\Sigma^{(i)}$ be block diagonal. We move the proof to the next subsection where we also construct an explicit form of the optimizing $Q^{(i)}$. Now we use this intermediary step repeatedly in reducing (5.6) to (5.4). First we can use Lemma 5.2 with $i = k = 1$ and $S_1 = I + \sum_{j=2}^{K} H^{(j)} \Sigma^{(j)} H^{(j)\dagger}$ allowing us to write (5.6) as

$$\min_{Q^{(i)}, i=2,\ldots,K} \max_{\sum_{i=2}^{K} \mathrm{tr}\left[\Sigma^{(i)} Q^{(i)}\right] + p_1 \le P} \max_{\sum_{j=1}^{K} \mathrm{tr}\left[\Sigma_j^{(1)}\right] = p_1}$$

$$a_1 \log\det\left(I + \sum_{j=1}^{K} H_j \Sigma_j^{(1)} H_j^\dagger + \sum_{j=2}^{K} H^{(j)} \Sigma^{(j)} H^{(j)\dagger}\right)$$

$$+ \sum_{i=2}^{K} (a_i - a_{i-1}) \log\det\left(I + \sum_{j=i}^{K} H^{(j)} \Sigma^{(j)} H^{(j)\dagger}\right). \tag{5.9}$$

Using the saddle function property of the objective function, we can move the minimization on $Q^{(2)}$ in (5.9) inside to arrive at

$$\min_{Q^{(i)},i=3,\dots,K} \max_{\sum_{i=3}^K \operatorname{tr}\left[\Sigma^{(i)}Q^{(i)}\right]+\sum_{j=1}^K \operatorname{tr}\left[\Sigma_j^{(1)}\right]+p_2\le P} \min_{Q^{(2)}} \max_{\operatorname{tr}\left[\Sigma^{(2)}Q^{(2)}\right]=p_2}$$

$$a_1 \log\det\left(I+\sum_{j=1}^K H_j\Sigma_j^{(1)}H_j^\dagger+\sum_{j=2}^K H^{(j)}\Sigma^{(j)}H^{(j)\dagger}\right)$$

$$+\sum_{i=2}^K (a_i-a_{i-1})\log\det\left(I+\sum_{j=i}^K H^{(j)}\Sigma^{(j)}H^{(j)\dagger}\right). \tag{5.10}$$

We now reuse Lemma 5.2 with $i=k=2$ and

$$\begin{aligned} S_1 &= I+\sum_{j=1}^K H_j\Sigma_j^{(1)}H_j^\dagger+\sum_{j=3}^K H^{(j)}\Sigma^{(j)}H^{(j)\dagger}, \\ S_2 &= I+\sum_{j=3}^K H^{(j)}\Sigma^{(j)}H^{(j)\dagger}, \end{aligned}$$

and $b_1=a_1$ and $b_2=a_2-a_1$. This allows us to write (5.10) as

$$\min_{Q^{(i)},i=3,\dots,K} \max_{\sum_{i=3}^K \operatorname{tr}\left[\Sigma^{(i)}Q^{(i)}\right]+\sum_{i=1}^2\sum_{j=i}^K \operatorname{tr}\left[\Sigma_j^{(i)}\right]\le P}$$

$$\sum_{i=1}^2 (a_i-a_{i-1})\log\det\left(I+\sum_{j=i}^K H_j\Sigma_j^{(i)}H_j^\dagger+\sum_{j=3}^K H^{(j)}\Sigma^{(j)}H^{(j)\dagger}\right)$$

$$+\sum_{i=2}^K (a_i-a_{i-1})\log\det\left(I+\sum_{j=i}^K H^{(j)}\Sigma^{(j)}H^{(j)\dagger}\right). \tag{5.11}$$

Continuing this process up to $i=K$, we arrive from (5.11) at

$$\max_{\sum_{i=1}^K\sum_{j=i}^K \operatorname{tr}\left[\Sigma_j^{(i)}\right]\le P} \sum_{i=1}^K (a_i-a_{i-1})\log\det\left(I+\sum_{l=i}^K\sum_{j=l}^K H_j\Sigma_j^{(l)}H_j^\dagger\right). \tag{5.12}$$

We now claim that the optimizing solution must have the property

$$\Sigma_j^{(i)}=0,\quad i=1\dots K,\ j=i+1,\dots,K. \tag{5.13}$$

To see this, we can rewrite (5.12) as

$$\max_{\sum_{i=1}^K\sum_{j=i}^K \operatorname{tr}\left[\Sigma_j^{(i)}\right]\le P} \sum_{i=1}^K (a_i-a_{i-1})\log\det\left(I+\sum_{j=i}^K H_j\left(\sum_{l=j}^K \Sigma_j^{(l)}\right)H_j^\dagger\right). \tag{5.14}$$

From (5.14), we can make the following observation. Fix $l_1>l_2$. Then the input pair

$$\Sigma_{l_1}^{(l_1)}=\Sigma_{l_1}^{(l_1)*},\quad\text{and}\quad \Sigma_{l_1}^{(l_2)}=\Sigma_{l_1}^{(l_2)*}$$

provides a smaller value of the objective function than the input pair

$$\Sigma_{l_1}^{(l_1)}=\Sigma_{l_1}^{(l_1)*}+\Sigma_{l_1}^{(l_2)*},\quad\text{and}\quad \Sigma_{l_1}^{(l_2)}=0.$$

This allows us to write (5.14) as

$$\max_{\sum_{i=1}^K \operatorname{tr}\left[\Sigma_i^{(i)}\right] \le P} \sum_{i=1}^K (a_i - a_{i-1}) \log\det\left(I + \sum_{j=i}^K H_j \Sigma_j^{(j)} H_j^\dagger\right).$$

Comparing (5.4) with this value, we have completed the proof of the theorem. Lemma 5.2 is proved next. □

5.2.2. *Proof of Lemma 5.2.* Fix $1 \le i \le K$. We first note that $C_1 \ge C_2$. To see this, observe that any valid input to the optimization problem in (5.8), $\Sigma^{(i)} \stackrel{\text{def}}{=} \operatorname{diag}\{\Sigma_i, \ldots, \Sigma_K\}$, is also a valid input to the optimization problem in (5.7). This follows due to the constraint $Q^{(i)} \preceq^{(i)} I$. We now find a $Q^{(i)*}$ satisfying the constraint $Q^{(i)*} \preceq^{(i)} I$ such that the value of the maximization in (5.7) with $Q^{(i)} = Q^{(i)*}$ is equal to C_2. This means that the optimal input to the maximization problem in (5.7), $\Sigma^{(i)*}$ is equal to the optimal input to the maximization problem in (5.8), i.e., $\Sigma^{(i)*}$ is block diagonal and the block diagonal matrix elements are $\Sigma_i^*, \ldots, \Sigma_K^*$, the optimal inputs to (5.8). A *sufficient* Kuhn-Tucker condition for $\Sigma^{(i)} = \operatorname{diag}\{\Sigma_i^*, \ldots, \Sigma_K^*\}$ to be optimal for the maximization problem in (5.7) for a fixed $Q^{(i)}$ is that:

$$\operatorname{tr}\left[Q^{(i)}\Sigma^{(i)}\right] \le P, \tag{5.15}$$

$$\sum_{j=1}^k b_j \nabla_{\Sigma^{(i)}} \log\det\left(S_j + H^{(i)}\Sigma^{(i)}H^{(i)\dagger}\right) = \lambda Q^{(i)}, \quad \text{for some } \lambda > 0. \tag{5.16}$$

Hence, if we set

$$\begin{aligned} Q^{(i)*} &= \frac{1}{\lambda}\sum_{j=1}^k b_j \nabla_{\Sigma^{(i)}} \log\det\left(S_j + H^{(i)}\Sigma^{(i)}H^\dagger\right)\Big|_{\Sigma^{(i)}=\operatorname{diag}\{\Sigma_i^*,\ldots,\Sigma_K^*\}}, \\ &= \frac{1}{\lambda}\sum_{j=1}^k b_j H^{(i)\dagger}\left(S_j + \sum_{l=i}^K H_l \Sigma_l^* H_l^\dagger\right)^{-1} H^{(i)}, \end{aligned} \tag{5.17}$$

then condition (5.16) is satisfied. To show that there is a λ such that condition (5.15) is satisfied as well, we observe that since $\Sigma^{(i)*} = \operatorname{diag}\{\Sigma_i^*, \ldots, \Sigma_K^*\}$ solves the optimization problem (5.8), it satisfies the Kuhn-Tucker conditions, which can be easily derived: for all $l = i, \ldots, K$,

$$\sum_{j=1}^k b_j H_l^\dagger\left(S_j + \sum_{m=i}^K H_m \Sigma_m^* H_m^\dagger\right)^{-1} H_l = \lambda^* I \qquad \text{if } \Sigma_j^* \succ 0, \tag{5.18}$$

$$\sum_{j=1}^k b_j H_l^\dagger\left(S_j + \sum_{m=i}^K H_m \Sigma_m^* H_m^\dagger\right)^{-1} H_l \preceq \lambda^* I \qquad \text{else}, \tag{5.19}$$

for some $\lambda^* > 0$. Thus, if we now set $\lambda = \lambda^*$ in (5.17), then $Q^{(i)} \preceq^{(i)} I$ and further

$$\begin{aligned}
\operatorname{tr}\left[Q^{(i)}\Sigma^{(i)*}\right] &= \frac{1}{\lambda^*}\sum_{l=i}^{K}\sum_{j=1}^{k} b_j \operatorname{tr}\left[H_l^\dagger\left(S_j + \sum_{m=i}^{K} H_m \Sigma_m^* H_m^\dagger\right)^{-1} H_l \Sigma_l^*\right], \text{ from (5.17)},\\
&\le \sum_{l=i}^{K} \operatorname{tr}\left[\Sigma_l^*\right], \text{ from (5.18) and (5.19)},\\
&\le P, \text{ since } \Sigma_1^*, \ldots, \Sigma_K^* \text{ are valid inputs to the problem in (5.8)},
\end{aligned}$$

i.e., condition (5.15) is satisfied as well. Hence $C_1 = C_2$ and the lemma is proved. □

References

[1] P. P. Bergmans, "A simple converse for broadcast channels with white Gaussian noise", *IEEE Trans. on Information Theory*, Vol. 20, pp. 279-281, Mar 1974.

[2] S. Boyd and L. Vandenberghe, "Convex Optimization", *course reader for EE364 (Stanford) / EE236B (UCLA).*

[3] G. Caire and S. Shamai, "On the Achievable throughput in Multiple Antenna Gaussian Broadcast Channel", submitted to *IEEE Transactions on Information Theory.*

[4] T. Cover and J. Thomas, *Elements of Information Theory*, John Wiley, 1991.

[5] M. H. M. Costa, "Writing on dirty paper", *IEEE Transactions on Information Theory*, pp. 439 - 441, May 1983.

[6] S. N. Diggavi and T. Cover, "The worst additive noise under a covariance constraint", *IEEE Trans. on Information Theory*, Vol. 47(7), pp. 3072-3081, Nov. 2001.

[7] M. Gastpar, B. Rimoldi and M. Vetterli, "To code, or not to code: On the optimality of symbol-by-symbol communication," Technical Report, EPFL, Switzerland, May 2001. (Also submitted to IEEE Trans Info Theory.)

[8] S. Hanly and D. Tse, "Power control and capacity of spread spectrum wireless networks", *Automatica*, Vol. 35(12), Dec. 1999.

[9] N. Jindal, S. Vishwanath and A. Goldsmith, "On the duality between multiple access and broadcast channels", *Annual Allerton Conference*, 2001.

[10] K. Marton, "A coding theorem for the discrete memoryless broadcast channel", *IEEE Transactions on Information Theory*, pp. 306 - 311, May 1979.

[11] F. Rashid-Farrokhi, L. Tassiulas and K. J. R. Liu, "Joint optimal power control and beamforming in wireless networks using antenna arrays", *IEEE Trans. on Comm.*, Vol. 46(10), pp. 1313-1323, Oct. 1998.

[12] R. T. Rockafellar, *Convex Analysis*, Princeton University Press, Princeton, 1970.

[13] H. Sato, "An outer bound to the capacity region of broadcast channels", *IEEE Trans. on Information Theory*, Vol. 24, No. 3, May 1978, pp. 374-377.

[14] M. Schubert and H. Boche, "A unifying theory for uplink and downlink multi-user beamforming." In *Proc. International Zurich Seminar*, Zurich, Switzerland, February 2002.

[15] E. Telatar, "Capacity of the Multiple antenna Gaussian channel", *European Transactions on Telecommunications*, vol. 10(6), pp. 585-595, Nov/Dec 1999.

[16] D. Tse and S. Hanly, "Multi-Access Fading Channels: Part I: Polymatroidal Structure, Optimal Resource Allocation and Throughput Capacities", *IEEE Transactions on Information Theory*, Vol. 44, No. 7, Nov 1998, pp 2796-2815.

[17] E. Visotsky and U. Madhow, "Optimal beamforming using tranmit antenna arrays", *Proc. of VTC*, 1999.

[18] S. Vishwanath, N. Jindal, and A. Goldsmith, "On the Capacity of Multiple Input Multiple Output Broadcast Channels", *International Conference on Communications*, 2002.

[19] S. Vishwanath, G. Kramer, S. Shamai, S. A. Jafar and A. Goldsmith, to appear in Proceedings of the *DIMACS Workshop in Signal processing in Wireless Communications*, Oct. 2002.

[20] P. Viswanath, D. Tse and V. Anantharam, "Asymptotically optimal waterfilling in vector multiple access channels", *IEEE Transactions on Information Theory*, Vol. 47(1), Jan 2001, pp. 241-267.

[21] P. Viswanath and D. Tse, "Sum Capacity of the Multiple Antenna Gaussian Broadcast Channel", *ISIT* 2001, also submitted to *IEEE Trans. on Information Theory.*

[22] W. Yu, and J. Cioffi "Sum Capacity of Gaussian Vector Broadcast Channels", submitted to IEEE Transactions on Information Theory, Nov., 2001.

DEPARTMENT OF EECS, UNIVERSITY OF CALIFORNIA AT BERKELEY
E-mail address: dtse@eecs.berkeley.edu

DEPARTMENT OF ECE, UNIVERSITY OF ILLINOIS AT URBANA-CHAMPAIGN
E-mail address: pramodv@uiuc.edu

DIMACS Series in Discrete Mathematics
and Theoretical Computer Science
Volume **62**, 2003

Capacity Bounds for Gaussian Vector Broadcast Channels

S. Vishwanath, G. Kramer, S. Shamai, S. Jafar, and A. Goldsmith

ABSTRACT. An outer bound on the capacity region of broadcast channels is presented called the degraded, same marginals (DSM) bound. This bound includes and improves on Sato's sum-rate bound. The DSM bound is applied to Gaussian vector broadcast channels, and it is found that the Gel'fand-Pinsker/Costa dirty-paper coding technique achieves all points inside the bound if only Gaussian code books are permitted. This result suggests that dirty paper coding gives the capacity region of Gaussian vector broadcast channels, but the optimality of Gaussian code books remains to be verified.*

1. Introduction

A broadcast channel [1] with K receivers is defined by an input alphabet $\mathcal{X}$, K output alphabets $\mathcal{Y}_k$ for $1 \leq k \leq K$, and a conditional probability distribution

$$P(y_1, y_2, \ldots, y_K|x) \tag{1.1}$$

where $x \in \mathcal{X}$ and $y_k \in \mathcal{Y}_k$ for $1 \leq k \leq K$. The channel is said to be *physically degraded* if (1.1) factors as

$$P(y_{\pi(1)}|x)\, P(y_{\pi(2)}|y_{\pi(1)}) \cdots P(y_{\pi(K)}|y_{\pi(K-1)}) \tag{1.2}$$

for some permutation $\pi(\cdot)$ on the integers $1, 2, \ldots, K$. For simplicity, we assume the random variables Y_k to be labeled so that we can ignore the permutation, i.e., $\pi(k) = k$. Thus, the channel is physically degraded if

$$X \to Y_1 \to Y_2 \to \ldots \to Y_K$$

forms a Markov chain.

The capacity region of the physically degraded broadcast channel is known and was determined by Bergmans [2] (coding theorem) and Gallager [3] (converse). These two papers were concerned mainly with giving an information-theoretic *expression* for the capacity region, and did not attempt to find a channel input distribution that optimizes this expression. For the Gaussian scalar channel this optimization was done by Bergmans [4] who showed that Gaussian code books are optimal. After this early work there have been many papers addressing general

Key words and phrases. broadcast channels, multiple antennas.

* At the time of the workshop we thought we had a proof that Gaussian input distributions are optimal, and we presented the main arguments of this purported "proof". Unfortunately, in the time after the workshop and before this document was due, we found a step that we could not verify. Thus, despite the progress reported here showing that dirty-paper coding is optimal for Gaussian code books, the problem of determining the *capacity region* remains open.

broadcast channels. A review of the known results until 1977 is given in [5, Sec. V]. A recent review can be found in [6].

We are interested primarily in the recent work of Caire and Shamai [7] on Gaussian *vector* broadcast channels. Our interest stems in part from the practical importance of such channels for wireless communication, and in part because they are not degraded. The paper [7] exploits a coding technique of Gel'fand and Pinsker [8] that was applied to Gaussian scalar channels by Costa [9] and to Gaussian vector channels by Yu [10]. Costa called this coding scheme *writing on dirty paper* and we will call it *dirty paper coding.* It turns out that dirty paper coding, as applied to broadcast channels, is a special case of Marton's celebrated coding theorem [11] (see [8,12]).

It was found in [7] that dirty paper coding achieves the *sum-rate* capacity of a Gaussian broadcast channel. The paper [7] considers only the two-receiver problem where the transmitter has two antennas and the receivers each have one antenna. This work was subsequently generalized to any number of receivers with any number of antennas; [12] addresses the problem with receivers having one antenna and [13,14] address the general problem.

We are here concerned with the problem of determining the entire capacity *region* of Gaussian vector broadcast channels. Our approach will be to apply a genie-aided bound we call the *degraded, same-marginals* (DSM) bound that converts a non-degraded channel into a (generally better) degraded channel.[1] We find that this bound gives the best possible region *if the code books are restricted to be Gaussian.* It now seems natural to suspect that the capacity region must be the achievable region we have given here. However, until now we have been unable to prove that Gaussian code books are optimal.

This paper is organized as follows. Section 2 reviews the information-theoretic expression for the capacity region of a degraded broadcast channel. Section 3 reviews known outer bounds on the capacity region and introduces the DSM bound. Section 4 describes the Gaussian models we consider and gives achievable regions when one uses Gaussian code books. Section 5 shows that the dirty paper coding region is the same as the DSM outer bound when using Gaussian distributions.

2. A Capacity Expression

Suppose the transmitter wants to communicate a message W_k to receiver k for $k = 1, 2, \dots, K$, and that these K messages are statistically independent. If W_k has B_k bits and the channel is used N times, then the rate for receiver k is $R_k = B_k/N$ bits per channel use. The *capacity region* $\mathcal{C}$ of the broadcast channel is the closure of the set of K-tuples $(R_1, R_2, \dots, R_K)$ at which the communication can be made *reliable*, i.e., receiver k's estimate $\widehat{W}_k$ can be made to satisfy $\Pr(\widehat{W}_k \neq W_k) \leq \epsilon$ for any $\epsilon > 0$ and all k.

Broadcast channels turn out to have the following exceedingly useful property.

LEMMA 2.1 (Cover [1]). *Broadcast channels that have the same marginal distributions $P(y_k|x)$ for $k = 1, 2, \dots, K$ have the same capacity region $\mathcal{C}$.*

The import of this lemma is that we can consider any joint distribution $P(y_1, y_2, \dots, y_K|x)$ for which the marginals are the right ones. We can now design

[1] The DSM bound was derived independently by Viswanath and Tse [16], who also presented this bound at the workshop.

encoding schemes for this new channel and they will perform the same (in terms of rate) as for our original channel. Furthermore, any outer bound on the capacity region of the new channel will be an outer bound on the capacity region of the original channel. We will here consider the class of *degraded* (or stochastically degraded) broadcast channels which have the same marginal distribution as the physically degraded broadcast channel [1, p. 4].

The capacity region of the physically degraded broadcast channel is known to be the convex hull of the closure of the $(R_1, R_2, \dots, R_K)$ satisfying

$$0 \le R_k \le I(U_k; Y_k | U_{k+1} U_{k+2} \dots U_K) \tag{2.1}$$

for $k = 1, 2, \dots, K$, where $P(u_1, \dots, u_K, x, y_1, \dots, y_K)$ factors as

$$P(u_1, \dots, u_K, x)\, P(y_1|x) \prod_{k=2}^{K} P(y_k|y_{k-1}). \tag{2.2}$$

The above region was described by Bergmans [2, Sec. III.B] and is identical to his expression if we set $X_k = [U_k\, U_{k+1}\, \dots\, U_K]$. Observe that we have

$$I(U_1; Y_1 | U_2 \dots U_K) = I(X_1; Y_1 | X_2) \le I(X; Y_1 | X_2) \tag{2.3}$$

where the inequality follows by (2.2). This means that we can remove the random variable U_1 in (2.1) and (2.2) if we so desire. For example, for $K = 2$ the capacity region is the convex hull of the closure of the set of (R_1, R_2) satisfying

$$0 \le R_1 \le I(X; Y_1 | U_2) \tag{2.4}$$

$$0 \le R_2 \le I(U_2; Y_2) \tag{2.5}$$

where $P(u_2, x, y_1, y_2) = P(u_2, x) P(y_1|x) P(y_2|y_1)$. However, we will sometimes find it useful to keep U_1 because it represents the first receiver's message.

3. Capacity Outer Bounds

3.1. Basic Bounds. The first capacity outer bound for broadcast channels was given in the paper that introduced such channels.

LEMMA 3.1 (Cover [1, Sec. VIII], Bergmans [2, Sec. IV]).
The capacity region of a broadcast channel $P(y_1, \dots, y_K|x)$ satisfies, for all k,

$$R_k \le \max_{P_X} I(X; Y_k). \tag{3.1}$$

A simple extension of this bound for degraded channels appeared soon after [1], and we include it here for completeness.

LEMMA 3.2 (Wyner [2, Ack.]). *The capacity region of a degraded broadcast channel satisfies, for all k,*

$$\sum_{i=k}^{K} R_i \le \max_{P_X} I(X; Y_k). \tag{3.2}$$

Of course, this bound was soon superseded by Gallager's capacity characterization in [3]. An outer bound for *general* broadcast channels was given by Marton in [11].

LEMMA 3.3 (Körner-Marton [11]). *The capacity region of a two receiver broadcast channel $P(y_1, y_2|x)$ satisfies*

$$\begin{aligned} R_1 &\leq I(X;Y_1) \\ R_2 &\leq I(U;Y_2) \\ R_1 + R_2 &\leq I(X;Y_1|U) + I(U;Y_2) \end{aligned} \tag{3.3}$$

for some $P(u,x)$ where $P(u,x,y_1,y_2) = P(u,x)P(y_1,y_2|x)$. One can further add the bound obtained by swapping Y_1 and Y_2.

REMARK 3.4. Lemma 3.3 is currently the best known outer bound on the capacity of general broadcast channels. The bounds we give in the next section are in fact *included* in Lemma 3.3. However, there are reasons for using these other bounds, as explained below.

3.2. Bounds That Exploit Lemma 2.1. We now consider outer bounds that exploit Lemma 2.1 in the sense that *classes* of channels are considered. The first bound of this type was due to Sato [17], and the bound's two main steps are

1. let the receivers co-operate to get a point-to-point channel, and
2. minimize the capacity of this point-to-point channel over all joint channel distributions with the same marginals.

A descriptive name for Sato's bound might be the *co-operative, same marginals* (CSM) bound because of the two main ingredients involved in this bound. A formal statement of Sato's bound is the following.

LEMMA 3.5 (Sato [17]). *The capacity region $\mathcal{C}$ of a broadcast channel $P(y_1, \ldots, y_K|x)$ satisfies*

$$\sum_{k=1}^{K} R_k \leq \min_{P_{\widetilde{Y}_1 \ldots \widetilde{Y}_K|X} \in \mathcal{P}} \max_{P_X} I(X; \widetilde{Y}_1 \widetilde{Y}_2 \ldots \widetilde{Y}_K) \tag{3.4}$$

where $\mathcal{P}$ is the set of channels $P(\widetilde{y}_1 \ldots \widetilde{y}_K|x)$ that have the same marginals as $P(y_1, \ldots, y_K|x)$, i.e., $P_{\widetilde{Y}_k|X}(y|x) = P_{Y_k|X}(y|x)$ for all k, x and y.

Note that Sato's bound can be used for any broadcast channel, but is weaker than Lemma 3.3.[2] Sato applied this bound to the Gaussian scalar channel in [17] where he found that it is tight for the sum-rate. However, this result was hardly surprising since it was known that maximizing the sum rate for degraded channels is the same as sending to the better user only. Much more surprising was the discovery of Caire and Shamai [7] that this bound is tight for certain Gaussian *vector* broadcast channels that are *not* degraded. This discovery led to a number of papers that showed Sato's bound to be tight in the general Gaussian setting, i.e., any number of receivers and any number of transmit and receive antennas [12–14]. A key to proving these results was the multiple-access/broadcast channel duality of [15].

We here wish to make some progress on finding the entire capacity region $\mathcal{C}$. To this end, we combine Sato's "same marginals" insight with a "degraded" technique first used by Ozarow and Leung-Yan-Cheong in the context of broadcast channels with feedback [18]. The resulting bound is as follows.

[2] We have used "min" and "max" in Lemma 3.5, which suffices for the channels and distributions we consider. A bound for general channels and distributions would instead require an "inf" and a "sup", respectively.

1. Create a degraded broadcast channel by giving receiver k the channel outputs of receivers $k+1, k+2, \ldots, K$.
2. Minimize the capacity of this degraded channel over all joint channel distributions with the same marginals.

For obvious reasons, we call this bound the *degraded, same marginals* (DSM) outer bound. Of course, the first step could have been done in any order, so that one can further take the intersection of regions obtained with different degraded orderings. A formal statement of the DSM bound is the following lemma. For this lemma, we consider a broadcast channel $P(y_1, \ldots, y_K|x)$ and let $\mathcal{R}_D(\pi, P(y_1, \ldots, y_K|x))$ be the capacity region of the degraded broadcast channel created by giving receiver $\pi(k)$ the outputs $Y_{\pi(k)}, Y_{\pi(k+1)}, \ldots, Y_{\pi(K)}$. This capacity region is given by (2.1) and (2.2).

LEMMA 3.6 (DSM Bound). *The capacity region $\mathcal{C}$ of a broadcast channel $P(y_1, \ldots, y_K|x)$ satisfies*

$$\mathcal{C} \subseteq \bigcap_{\pi} \left\{ \bigcap_{P_{\widetilde{Y}_1 \ldots \widetilde{Y}_K|X} \in \mathcal{P}} \mathcal{R}_D(\pi, P_{\widetilde{Y}_1 \ldots \widetilde{Y}_K|X}) \right\}. \tag{3.5}$$

where $\mathcal{P}$ is the set of channels $P(\widetilde{y}_1 \ldots \widetilde{y}_K|x)$ that have the same marginals as $P(y_1, \ldots, y_K|x)$, i.e., $P_{\widetilde{Y}_k|X}(y|x) = P_{Y_k|X}(y|x)$ for all k, x and y.

REMARK 3.7. The DSM bound applies to *general* broadcast channels.

REMARK 3.8. The DSM bound is *stronger* than Sato's bound because the sum rate cannot be more than the capacity of the best channel.

REMARK 3.9. Lemma 3.3 includes the DSM bound, as can be seen by weakening (3.3) by replacing Y_1 with (Y_1, Y_2). However, just as Sato's bound seems easier to use than Lemma 3.3, so it seems that Lemma 3.6 is easier to use than Lemma 3.3. It is unclear whether one can obtain the results we present in Section 5 without going through Lemma 3.6.

4. Gaussian Scalar and Vector Channels

This section introduces the Gaussian channel models we wish to consider. The DSM bound is later applied to such channels.

4.1. Gaussian Scalar Channels. The Gaussian scalar broadcast channel has

$$Y_k = X + Z_k \tag{4.1}$$

for $1 \leq k \leq K$, where X is a complex random variable and Z_k is zero-mean, complex, Gaussian noise that is *circularly symmetric* in the sense that its real and imaginary parts are independent and have the same variance $N_k/2$. There is further a power constraint $\mathrm{E}[|X|^2] \leq P$ on the input.

Consider the region defined by (2.1) and (2.2). We choose the U_k to be independent, zero-mean, complex, circularly symmetric, Gaussian random variables with variance $\mathrm{E}[|U_k|^2] = Q_k$ and set $X = \sum_{k=1}^{K} U_k$. We thus have $\sum_{k=1}^{K} Q_k \leq P$ and obtain as our achievable region the set of rate-tuples satisfying

$$0 \leq R_k \leq \log\left(N_k + \sum_{i=1}^{k} Q_i\right) - \log\left(N_k + \sum_{i=1}^{k-1} Q_i\right) \tag{4.2}$$

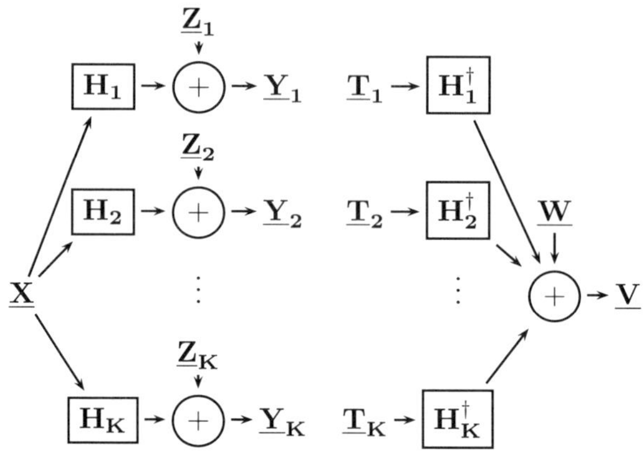

FIGURE 1. System models for the vector Gaussian broadcast channel (left) and its *dual* vector Gaussian multiple access channels (right).

where the units are nats per channel use (1 nat is $\log_2(e)$ bits). Bergmans [4] showed that (4.2) gives the capacity region by varying over all $(Q_1, \ldots, Q_K)$ with $\sum_{k=1}^{K} Q_k \leq P$. This implies that Gaussian input distributions are optimal.

4.2. Gaussian Vector Channels. The Gaussian vector broadcast channel has vectors as inputs and outputs. The output of receiver k is

$$\underline{Y}_k = H_k \, \underline{X} + \underline{Z}_k \tag{4.3}$$

where $\underline{X}$ is a complex column vector of length t and $\underline{Y}_k$ is a complex column vector of length r_k. The channel matrix H_k is a $r_k \times t$ complex matrix and $\underline{Z}_k$ is a vector of r_k zero-mean, complex, circularly symmetric, Gaussian random variables. Let $Q_{\underline{X}} = \mathrm{E}[\underline{X}_k \underline{X}_k^\dagger]$ be the covariance matrix of $\underline{X}$, where $\dagger$ denotes complex conjugation and transposition. The transmitter power constraint is

$$\mathrm{Tr}(Q_{\underline{X}}) \leq P \tag{4.4}$$

where $\mathrm{Tr}(\cdot)$ is the trace operator. The Gaussian vector broadcast channel is *not* degraded in general, as can be seen for $K = 2$, $t = 2$, $r_1 = r_2 = 1$, $H_1 = [1 \ 0]$ and $H_2 = [0 \ 1]$.

The dirty paper coding region where the messages are encoded in the order $K, K-1, \ldots, 1$ is the convex hull of

$$C_{BC}^{DP}(P) = \bigcup_{Q_1, \ldots, Q_k} \left\{ \begin{array}{l} (R_1, \ldots, R_K) : 0 \leq R_k \leq \\ \log \left| N_k + H_k \left(\sum_{i=1}^{k} Q_i \right) H_k^\dagger \right| \\ - \log \left| N_k + H_k \left(\sum_{i=1}^{k-1} Q_i \right) H_k^\dagger \right|, \text{ all } k \end{array} \right\} \tag{4.5}$$

where $|A|$ is the determinant of A, Q_k is now the covariance matrix of User k and $\sum_k Q_k = Q_{\underline{X}}$. Of course, one can perform the encoding in any of the $K!$ orderings.

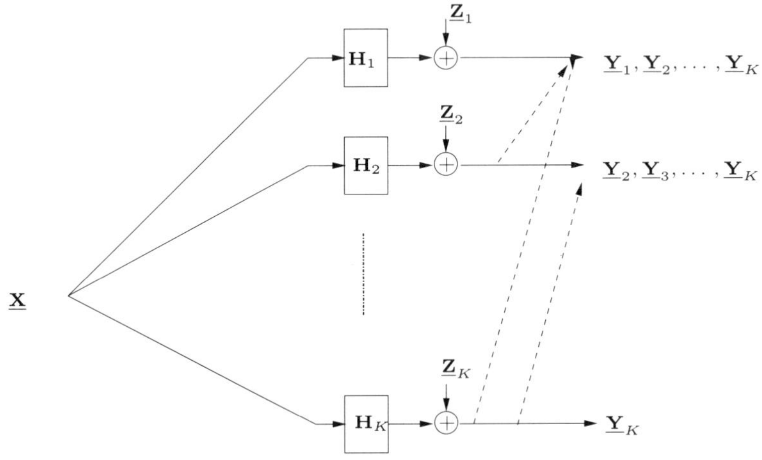

FIGURE 2. Model for the DSM bound.

4.3. The DSM Bound for Gaussian Vector Channels. We set $H_{Dk} = [H_k^\dagger \dots H_K^\dagger]^\dagger$ and $\underline{Z}_{Dk} = [\underline{Z}_k^\dagger \dots \underline{Z}_K^\dagger]^\dagger$ for all k. The channel outputs after step 1 of the DSM bound are thus

$$\underline{Y}_{Dk} = H_{Dk}\,\underline{X} + \underline{Z}_{Dk} \tag{4.6}$$

so that receiver k sees vectors of length $\sum_{i=k}^{K} r_i$. The second step of the DSM bound permits $\underline{Z}_{DK}$ to have *any* covariance matrix as long as the covariance matrices of the sub-vectors $\underline{Z}_k$ are the same as before.

Consider next the capacity region for the DSM bound given by (2.1) and (2.2). Suppose we choose the U_k, which we will now write as $\underline{U}_k$, to be independent, zero-mean, complex, Gaussian random vectors with covariance matrices $\mathrm{E}[\underline{U}_k \underline{U}_k^\dagger] = T_k$. We set $\underline{X} = \sum_{k=1}^{K} \underline{U}_k$ so that $\sum_{k=1}^{K} \mathrm{Tr}(T_k) \le P$. Thus, the DSM bound restricted to Gaussian inputs and the degraded ordering given in (4.6) is

$$C_{BC}^{DSM}(P) = \bigcap_{N_{DK}} \bigcup_{T_k} \left\{ \begin{array}{l} (R_1, \dots, R_K): \\ 0 \le R_k \le \log\left|N_{Dk} + H_{Dk}(\sum_{i=1}^{k} T_i) H_{Dk}^\dagger\right| \\ -\log\left|N_{Dk} + H_{Dk}\left(\sum_{i=1}^{k-1} T_i\right) H_{Dk}^\dagger\right|, \text{ all } k \end{array} \right\} \tag{4.7}$$

where $N_{Dk} = \mathrm{E}[\underline{Z}_{Dk}\underline{Z}_{Dk}^\dagger]$. The notation $\cap_{N_{DK}}$ is used as a shorthand for the intersection over all N_{DK} that keep the covariance matrices of $\underline{Z}_k$ the same as the original problem. Similarly, the notation $\cup_{T_k}$ is used as a shorthand for $\cup_{T_1,\dots,T_K}$. The next section shows that (4.7) is the same as (4.5).

5. Optimization of the DSM Bound

The region (4.7) can be rewritten as

$$\bigcap_{N_{DK}} \bigcup_{T_k} \left\{ \begin{array}{l} (R_1, \dots, R_K): 0 \le R_k \le \log |I + \\ \left(I + N_{Dk}^{-1/2} H_{Dk}\left(\sum_{i=1}^{k-1} T_i\right) H_{Dk}^\dagger N_{Dk}^{-1/2}\right)^{-1} \\ N_{Dk}^{-1/2} H_{Dk} T_k H_{Dk}^\dagger N_{Dk}^{-1/2}\Big|, \text{ all } k \end{array} \right\} \tag{5.1}$$

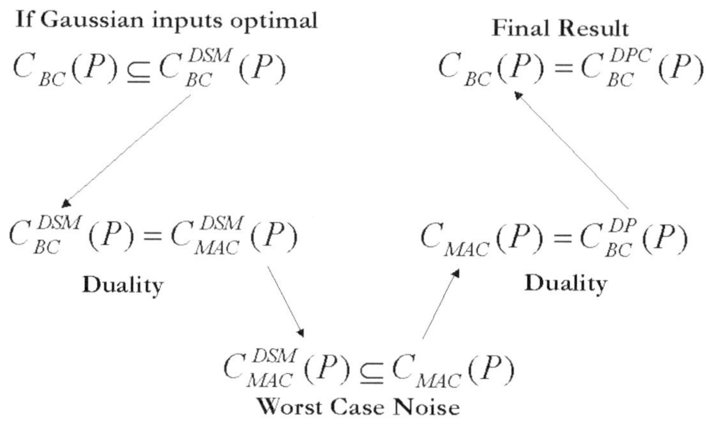

FIGURE 3. Pictorial representation of the steps of the proof.

Note that (5.1) is the dirty paper coding region of a broadcast channel with channel matrices $N_{Dk}^{-1/2} H_{Dk}$ and power constraint P. We use the duality result of [14] and replace (5.1) with the capacity region of the dual vector multiple access (MAC) with channel $H_{Dk}^{\dagger} N_{Dk}^{-1/2}$. The result is

$$C_{MAC}^{DSM}(P) =$$

$$\bigcap_{N_{DK}} \bigcup_{C_k} \left\{ \begin{array}{l} (R_1, \ldots, R_K) : 0 \leq \sum_{i=k}^{K} R_i \leq \\ \log |I + \sum_{i=k}^{K} H_{Di} N_{Di}^{-1/2} C_i N_{Di}^{-1/2} H_{Di}^{\dagger}|, \\ \text{all } k \end{array} \right\} \tag{5.2}$$

where C_k is positive semidefinite for all k and $\sum_k \mathrm{Tr}(C_k) \leq P$. Note that the numbering of users $\{1, \ldots, K\}$ is arbitrary, and we can obtain a tighter upper bound by taking an intersection over all $K!$ outer bounds of the form (5.2). In subsection 5.1, we show that this intersection is inside the capacity region of the K user vector multiple access (MAC) with channel $H_k^{\dagger}$ and *joint* power constraint P. This region is given by:

$$C_{MAC}(P) =$$

$$\bigcup_{S_k} \left\{ \begin{array}{l} (R_1, \ldots, R_K) : \\ 0 \leq \sum_{i \in E} R_i \leq \log \left| I + \sum_{i \in E} H_i S_i H_i^{\dagger} \right|, \end{array} \right\} \tag{5.3}$$

where E is any subset of $\{1, \ldots, K\}$, S_k is positive semidefinite for all k and $\sum_k \mathrm{Tr}(S_k) \leq P$. Employing duality [14] once again, we find that (5.3) is equivalent to the dirty paper coding region of the K user BC with channels H_k and power constraint P. Thus, after taking the intersection over all possible orderings, the DSM upper bound is the dirty paper coding region. The steps of our proof are depicted in Fig. 3, where $C_{BC}(P)$ is the "capacity" region of the vector Gaussian broadcast channel with Gaussian input distributions.

REMARK 5.1. One can show that there are interesting special cases where Gaussian input distributions are optimal, e.g., if the covariance matrices of the $\underline{U}_k$ are restricted to have unit rank.

5.1. Optimization Problems. Consider the K user MAC with channels $H_k^\dagger$ whose capacity region is (5.3). We later extend the analysis to the DSM upper bound, and finally show that (5.3) includes (5.2). Note that any point on the boundary of (5.3) is the solution of an optimization problem parameterized by a real nonnegative vector $(\mu_0 = 0, \mu_1 = 1, \ldots, \mu_K)$ [19]. Since the ordering of users $1, \ldots, K$ is arbitrary, it is sufficient to consider the case when $\mu_0 = 0, \mu_1 = 1 \leq \mu_2 \leq \mu_3 \leq \ldots \leq \mu_K$. The optimization problem is

$$\max_{P_1,\ldots,P_K} \max_{S_1,\ldots,S_K} \sum_k (\mu_k - \mu_{k-1}) \log \left| I + \sum_{i=k}^{K} H_i^\dagger S_i H_i \right| \tag{5.4}$$

subject to

$$\begin{aligned} &\sum_k P_k = P \\ &\mathrm{Tr}(S_k) \leq P_k \ \forall\ k \\ &P_k, S_k \geq 0 \ \forall\ k. \end{aligned}$$

This optimization problem is convex. We can therefore formulate a *Lagrangian* dual problem for which the dual and primal optimum values are the same [21].

Consider first S_1 and note that only the first term of the outer sum in (5.4) is a function of S_1. We thus begin with the dual problem of

$$\max_{S_1} \log \left| I + \sum_{k=1}^{K} H_k^\dagger S_k H_k \right| \tag{5.5}$$

subject to

$$\begin{aligned} &\mathrm{Tr}(S_1) \leq P_1 \\ &S_1 \geq 0. \end{aligned}$$

Using results from the Appendix, the dual problem to (5.5) is

$$\min_{\Gamma_1,\lambda_1} -\log|\Gamma_1| - t + \mathrm{Tr}\left(\Gamma_1 (I + \sum_{k=2}^{K} H_k^\dagger S_k H_k) \right) + \lambda_1 P_1 \tag{5.6}$$

subject to

$$\begin{aligned} &\lambda_1 I \geq H_1 \Gamma_1 H_1^\dagger \\ &\Gamma_1 \geq 0. \end{aligned}$$

Substituting (5.6) into (5.4), the boundary point maximization problem becomes

$$\begin{aligned} \max_{P_1,\ldots,P_K} \max_{S_2,\ldots,S_K} \min_{\Gamma_1,\lambda_1} & -\log|\Gamma_1| - t + \mathrm{Tr}\left(\Gamma_1 (I + \sum_{k=2}^{K} H_k^\dagger S_k H_k) \right) + \lambda_1 P_1 \\ & + \sum_{k=2}^{K} (\mu_k - \mu_{k-1}) \log \left| I + \sum_{i=k}^{K} H_i^\dagger S_i H_i \right| \end{aligned} \tag{5.7}$$

subject to

$$\sum_k P_k = P$$
$$\mathrm{Tr}(S_k) \leq P_k \text{ for } k = 2, \dots, K$$
$$\lambda_1 I \geq H_1 \Gamma_1 H_1^\dagger$$
$$P_k, S_k, \Gamma_1 \geq 0 \ \forall \ k.$$

Using Ky Fan's min-max switching theorem [20], the objective function in (5.7) can be rewritten as

$$\min_{\Gamma_1, \lambda_1} \max_{P_1, \dots, P_K} \max_{S_2, \dots, S_K} -\log|\Gamma_1| - t + \mathrm{Tr}\left(\Gamma_1 (I + \sum_{k=2}^{K} H_k^\dagger S_k H_k)\right) + \lambda_1 P_1$$
$$+ \sum_{k=2}^{K} (\mu_k - \mu_{k-1}) \log \left| I + \sum_{i=k}^{K} H_i^\dagger S_i H_i \right|. \tag{5.8}$$

Now observe that in (5.8) there are two terms that depend on S_2. Isolating these terms, we formulate a primal problem as

$$\max_{S_2} \log \left| I + \sum_{k=2}^{K} H_k^\dagger S_k H_k \right| + \frac{1}{\mu_2 - 1} \mathrm{Tr}\left(\Gamma_1 (I + \sum_{k=2}^{K} H_k^\dagger S_k H_k)\right) \tag{5.9}$$

subject to

$$\mathrm{Tr}(S_2) \leq P_2.$$

Again using results from the Appendix, the dual problem to (5.9) is

$$\min_{\Gamma_2, \lambda_2} -\log \left| \Gamma_2 - \frac{\Gamma_1}{\mu_2 - 1} \right| - t + \mathrm{Tr}\left(\Gamma_2 (I + \sum_{k=3}^{K} H_k^\dagger S_k H_k)\right) + \lambda_2 P_2 \tag{5.10}$$

subject to

$$\lambda_2 I \geq H_2 \Gamma_2 H_2^\dagger$$
$$\Gamma_2 \geq 0.$$

Continuing in this fashion, at the jth step we find that the optimization problem in terms of S_j is

$$\max_{S_j} \log \left| I + \sum_{k=j}^{K} H_k^\dagger S_k H_k \right| + \frac{\mu_{j-1} - \mu_{j-2}}{\mu_j - \mu_{j-1}} \mathrm{Tr}\left(\Gamma_{j-1} (I + \sum_{k=j}^{K} H_k^\dagger S_k H_k)\right) \tag{5.11}$$

subject to

$$\mathrm{Tr}(S_j) \leq P_j.$$

The corresponding dual problem to (5.11) is

$$\min_{\Gamma_j, \lambda_j} -\log \left| \Gamma_j - \frac{\mu_{j-1} - \mu_{j-2}}{\mu_j - \mu_{j-1}} \Gamma_{j-1} \right| - t + \mathrm{Tr}\left(\Gamma_j (I + \sum_{k=j+1}^{K} H_k^\dagger S_k H_k)\right) + \lambda_j P_j$$

subject to

$$\lambda_j I \geq H_j \Gamma_j H_j^\dagger$$
$$\Gamma_j \geq 0.$$

After the final (K^{th}) step, the overall dual problem is (defining $\Gamma_0 = 0$)

$$\min_{\Gamma_1,\dots,\Gamma_K} \min_{\lambda_1,\dots,\lambda_K} \max_{P_1,\dots,P_K} - \left[\sum_{j=1}^{K} (\mu_j - \mu_{j-1}) \log \left| \Gamma_j - \frac{\mu_{j-1} - \mu_{j-2}}{\mu_j - \mu_{j-1}} \Gamma_{j-1} \right| \right]$$

$$- \mu_K t + (\mu_K - \mu_{K-1}) \mathrm{Tr}(\Gamma_K) + \sum_k (\mu_k - \mu_{k-1}) \lambda_k P_k \tag{5.12}$$

subject to

$$\sum_k P_k = P$$
$$\lambda_k I \geq H_k \Gamma_k H_k^\dagger \ \forall\ k$$
$$P_k, \Gamma_k \geq 0 \ \forall\ k.$$

Note that the dual problem (5.12) is linear in $\{P_1, \dots, P_K\}$ and in $\{\lambda_1, \dots, \lambda_K\}$.

The same analysis as for the K user MAC can be repeated to obtain the dual problem for (5.2) for a point with priorities $\mu_0 = 0, \mu_1 = 1 \leq \mu_2 \leq \mu_3 \dots \leq \mu_K$ by replacing H_k with $N_{Dk}^{-1/2} H_{Dk}$. This dual problem is

$$\min_{N_{D1},\dots,N_{DK}} \min_{\lambda_1,\dots,\lambda_K} \min_{\Gamma_1,\dots,\Gamma_K} \max_{P_1,\dots,P_K} - \left[\sum_{j=1}^{K} (\mu_j - \mu_{j-1}) \log \left| \Gamma_j - \frac{\mu_{j-1} - \mu_{j-2}}{\mu_j - \mu_{j-1}} \Gamma_{j-1} \right| \right]$$

$$- \mu_K t + (\mu_K - \mu_{K-1}) \mathrm{Tr}(\Gamma_K) + \sum_k (\mu_k - \mu_{k-1}) \lambda_k P_k \tag{5.13}$$

subject to

$$\sum_k P_k = P$$
$$\lambda_k N_{Dk} \geq H_{Dk} \Gamma_k H_{Dk}^\dagger \ \forall\ k$$
$$P_k, \Gamma_k \geq 0 \ \forall\ k.$$

The above problem (5.13) is linear in $\{N_{D1}, \dots, N_{DK}\}$, so we can use Ky-Fan's min-max switching theorem [20] to obtain

$$\min_{\lambda_1,\dots,\lambda_K} \min_{\Gamma_1,\dots,\Gamma_K} \max_{P_1,\dots,P_K} \min_{N_{D1},\dots,N_{DK}} - \left[\sum_{j=1}^{K} (\mu_j - \mu_{j-1}) \log \left| \Gamma_j - \frac{\mu_{j-1} - \mu_{j-2}}{\mu_j - \mu_{j-1}} \Gamma_{j-1} \right| \right]$$

$$- \mu_K t + (\mu_K - \mu_{K-1}) \mathrm{Tr}(\Gamma_K) + \sum_k (\mu_k - \mu_{k-1}) \lambda_k P_k \tag{5.14}$$

subject to

$$\sum_k P_k = P$$
$$\lambda_k N_{Dk} \geq H_{Dk} \Gamma_k H_{Dk}^\dagger \ \forall\ k$$
$$P_k, \Gamma_k \geq 0 \ \forall\ k.$$

To obtain the final result, we show that (5.14) is at most (5.12). We show this by constructing a set of K MACs having $K+1$ users each. The first MAC has channel H_1 for user 1, channels $N_{Dj}^{-1/2} H_{Dj}$ for user j, $j \in \{2, \dots, K\}$, and channel $N_{D2}^{-1/2} H_{D2}$ for user $K+1$. Let us consider a point on the boundary of the capacity region of this channel, where users 1 and $K+1$ have priorities $\mu_1 = \mu_{K+1} = 1$, and user j has priority μ_j, $j \in \{2, \dots, K\}$. Calling $B_{Di} \triangleq N_{Di}^{-1/2} H_{Di}$, the primal problem for the point $(\mu_0, \mu_1, \mu_2, \dots, \mu_K, \mu_{K+1}) = (0, 1, \mu_2, \dots, \mu_K, 1)$ is given by:

$$\begin{aligned} \max_{P_1,\dots,P_K} \max_{S_1,\dots,S_{K+1}} & \log \left| I + H_1^\dagger S_1 H_1 + \sum_{i=2}^{K} B_{Di}^\dagger S_i B_{Di} + B_{D2}^\dagger S_{K+1} B_{D2} \right| \\ & + \sum_{k=2}^{K} (\mu_k - \mu_{k-1}) \log \left| I + \sum_{i=k}^{K} B_{Di}^\dagger S_i B_{Di} \right| \end{aligned} \tag{5.15}$$

subject to

$$\begin{aligned} & \sum_{k}^{K} P_k = P \\ & \mathrm{Tr}(S_1) + \mathrm{Tr}(S_{K+1}) = P_1 \\ & \mathrm{Tr}(S_k) \leq P_k \text{ for } k = \{2, \dots, K\} \\ & P_k, S_k \geq 0 \ \forall\ k. \end{aligned}$$

Extending the argument presented earlier, one can show that the dual problem to (5.15) is given by

$$\begin{aligned} \min_{\lambda_1,\dots,\lambda_K} \min_{\Gamma_1,\dots,\Gamma_K} \max_{P_1,\dots,P_K} \min_{N_{D2},\dots,N_{DK}} & - \left[\sum_{j=1}^{K} (\mu_j - \mu_{j-1}) \log \left| \Gamma_j - \frac{\mu_{j-1} - \mu_{j-2}}{\mu_j - \mu_{j-1}} \Gamma_{j-1} \right| \right] \\ & - \mu_K t + (\mu_K - \mu_{K-1}) \mathrm{Tr}(\Gamma_K) + \sum_{k} (\mu_k - \mu_{k-1}) \lambda_k P_k \end{aligned} \tag{5.16}$$

subject to

$$\begin{aligned} & \sum_{k} P_k = P \\ & \lambda_k N_{Dk} \geq H_{Dk} \Gamma_k H_{Dk}^\dagger && \text{for } k = 2, \dots, K \\ & \lambda_1 I \geq H_1 \Gamma_1 H_1^\dagger && \text{for user 1} \\ & \lambda_1 N_{D2} \geq H_{D2} \Gamma_1 H_{D2}^\dagger && \text{for user } K+1 \\ & P_k, \Gamma_k \geq 0 \ \forall\ k. \end{aligned}$$

Note that the objective functions of (5.14) and (5.16) are the same. The only difference between them lies in the constraints.

We now show that (5.14) is at most (5.16). Given that Γ_1, λ_1 and N_{D2} satisfy the constraints of (5.16), we choose N_{D1} for (5.14) to be

$$N_{D1} = \begin{bmatrix} I & \frac{H_1 \Gamma_1 H_{D2}^\dagger}{\lambda_1} \\ \frac{H_{D2} \Gamma_1 H_1^\dagger}{\lambda_1} & N_{D2} \end{bmatrix} \tag{5.17}$$

which satisfies the constraints on N_{D1} in (5.14). Since (5.14) is a minimization over all N_{D1}, the optimal value of (5.14) is at most that of (5.16).

An important point to note here is that, for the point under consideration in the $K+1$ user MAC (5.15), the optimal value of S_{K+1} equals 0. Since user 2 and user $K+1$ have the same channels, but user 2 has a higher priority (μ_2) than user $K+1$ (μ_1), it is optimal for user $K+1$ to not transmit.

For the $(j-1)^{\text{th}}$ step, we consider a $K+1$ user MAC where users $1,\ldots,j-1$ have channels $H_1,\ldots,H_{j-1}$, users $j,\ldots,K$ have channels $N_{Dj}^{-1/2}H_{Dj},\ldots,N_{DK}^{-1/2}H_{DK}$ and user $K+1$ has channel $N_{Dj}^{-1/2}H_{Dj}$. Suppose the priorities are μ_k for user k, $k\in\{1,\ldots,K\}$ and μ_{j-1} for User $K+1$. It can again be shown that the dual problem to the primal problem corresponding to this point in the capacity region is

$$\min_{\lambda_1,\ldots,\lambda_K}\min_{\Gamma_1,\ldots,\Gamma_K}\max_{P_1,\ldots,P_K}\min_{N_{Dj},\ldots,N_{DK}} -\left[\sum_{j=1}^{K}(\mu_j-\mu_{j-1})\log\left|\Gamma_j-\frac{\mu_{j-1}-\mu_{j-2}}{\mu_j-\mu_{j-1}}\Gamma_{j-1}\right|\right]$$

$$-\mu_K t+(\mu_K-\mu_{K-1})\mathrm{Tr}(\Gamma_K)+\sum_k(\mu_k-\mu_{k-1})\lambda_k P_k \tag{5.18}$$

subject to

$$\begin{aligned}
&\sum_k P_k = P\\
&\lambda_k N_{Dk}\geq H_{Dk}\Gamma_k H_{Dk}^{\dagger} \text{ for } k=j,\ldots,K\\
&\lambda_k I\geq H_k\Gamma_k H_k^{\dagger} \text{ for } k=1,\ldots,j-1\\
&\lambda_{j-1}N_{Dj}\geq H_{Dj}\Gamma_{j-1}H_{Dj}^{\dagger} \text{ for user } K+1\\
&P_k,\Gamma_k\geq 0\ \forall\ k.
\end{aligned}$$

We choose

$$N_{D2}=\begin{bmatrix} I & \frac{H_2\Gamma_2 H_{D3}^{\dagger}}{\lambda_2}\\ \frac{H_{D3}\Gamma_2 H_2^{\dagger}}{\lambda_2} & N_{D3}\end{bmatrix} \tag{5.19}$$

where Γ_2, N_{D3} and λ_2 satisfy the constraints of (5.18) for $j=3$. This choice of N_{D2} sets (5.16) equal to (5.18) for $j=3$. Since (5.16) is a minimization over all N_{D2}, the optimal value of (5.18) for $j=3$ is at least (5.16). Note again that the optimal transmission strategy for user $K+1$ is to not transmit. In general, by using

$$N_{Dk}=\begin{bmatrix} I & \frac{H_k\Gamma_k H_{D(k+1)}^{\dagger}}{\lambda_k}\\ \frac{H_{D(k+1)}\Gamma_k H_k^{\dagger}}{\lambda_k} & N_{D(k+1)}\end{bmatrix} \tag{5.20}$$

where Γ_k, $N_{D(k+1)}$ and λ_k satisfy the constraints of (5.18) for $j=k+1$, we find that the optimum value of (5.18) for $j=k$ is at most that of (5.18) for $j=k+1$. In all these cases, since the priority of user $K+1$ (μ_k) is less than that of user $k+1$ (μ_{k+1}), but they have the same channels, it is optimal for user $K+1$ to not transmit ($S_{K+1}=0$).

Finally, we have the result that the optimum value of (5.14) is at most that of (5.18) for $j=K$. But the optimum value of (5.18) for $j=K$ equals that of (5.12). Hence, we have the result that (5.14) is at most (5.12). □

Corollary: The worst case noise is given by:

$$\begin{aligned}
N_{D1(i,i)} &= I\\
N_{D1(i,j)} &= \frac{H_i\Gamma_{i0}H_j^{\dagger}}{\lambda_{i0}} \text{ for } i<j
\end{aligned} \tag{5.21}$$

where $N_{D1(i,j)}$ denotes the $(i,j)^{\text{th}}$ matrix entry of N_{D1}, and $\{\Gamma_{10},\dots,\Gamma_{K0}\}$ and $\{\lambda_{10},\dots,\lambda_{K0}\}$ are values of $\{\Gamma_1,\dots,\Gamma_K\}$ and $\{\lambda_1,\dots,\lambda_K\}$ corresponding to an extremizing solution of (5.12).

6. Conclusions

A new outer bound, termed the DSM bound, is obtained for the broadcast channel. It is shown that, for the vector broadcast channel constrained to Gaussian inputs, the DSM outer bound equals the Marton achievable region.

Appendix A
A Primal Problem and its Dual

We wish to find the dual problem for

$$\max_{S_1,\dots,S_L} \log\left|A+\sum_l H_l^\dagger S_l H_l\right| + \operatorname{Tr}\left(B(A+\sum_l H_l^\dagger S_l H_l)\right) \tag{A.1}$$

subject to

$$\sum_l \operatorname{Tr}(S_l) \le P$$
$$S_l \ge 0 \ \forall\, l \in \{1,\dots,L\}.$$

where L is a positive integer. Consider first the optimization problem

$$\min_T -\log|T| - \operatorname{Tr}(BT) \tag{A.2}$$

subject to

$$T \le A + \sum_l H_l^\dagger S_l H_l$$
$$\operatorname{Tr}(S_l) \le P$$
$$S_l \ge 0 \ \forall\, l.$$

Note that (A.2) is obtained by negating the objective of (A.1). A Lagrangian for (A.2) is

$$\begin{aligned} L = &-\log|T| - \operatorname{Tr}(BT) + \operatorname{Tr}\left(\Gamma(T-A-\sum_l H_l^\dagger S_l H_l)\right) \\ &+\lambda\left(\sum_l \operatorname{Tr}(S_l) - P\right) - \sum_l \operatorname{Tr}(\Psi_l S_l). \end{aligned} \tag{A.3}$$

The saddle point of this Lagrangian gives the optimum value of (A.2), and extremizing with respect to T and S_l gives us the dual problem. Rewriting (A.3), we have

$$\begin{aligned} L = &-\log|T| - \operatorname{Tr}(BT) + \operatorname{Tr}(\Gamma T) - \operatorname{Tr}(\Gamma A) - \lambda P \\ &-\sum_l \operatorname{Tr}\left(S_l(\Psi_l - \lambda I + H_l \Gamma H_l^\dagger)\right). \end{aligned} \tag{A.4}$$

Taking the derivative with respect to S_l and setting it to zero, we have $\Psi_l - \lambda I + H_l\Gamma H_l^\dagger = 0$. Taking the derivative with respect to T and setting it to zero, we have

$$-T^{-1} - B + \Gamma = 0.$$

Substituting the resulting Ψ_l and T into (A.4) we find that

$$L = \log|\Gamma - B| + t - \mathrm{Tr}(\Gamma A) - \lambda P. \quad \text{(A.5)}$$

The slackness variable Ψ_l is semidefinite, so the dual problem for (A.2) is

$$\max_{\Gamma,\lambda} \log|\Gamma - B| + t - \mathrm{Tr}(\Gamma A) - \lambda P \quad \text{(A.6)}$$

subject to

$$\lambda I \geq H_l \Gamma H_l^\dagger \ \forall\, l$$
$$\lambda, \Gamma \geq 0.$$

Hence, the dual problem for (A.1) is

$$\min_{\Gamma,\lambda} -\log|\Gamma - B| - t + \mathrm{Tr}(\Gamma A) + \lambda P$$

subject to

$$\lambda I \geq H_l \Gamma H_l^\dagger \ \forall\, l$$
$$\lambda, \Gamma \geq 0.$$

References

[1] T.M. Cover, "Broadcast channels," *IEEE Trans. Inform. Theory*, vol. 18, no. 1, pp. 2-14, Jan. 1972.

[2] P.P. Bergmans, "Random coding theorem for broadcast channels with degraded components," *IEEE Trans. Inform. Theory*, vol. 19, no. 2, pp. 197-207, March 1973.

[3] R.G. Gallager, "Capacity and coding for degraded broadcast channels," *Problemy Peredachi Informatsii*, vol. 10, no. 3, pp. 3-14, July-Sept. 1974.

[4] P.P. Bergmans, "A simple converse for broadcast channels with additive white Gaussian noise," *IEEE Trans. Inform. Theory*, vol. 20, no. 2, pp. 279-280, March 1974.

[5] E.C. van der Meulen, "A survey of multi-way channels in information theory: 1961-1976," *IEEE Trans. Inform. Theory*, vol. 23, no. 1, pp. 1-37, Jan. 1977.

[6] T.M. Cover, "Comments on broadcast channels," *IEEE Trans. Inform. Theory*, vol. 44, no. 6, pp. 2524-2530, Oct. 1998.

[7] G. Caire and S. Shamai, "On the achievable throughput of a multi-antenna Gaussian broadcast channel," submitted to *IEEE Trans. Inform. Theory*, July 2001.

[8] S.I. Gel'fand and M.S. Pinsker, "Coding for channel with random parameters," *Problems of Control and Information Theory*, vol. 9, no. 1, pp. 19-31, 1980.

[9] M.H.M. Costa, "Writing on dirty paper," *IEEE Trans. Inform. Theory*, vol. 29, no. 3, pp. 439-441, May 1983.

[10] W. Yu and J. Cioffi, "Trellis precoding for the broadcast channel," *Proc. IEEE Global Telecomm. Conf. (Globecom)*, pp. 1338-1344, San Antonio. TX, USA Nov. 25-29, 2001.

[11] K. Marton, "A coding theorem for the discrete memoryless broadcast channel," *IEEE Trans. Inform. Theory*, vol. 25, no. 3, pp. 306-311, May 1979.

[12] P. Viswanath and D. Tse, "Sum capacity of the multi-antenna broadcast channel," *Proc. IEEE Int. Symp. Inform. Theory*, pp. 497, Lausanne, Switzerland, June 30 - July 5, 2002.

[13] W. Yu and J. Cioffi, "Sum capacity of a Gaussian vector broadcast channel," *Proc. IEEE Int. Symp. Inform. Theory*, pp. 498, Lausanne, Switzerland, June 30 - July 5, 2002.

[14] S. Viswanath, N. Jindal, and A. Goldsmith, "Duality, achievable rates and sum capacity of Gaussian MIMO broadcast channels," submitted to *IEEE Trans. on Inform. Theory*, Aug. 2002.

[15] N. Jindal, S. Vishwanath, and A. Goldsmith, "On the duality of Gaussian multiple-access and broadcast channels," submitted to *IEEE Trans. on Inform. Theory*, Aug. 2002.

[16] P. Viswanath and D. N. C. Tse, "Capacity of the Multiantenna Gaussian Broadcast Channel" *Proc. DIMACS workshop on Signal Processing for Wireless Transmission*, Rutgers, NJ, Oct. 7-9 2002.

[17] H. Sato, "An outer bound to the capacity region of broadcast channels," *IEEE Trans. Inform. Theory*, vol. 24, no. 3, pp. 374-377, May 1978.

[18] L.H. Ozarow and S.K. Leung-Yan-Cheong, "An achievable region and outer bound for the Gaussian broadcast channel with feedback," *IEEE Trans. Inform. Theory*, vol. 30, no. 4, pp. 667-671, July 1984.

[19] D. Tse and S. Hanly, "Multi-Access Fading Channels: Part I: Polymatroid Structure, Optimal Resource Allocation and Throughput Capacities", IEEE Trans. Inf. Theory, Nov. 1998, pp.2796-2815.

[20] K. Fan, " Minimax Theorems", Proc. Nat. Acad. Sci, 39, 1953, pp.42-47.

[21] S. Boyd, " Convex Optimization", Lecture Notes, Stanford University.

Stanford University, Stanford, CA USA.
E-mail address: sriram@wsl.stanford.edu

Bell Laboratories, Lucent Technologies, Murray Hill, NJ 07974 USA
E-mail address: gkr@bell-labs.com

Technion–Israel Institute of Technology, Haifa, Israel
E-mail address: sshlomo@ee.technion.ac.il

Stanford University, Stanford, CA USA.
E-mail address: syed@wsl.stanford.edu

Stanford University, Stanford, CA USA.
E-mail address: andrea@wsl.stanford.edu

DIMACS Series in Discrete Mathematics
and Theoretical Computer Science
Volume **62**, 2003

Writing on Dirty Tape with LDPC Codes

Giuseppe Caire and Shlomo Shamai (Shitz)

ABSTRACT. We consider channels with interference known to the transmitter but unknown to the receiver. A precoding technique based on lattice quantization is capable of achieving Costa's "Dirty-Paper" limit for arbitrary interference signals, when interference is known non-causally. The same technique can be applied when interference is known with given anticipation k, fixed and independent of the code blocklength n. We review the main information-theoretic results in the non-causal $k = n$ and causal ($k = 1$) settings, and provide specific examples for a modulo-2 additive-noise binary channel with binary interference and for an AWGN channel with additive interference.

Then, we turn to coding design for the AWGN case with causally-known interference (nicknamed "Dirty-Tape" channel, in analogy with Costa's "Dirty-Paper"). We provide explicit code constructions based on Low-Density Parity-Check codes and on M-PAM modulation, able to approach the rates achievable by lattice precoding for the AWGN Dirty-Tape channel. Finally, we point out some research problems such as the wideband limit of the causally-known interference channel (and the related optimal signaling) and the extension of our coding technique to the case of non-causally known interference (AWGN Dirty-Paper channel).

1. Introduction

Memoryless channels with input X, output Y and state-dependent transition probability $P_{Y|X,S}$ where the channel state S is i.i.d., known to the transmitter and unknown to the receiver, date back to Shannon [**1**], who considered the case of state sequence known causally, and to Kusnetsov and Tsybakov [**2**], who considered the case of state sequence known non-causally. Gel'fand and Pinsker [**3**] proved the capacity formula

$$C = \sup_{P_{T|S}} \{I(T;Y) - I(T;S)\} \tag{1.1}$$

for the non-causal case, where T is an auxiliary random variable with conditional distribution $P_{T|S}$ and X is a deterministic function of S and T. This yields Shannon's capacity formula [**1**]

$$C = \sup_{P_T} I(T;Y) \tag{1.2}$$

1991 *Mathematics Subject Classification.* Primary 94B99, 94A15.

Key words and phrases. Low-Density Parity-Check codes, Channels with Side Information at the Transmitter.

The work of G. Caire has been partially supported by the French *Réseau national de Recherche en Télécommunications.*

The work of S. Shamai has been supported by the Israel Sience Foundation (ISF)..

for the causal case with i.i.d. state sequence, by restricting the supremization in (1.1) to T independent of S [**6**].

In the case where the channel is $Y = X+S+Z$, with $Z \sim \mathcal{N}(0, \sigma^2)$, $E[|X|^2] \leq \mathcal{E}$ and the interference S is also Gaussian and known non-causally at the transmitter, Costa [**4**] proved that the capacity in (1.1) is equal to the standard AWGN capacity $\frac{1}{2}\log(1+\mathcal{E}/\sigma^2)$, as if interference was not present. From the title of Costa's paper, coding strategies for the non-causal known interference case are nicknamed "Dirty-Paper" coding and, by analogy, coding strategies for the causal case are nicknamed "Dirty-Tape" coding [**5, 6**].

While in early works such problems were motivated by data storage on defective media [**2**], more recently "Dirty-Paper" coding gained renewed attention because it arises as the main tool in several important settings such as broadcast vector Gaussian channels [**7, 8, 9, 10, 11, 12**], precoding for ISI channels [**13, 6**] and data hiding [**14, 15, 16**]. A widescope recent survey of applications of Dirty-Paper coding is provided in [**6**].

From the information theoretic point of view, Gel'fand, Pinsker and Costa's results were generalized in many ways (see [**6**] and references therein). However, efficient coding strategies able to approach the theoretical results are still not a common practice (preliminary results can be found in [**14, 15**]) although, based on random coding arguments, it can be shown that sequences of good lattice codes, mimicking Gel'fand and Pinsker random binning scheme, do exist and can approach Costa's result in the additive interference AWGN case [**17**].

The reminder of this paper is organized as follows. In Sections 2 and 3 we illustrate the capacity formulas (1.1) and (1.2) through two well-known examples: an input-constrained Binary-Symmetric Channel (BSC) with binary equiprobable interference, and Costa's AWGN channel with arbitrary interference. Then, Section 4 is devoted to the construction of coded modulation schemes based on Low-Density Parity-Check codes (LDPCs) for the AWGN Dirty-Tape problem (the example in Section 3 with causal interference knowledge). Conclusions and discussion are pointed out in Section 5.

2. BSC with known binary interference

Suppose that we wish to hide data into the least significant bits (LSBs) of the pixels of a gray-scale image (host signal) with a Hamming distortion constraint W, and assume that the host signal goes through some transformation that, acting on the LSBs, can be modeled as a BSC with transition probability p.

This problem can be modeled by the binary modulo-2 adder channel $Y = X + S + Z$ where S is the host signal LSB sequence, Z is the BSC noise and the input X is subject to the constraint

$$\frac{1}{n}\sum_{i=1}^{n} d_H(x_i, 0) \leq W \tag{2.1}$$

where $d_H(a,b)$ denotes Hamming distance.

In the following we shall assume that S is i.i.d. with $P(S=0) = P(S=1) = 1/2$. If the encoder has available the whole interference sequence realization s non-causally (i.e., before encoding), it can be shown (see [**18**] and references therein) that the maximum achievable rate is given by

$$C = \begin{cases} K(W) & \text{for } W_c \leq W \leq 1/2 \\ \alpha W & \text{for } 0 \leq W \leq W_c \end{cases} \tag{2.2}$$

where the function $K(w)$, defined for $w \in [0, 1/2]$, is given by

$$K(w) = \begin{cases} h(w) - h(p) & \text{for } p \leq w \leq 1/2 \\ 0 & \text{for } 0 \leq w \leq p \end{cases} \tag{2.3}$$

where $h(p)$ denotes the binary entropy function and where we let $W_c = 1 - \exp(-h(p))$ and $\alpha = \log((1 - W_c)/W_c)$.[1]

The rate $K(W)$ (for $W > p$) is achieved by Dirty-Paper coding. Namely, we construct a binary random codebook $\mathcal{C}$ of size $|\mathcal{C}| = \exp(n(1 - h(p) - \epsilon/2))$ and make a random partition of $\mathcal{C}$ into subsets $\{\mathcal{C}_m : m = 1, \ldots, M\}$ of size $|\mathcal{C}_m| = \exp(n(1-h(W)+\epsilon/2))$. For large n, the number of subsets is given by $M = \exp(n(K(W) - \epsilon))$. Both $\mathcal{C}$ and the partition $\{\mathcal{C}_m\}$ are revealed to encoder and decoder.

Let $\mathbf{s}$ be the interference sequence and m be the information message to be sent. The encoder finds a codeword $\mathbf{c} \in \mathcal{C}_m$ such that

$$\frac{1}{n}\sum_{i=1}^{n} d_H(s_i, c_i) \leq W$$

Since each subset is "large enough", such sequence $\mathbf{c}$ can be found with probability $1-\epsilon/2$ for sufficiently large n. Then, the encoder sends $\mathbf{x} = \mathbf{c} + \mathbf{s}$. The decoder observes the standard BSC output $\mathbf{y} = \mathbf{x} + \mathbf{s} + \mathbf{z} = \mathbf{c} + \mathbf{z}$, where $\mathbf{z}$ is the noise realization, finds the minimum distance codeword

$$\widehat{\mathbf{c}} = \arg\min_{\mathbf{c} \in \mathcal{C}} d_H(\mathbf{c}, \mathbf{y})$$

and outputs the message $\widehat{m}$ as the index of the subset containing $\widehat{\mathbf{c}}$. Since the codebook $\mathcal{C}$ is "small enough", $\Pr(\widehat{m} \neq m) \leq \epsilon/2$ for sufficiently large n. Eventually, the rate $K(W) - \epsilon$ can be transmitter with error probability not larger than ϵ for sufficiently large n.

Notice the different roles of the $\mathcal{C}$ and its subsets $\mathcal{C}_m$: the codebook $\mathcal{C}$ must be a good channel code for the BSC with transition probability p while its bins must be good Hamming quantizers for the interference sequence $\mathbf{s}$ (a Bernoulli i.i.d. unbiased source) with Hamming distortion W.

In the range $0 \leq W \leq W_c$, capacity is achieved by time-sharing with duty-cycle $\theta = W/W_c$ the above scheme for Hamming distortion equal to W_c and "silence" (zero rate and zero Hamming distortion).

The above intuitive argument can be made formal and proves achievability of (2.2). A converse is provided, for example, in [**18**].

Let's consider now the Dirty-Tape case, where the transmitter at time i knows only $(s_1, \ldots, s_i)$. In this case, Shannon's formula (1.2) is given explicitly by

$$C = 2W(\log 2 - h(p)) \tag{2.4}$$

The proof of the above capacity formula serves as an example of the rather abstract idea of "coding over strategies" underlying the general capacity formula (1.2). The auxiliary variable T in (1.2) takes on values in the set $\mathcal{T}$ of memoryless functions (or "strategies") mapping the state S into the input X. Therefore, a code for the Dirty-Tape channel (with i.i.d. states) is given by a set $\mathcal{C} \subseteq \mathcal{T}^n$ of sequences $\mathbf{c}_m = (c_{m,1}, \ldots, c_{m,n})$ of mappings $c_{m,i} : S \to X$, such that the input sequence $\mathbf{x}$ corresponding to the information message m and the state sequence $\mathbf{s}$ is obtained as $x_i = c_{m,i}(s_i)$.

---

[1] When information rates are measured in bits, log and exp are base-2, when it is measured in nats, log and exp are base-e.

When both S and X are binary, then $\mathcal{T} = \{\mathbf{0}, \mathbf{1}, \mathbf{id}, \mathbf{not}\}$, i.e., the identically zero, identically 1 functions, identity and negation, respectively. The transition probability of the associated channel with input T and output Y, given by

$$P_{Y|T}(y|t) = \sum_s P_{Y|X,S}(y|t(s), s)P_S(s),$$

is given by the table

Y/T	**0**	**1**	**id**	**not**
0	$1/2$	$1/2$	$1-p$	p
1	$1/2$	$1/2$	p	$1-p$

We start with an upperbound to capacity. Due to the concavity of capacity as a function of the input average "cost" W and to the fact that the input symbol **0** has zero associated cost, we have [**21**]

$$C \leq W \sup_{t \in \mathcal{T}} \frac{D(P_{Y|T=t} \| P_{Y|T=\mathbf{0}})}{E[d_H(t(S), 0)]} = 2W(\log 2 - h(p)) \tag{2.5}$$

where the supremization is achieved by either $t = \mathbf{id}$ or by $t = \mathbf{not}$. In order to show that (2.5) is indeed the capacity, we find an input probability assignment P_T such that $I(T;Y)$ equals (2.5). The input constraint is given by

$$E[d_H(T(S), 0)] = P_T(\mathbf{0}) \cdot 0 + P_T(\mathbf{1}) \cdot 1 + \frac{1}{2}\left(P_T(\mathbf{id}) + P_T(\mathbf{not})\right) \leq W \tag{2.6}$$

By direct inspection, it is clear that the optimal input probability assignment must put zero probability on the input $\mathbf{1}$, since $P_T(\mathbf{1}) > 0$ would increase the input average Hamming weight without increasing mutual information ($I(T = \mathbf{1}; Y) = 0$). Moreover, by symmetry, it must be $P_T(\mathbf{id}) = P_T(\mathbf{not})$ (notice that this choice makes the output distribution uniform). Hence, we choose $P_T(\mathbf{id}) = P_T(\mathbf{not}) = \theta/2$, and $P_T(\mathbf{0}) = 1 - \theta$ for some $\theta \in [0, 1]$. The resulting mutual information is given by

$$I(T;Y) = \theta\left(\log 2 - h(p)\right) \tag{2.7}$$

By letting $\theta = 2W$, we obtain (2.5).

Fig. 1 shows the Dirty-Paper and the Dirty-Tape capacities vs. the input constraint W for $p = 0.1$. Causal knowledge of the interference sequence incurs a noticeable capacity loss with respect to non-causal knowledge. The capacity per unit cost [**21**], defined as the maximum number of information bits per input "one", yields

$$\begin{aligned} U^{\text{non-causal}} &= \log \frac{1}{1 - \exp(-h(p))} - h(p) \\ U^{\text{causal}} &= 2(\log 2 - h(p)) \end{aligned} \tag{2.8}$$

3. AWGN channel with known arbitrary interference

Consider the real additive noise channel $Y = X + S + Z$, where $Z \sim \mathcal{N}(0, \sigma^2)$ is white Gaussian noise, the input is constrained by $E[X^2] \leq \mathcal{E}$ and S is an interference signal arbitrarily distributed according to some P_S. This model differ from Costa's channel since P_S is not necessarily known, therefore, the random coding argument for achievability (outlined in Section 2 for the BSC) does not hold any longer, as it makes use of typicality and, implicitly, exploits the knowledge of P_S. Nevertheless, in [**5, 6**] an explicit coding

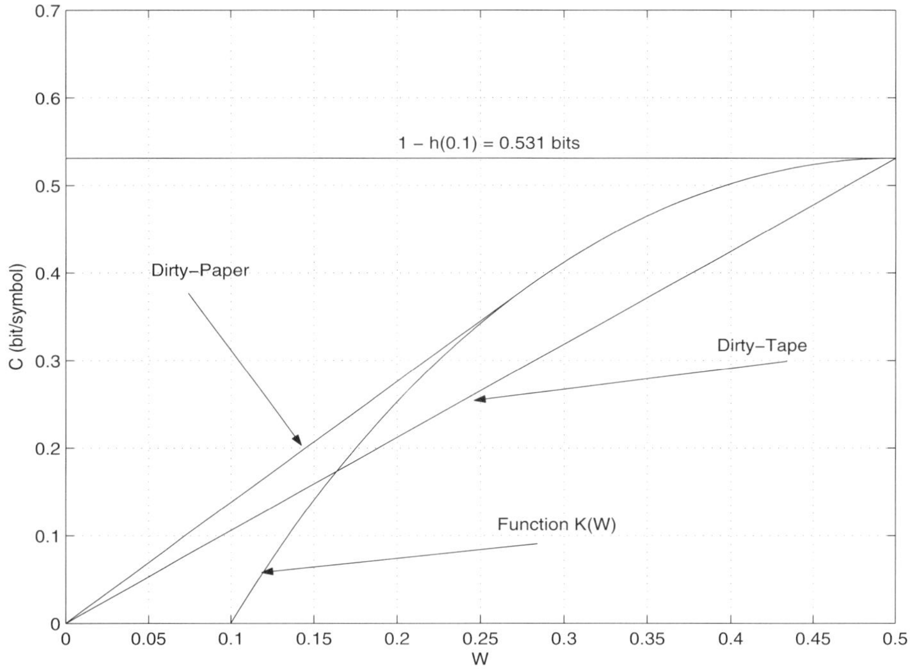

FIGURE 1. Capacity of the input-constrained BSC with causally and non-causally known interference, for $p = 0.1$.

strategy referred to as *inflated lattice precoding* is shown to achieve capacity $\frac{1}{2}\log(1 + \mathcal{E}/\sigma^2)$ when the interference signal *realization* is known non-causally.

Inflated lattice precoding is based on lattice dithered quantization and on the scaling of the received signal (inflation) followed by lattice decoding. Dithering requires that encoder and decoder share common randomness [2], and allows the extension of Costa's result to any interference statistics and even to arbitrary interference sequences [**6**], making it an efficient coding approach for the arbitrary varying channel with states known to the transmitter [**22**].

Moreover, the inflated lattice scheme can be applied to the case where interference is known non-causally with anticipation k, i.e., the encoder at time i knows the realization of the interference signal $(s_1, \dots, s_{\min\{i+k-1,n\}})$. In particular, for $k = 1$ is provides an effective coding strategy for the Dirty-Tape setting.

The inflated lattice precoding scheme is illustrated in Fig. 2. Let k be an integer dividing the block length n, and consider a k-dimensional lattice Λ with fundamental Voronoi cell $\mathcal{V}$, with normalized second-order moment $\mathcal{E}$ (i.e., such that $\frac{1}{V(\Lambda)}\int_{\mathcal{V}} |\mathbf{x}|^2 d\mathbf{x} = k\mathcal{E}$).

A code $\mathcal{C}$ of rate R and block length n is constructed according to a uniform distribution over $\mathcal{V}$, so that

$$\mathcal{C} \subseteq \mathcal{V}^{n/k} \triangleq \underbrace{\mathcal{V} \times \cdots \times \mathcal{V}}_{n/k \text{ times}}$$

[2]Notice that sharing randomness is common practice in wireless communications. For example, in standard randomly spread CDMA transmitter and receiver share the (pseudo-)random spreading code generator.

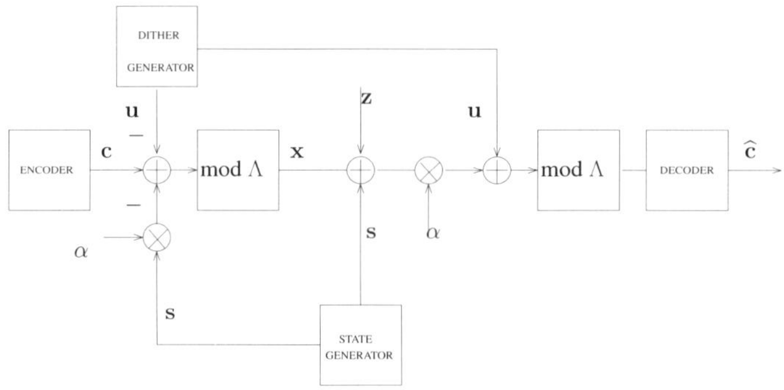

FIGURE 2. The inflated lattice precoding scheme.

Furthermore, the transmitter and the receiver generate the same *dither* signal $\mathbf{u}$, uniformly distributed over $\mathcal{V}^{n/k}$.

For an n-dimensional vector $\mathbf{v}$, the vector $\mathbf{v} \bmod \Lambda^{n/k}$ is defined as

$$\mathbf{v} \bmod \Lambda^{n/k} = \mathbf{v} - \boldsymbol{\lambda}(\mathbf{v}) \tag{3.1}$$

where

$$\boldsymbol{\lambda}(\mathbf{v}) \stackrel{\Delta}{=} \arg \min_{\boldsymbol{\lambda} \in \Lambda^{n/k}} |\mathbf{v} - \boldsymbol{\lambda}|^2 \tag{3.2}$$

In other words, reducing $\mathbf{v}$ modulo $\Lambda^{n/k}$ consists of quantizing $\mathbf{v}$ by using $\Lambda^{n/k}$ as a lattice quantizer, and computing the total quantization error vector.

Let $\mathbf{c} \in \mathcal{C}$ be the codeword to be transmitted. After observing the interference signal $\mathbf{s}$, the transmitter produces the channel input sequence

$$\mathbf{x} = [\mathbf{c} - \alpha \mathbf{s} - \mathbf{u}] \bmod \Lambda^{n/k} \tag{3.3}$$

where $\alpha \in [0, 1]$ is a scaling coefficient (to be optimized), and sends $\mathbf{x}$. Thanks to the dither signal $\mathbf{u}$, $\mathbf{x}$ is uniformly distributed on $\mathcal{V}^{n/k}$ and its average energy per symbol is $\mathcal{E}$, so that the power input constraint is satisfied.

After receiving $\mathbf{y} = \mathbf{x} + \mathbf{s} + \mathbf{z}$, the receiver computes

$$\mathbf{y}' = [\alpha \mathbf{y} + \mathbf{u}] \bmod \Lambda^{n/k} \tag{3.4}$$

It can be shown [**5, 6**] that the channel from the encoder output to the decoder input (see Fig. 2) is equivalent to the additive modulo$-\Lambda^{n/k}$ noise channel

$$\mathbf{y}' = [\mathbf{c} + \mathbf{z}'] \bmod \Lambda^{n/k} \tag{3.5}$$

where $\mathbf{z}'$ is distributed as $[(1-\alpha)\mathbf{u} + \alpha \mathbf{z}] \bmod \Lambda^{n/k}$. Finally, the decoder computes (or approximates) the ML decision

$$\widehat{\mathbf{c}} = \arg \max_{\mathbf{c} \in \mathcal{C}} p_{Z'}(\mathbf{y}' - \mathbf{c}) \tag{3.6}$$

where $p_{Z'}$ denotes the pdf of $\mathbf{z}'$ defined above.

As anticipated above, inflated lattice precoding for $k = n$ applies to the non-causal Dirty-Paper case and, by choosing a sequence of good n-dimensional lattices, it achieves the AWGN capacity $\frac{1}{2}\log_2(1 + \mathcal{E}/\sigma^2)$, where the optimized inflation factor is given by [**4**]

$$\alpha = \frac{\mathcal{E}}{\sigma^2 + \mathcal{E}} \tag{3.7}$$

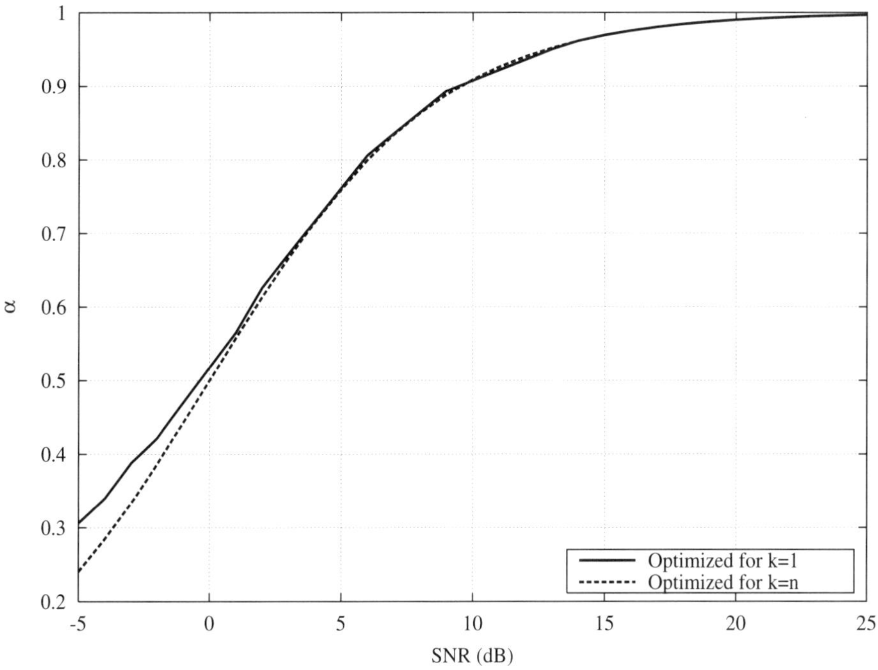

FIGURE 3. Optimal inflation factor for $k = 1$ and $k = n \to \infty$.

For the Dirty-Tape case ($k = 1$), we have $\Lambda = \Delta\mathbb{Z}$, with $\Delta = \sqrt{12\mathcal{E}}$, and the Voronoi region $\mathcal{V}$ is the interval $[-\frac{\Delta}{2}, \frac{\Delta}{2}]$. The scheme is a special case of Shannon's coding over strategies, where the code is *randomized* (via the dither signal) and the strategy alphabet is given by

$$\mathcal{T} = \{t_{v,u}(s) = [v - \alpha s - u] \bmod [-\Delta/2, \Delta/2] \ : \ v \in [-\Delta/2, \Delta/2]\} \tag{3.8}$$

Random coding over $\mathcal{T}$ where the code ensemble is generated according to a uniform distribution achieves the rate

$$R^{\text{prec.}} = \max_{\alpha \in [0,1]} \{\log \Delta - h(Z')\} \tag{3.9}$$

where the optimization of the inflation coefficient α is obtained numerically. Fig. 3 shows the optimal α as a function of the SNR $\triangleq \mathcal{E}/\sigma^2$. The optimal Dirty-Paper value (3.7) is shown for comparison.

By choosing α as in (3.7), we obtain the lower bound

$$R^{\text{prec.}} \geq \frac{1}{2}\log\left(1 + \frac{\mathcal{E}}{\sigma^2}\right) - \frac{1}{2}\log\frac{2\pi e}{12} \tag{3.10}$$

where $\frac{1}{2}\log_2 \frac{2\pi e}{12} = 0.254$ bits is the "shaping gain". Moreover, in [**5**] it is proved that the above rate is asymptotically optimal, i.e., it coincides with the capacity of the AWGN Dirty-Tape channel with arbitrary interference, in the limit of high SNR. It is interesting to notice that, for high SNR, the penalty incurred by causal versus non-causal interference knowledge is only the shaping gain.

4. LDPC coded-modulation for writing on Dirty-Tape

In this section we consider the AWGN Dirty-Tape channel and we construct inflated lattice coding schemes approaching the performance of random coding with uniform probability over the alphabet $\mathcal{T}$ defined in (3.8). The obtained codes can be seen either as "Dirty-Tape" codes, either as "Dirty-Paper" codes with a suboptimal choice of the lattice Λ in the general inflated lattice precoding scheme outlined in Section 3.

It is natural to construct the code $\mathcal{C}$ by concatenating a linear binary code with M-PAM signal set

$$\mathcal{A} = \left\{ \frac{\Delta}{2M}(2m - M + 1) \ : \ m = 0, \ldots, M - 1 \right\}$$

thus inducing an input marginal distribution close to the uniform distribution over $[-\Delta/2, \Delta/2]$, for large PAM alphabet size M.

The first-order transition pdf of the corresponding modulo-noise channel (defined in general by (3.5)) is given by

$$p_{Z'}(z) = \sum_{k \in \mathbb{Z}} \frac{P_Z\left(\frac{z + k\Delta + (1-\alpha)\Delta/2}{\alpha}\right) - P_Z\left(\frac{z + k\Delta - (1-\alpha)\Delta/2}{\alpha}\right)}{(1-\alpha)\Delta} \tag{4.1}$$

where $P_Z(z)$ denotes the cdf of the noise of the original channel. In this work we consider AWGN, therefore $P_Z(z)$ is the Gaussian distribution $\mathcal{N}(0, \sigma^2)$.

Next, we consider two code constructions for $\mathcal{C}$: LDPC multilevel coding and direct design of LDPC-coded modulation.

4.1. Multilevel coding. Multilevel coding (see [**23**] and references therein) is a general method for constructing coded modulation schemes. Fig. 4 shows a block diagram of the multilevel encoder, where m binary codes produce codewords $\mathbf{c}_1, \ldots, \mathbf{c}_m$ of length n. These are arranged as rows of a $m \times n$ matrix, and a binary labeling function $\phi : \mathbb{F}_2^m \to \mathbf{A}$ is applied columnwise, in order to form the corresponding codeword of $\mathcal{C}$.

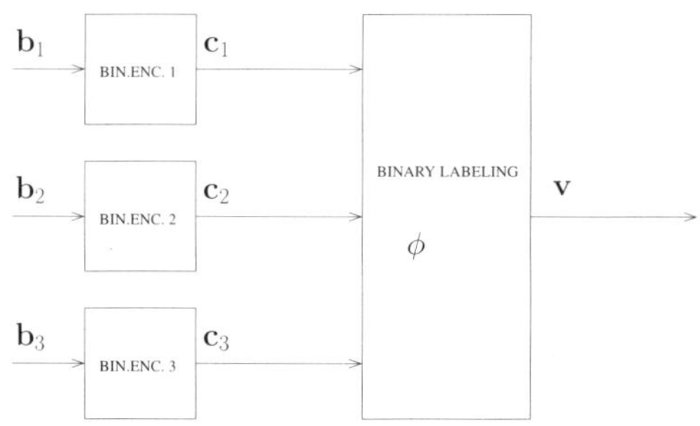

FIGURE 4. Multilevel coded modulator with three levels.

As binary component codes we choose LDPCs from the database of LDPC ensembles optimized for the binary-input AWGN channel provided in [**24**]. Following [**23**], the choice of the component code rates is dictated by the mutual information chain rule. Namely, let A be uniformly distributed on $\mathcal{A}$ and let $(b_1, \ldots, b_m)$ be binary uniform random variables, then

$$I(A; Y') = I(b_1, \ldots, b_m; Y') = \sum_{i=1}^{m} R_i$$

where we define the rate at level i as

$$
\begin{aligned}
R_i &= I(Y'; b_i | b_1, \ldots, b_{i-1}) \\
&= E\left[\log_2 \frac{\sum_{a \in \mathcal{A}(b_1,\ldots,b_i)} p_{Z'}(Y' - a)}{\sum_{a' \in \mathcal{A}(b_1,\ldots,b_{i-1})} p_{Z'}(Y' - a')}\right]
\end{aligned} \tag{4.2}
$$

where $\mathcal{A}(b_1, \ldots, b_i)$ denotes the subset of points of $\mathcal{A}$ whose label first i positions are given by $(b_1, \ldots, b_i)$. The rates R_i are achievable by a multistage decoder that considers the levels in sequence, by decoding each level i assuming that the decoding outcomes at previous levels $1, \ldots, i-1$ are correct, and by treating the levels $i+1, \ldots, m$ as a random nuisance. The multistage decoder is analogous to the well-known successive interference cancellation decoder used in Gaussian multiaccess and broadcast channels [**20**].

The binary labeling ϕ affects the level rates R_i but, as long as ϕ is a one-to-one mapping, the total mutual information is independent of ϕ. Fig. 5 shows an example of the set-partitioning labeling [**25**] considered in this work for 8-PAM codes and Fig. 6 shows the corresponding achievable rate of the 8-PAM modulo-noise channel, with the level rates R_i.

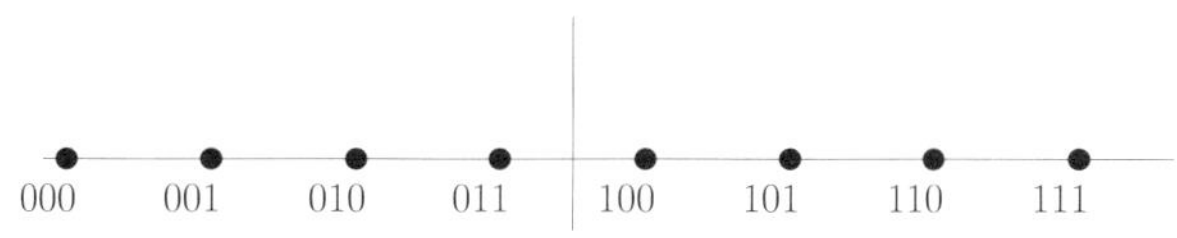

FIGURE 5. Set-partitioning labeling of 8-PAM.

Levels are decoded from right to left of the labels. We notice that for high SNR only the first level need coding at rate $R_1 < 1$, while the other two can be transmitted uncoded (i.e., $R_2 = R_3 = 1$). This provides a very simple scheme requiring a single decoding stage followed by symbol-by-symbol detection of the remaining stages. Hence, multilevel coding with set-partitioning labeling is particularly attractive in the high-SNR region. Incidentally, this is also the region where the inflated lattice precoding scheme with $k = 1$ pays the smallest relative penalty with respect to the full non-causal case (as already noticed, the shaping gain 0.254 bit/symbol).

4.2. Direct optimization. Our second approach consists of directly optimizing the ensemble of LDPC-coded modulation schemes constructed over the M-PAM alphabet $\mathcal{A}$, by exploiting the *density-evolution* method developed to analyze LDPC codes under message-passing decoding in the limit of infinite block length [**26, 27**]. Fig. 7 shows the Tanner graph of the code, where the bitnodes are partitioned into subsets of m nodes, each of which is associated to the m label positions of a M-PAM symbol. The super-nodes corresponding to modulation symbols will be referred to as "$\mathcal{A}$-nodes".

We say that a $\mathcal{A}$-node is of *type* $(d_1, \ldots, d_m)$ if its i-th label bitnode has degree d_i. We enumerate the $\mathcal{A}$-node types in lexicographic order, and let $d_{t,i}$ be the degree of the i-th bitnode in the $\mathcal{A}$-nodes of type t. We let $\lambda_{t,i}$ denote the fraction of edges connected with $\mathcal{A}$-nodes of type t in position i. For a graph with e edges, the number of $\mathcal{A}$-nodes of type t is given by $n_t = e\lambda_{t,i}/d_{t,i}$, therefore $\lambda_{t,i}/d_{t,i}$ must not depend on i.

As in standard LDPC notation, we let ρ_j be the fraction of edges connected to checknodes of degree j. The ensemble optimization consists of finding, for any given SNR, the set of $\mathcal{A}$-node types, the left degree distribution $\{\lambda_{t,i}\}$ and the right degree distribution $\{\rho_j\}$ such that the coding rate is maximized subject to the constraint that the density

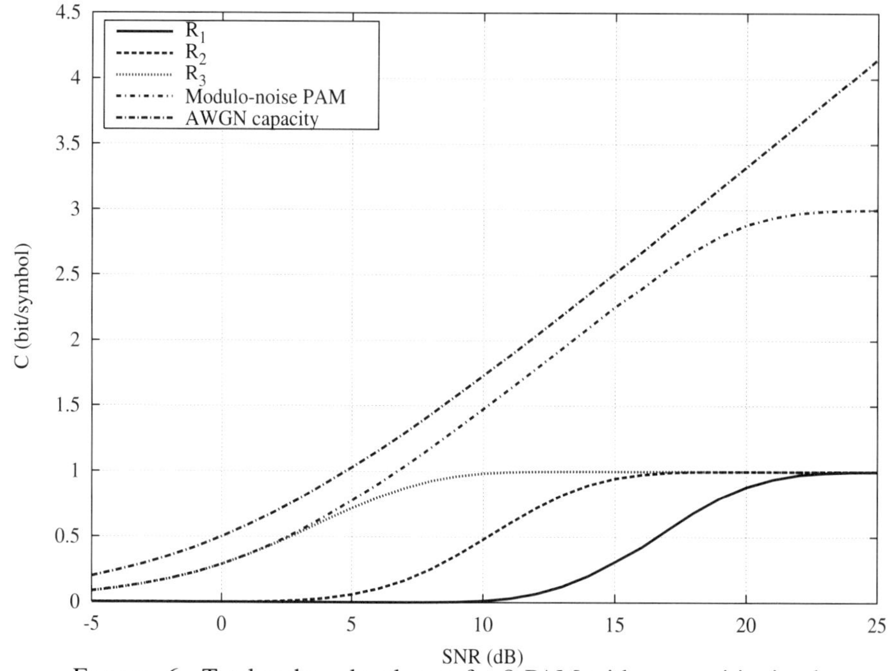

FIGURE 6. Total and per-level rates for 8-PAM with set-partitioning labeling in the modulo-noise channel.

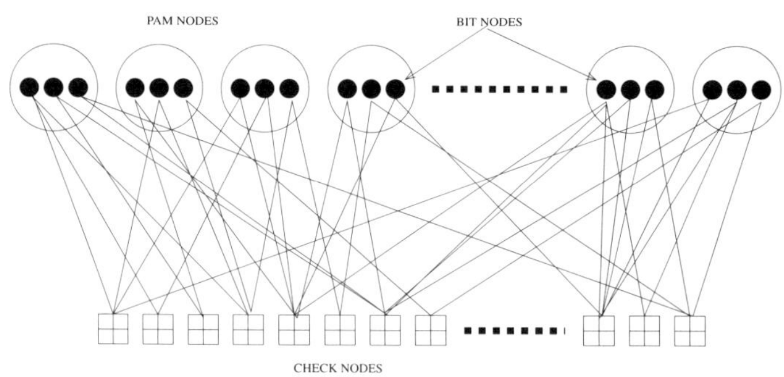

FIGURE 7. Tanner graph of an LDPC-coded modulation scheme.

evolution (DE) converges to zero bit-error probability as the number of iterations goes to infinity (see [**26, 28**] for the details). Since optimizing the degree distributions based on the exact DE is computationally very intensive, we propose a design method based on a one-dimensional approximation of DE, obtained by approximating the message densities by Gaussian pdfs (see [**27**] and especially [**29**] for similar approaches in different contexts).

From $\sum_{t,i} \lambda_{t,i} = 1$ we obtain the constraint $\sum_t \lambda_{t,1} \sum_{i=1}^m \frac{d_{t,i}}{d_{t,1}} = 1$. The design coding rate is given by

$$R = \log_2 M - \frac{\sum_j \rho_j / j}{\sum_t \lambda_{t,1}/d_{t,1}} \quad \text{bit/symbol} \tag{4.3}$$

For given right degree sequence $\{\rho_j\}$, we wish to obtain an optimization problem in the variables $\{\lambda_{t,1}\}$.

Consider an $\mathcal{A}$-node of type $(d_1, \ldots, d_m)$ and consider an edge o connected in position i. The message-passing transformation of the iterative belief-propagation decoder associated to the message output by the node onto edge o is given by

$$\mathcal{L}_{i,o}^{\text{out}} = \log \frac{\sum_{a \in \mathcal{A}_0^i} p_{Z'}(y' - a) \exp\left(-\sum_{j=1}^{m} b_j \sum_{u=1}^{d_j} \mathcal{L}_{j,u}^{\text{in}}\right)}{\sum_{a \in \mathcal{A}_1^i} p_{Z'}(y' - a) \exp\left(-\sum_{j=1}^{m} b_j \sum_{u=1}^{d_j} \mathcal{L}_{j,u}^{\text{in}}\right)} - \mathcal{L}_{i,o}^{\text{in}} \tag{4.4}$$

where $\mathcal{L}_{j,u}^{\text{in}}$ denotes the input message from edge u connected to the bitnode in position j, and $\mathcal{A}_0^i$ (resp. $\mathcal{A}_1^i$) denotes the signal subset of all points of $\mathcal{A}$ having symbol 0 (resp. 1) in label position i and where y' denotes the channel output corresponding to the given $\mathcal{A}$-node.

It has been shown in [**28**] that the message pdfs generated by the belief propagation algorithm at each iteration satisfy a *symmetry condition*. If the pdfs are Gaussian, the symmetry condition imposes that the variance must be twice the mean. Assuming that all input messages are Gaussian i.i.d. $\sim \mathcal{N}(\mu, 2\mu)$, we can obtain by Monte Carlo simulation of (4.4) the distribution of $\mathcal{L}_{i,o}^{\text{out}}$ for every i and node type t, parameterized in the input mean value μ.

Following [**29**], we shall replace the pdf $f_{\mathcal{L}}(z) \triangleq \frac{d}{dz} \Pr(\mathcal{L} \leq z | b = 0)$ of a message $\mathcal{L}$ relative to a bitnode b by the value of the mutual information functional $I(b; \mathcal{L})$ which, for symmetric pdfs, is given by

$$I(b; \mathcal{L}) = 1 - \int_{-\infty}^{\infty} \log_2\left(1 + e^{-z}\right) f_{\mathcal{L}}(z) dz \tag{4.5}$$

Then, from the Gaussian assumption of the input messages and the explicit mapping (4.4), for any $\mathcal{A}$-node type and bitnode position i we can find (numerically) a mutual information transfer function

$$\mathsf{y} = \Gamma_{t,i}(\mathsf{x}) \tag{4.6}$$

where $\mathsf{x} \triangleq I(b; \mathcal{L}^{\text{in}})$ and $\mathsf{y} \triangleq I(b; \mathcal{L}^{\text{out}})$.

Again as a consequence of the message pdf symmetry, the mutual information between a message $\mathcal{L}^{\text{out}}$ and its associated bitnode value b on a randomly selected graph edge, chosen with probability $\lambda_{t,i}$, is given by

$$\mathsf{y} = \sum_{t,i} \lambda_{t,i} \Gamma_{t,i}(\mathsf{x}) \tag{4.7}$$

Fig. 8 shows the functions $\Gamma_{t,i}$ for an 8-PAM-node of type $(2, 5, 10)$.

For the message-passing mapping at the checknodes we use the approximated duality property [**30**], stating that the mutual information transfer function of a checknode is closely approximated by the mutual information transfer function of a bitnode with the same degree, by applying the mapping $\mathsf{x} \mapsto 1 - \mathsf{x}$ to the input and $\mathsf{y} \mapsto 1 - \mathsf{y}$ to the output. Assuming the input messages i.i.d. Gaussian $\sim \mathcal{N}(\mu, 2\mu)$ and by defining the binary-input Gaussian mutual information functional

$$J(\mu) \triangleq 1 - \frac{1}{\sqrt{\pi}} \int_{-\infty}^{+\infty} e^{-z^2} \log_2\left(1 + e^{-2\sqrt{\mu}z - \mu}\right) dz, \tag{4.8}$$

the mutual information transfer function of a checknode of degree j is given by

$$\mathsf{y} = 1 - J\left((j-1)J^{-1}(1 - \mathsf{x})\right)$$

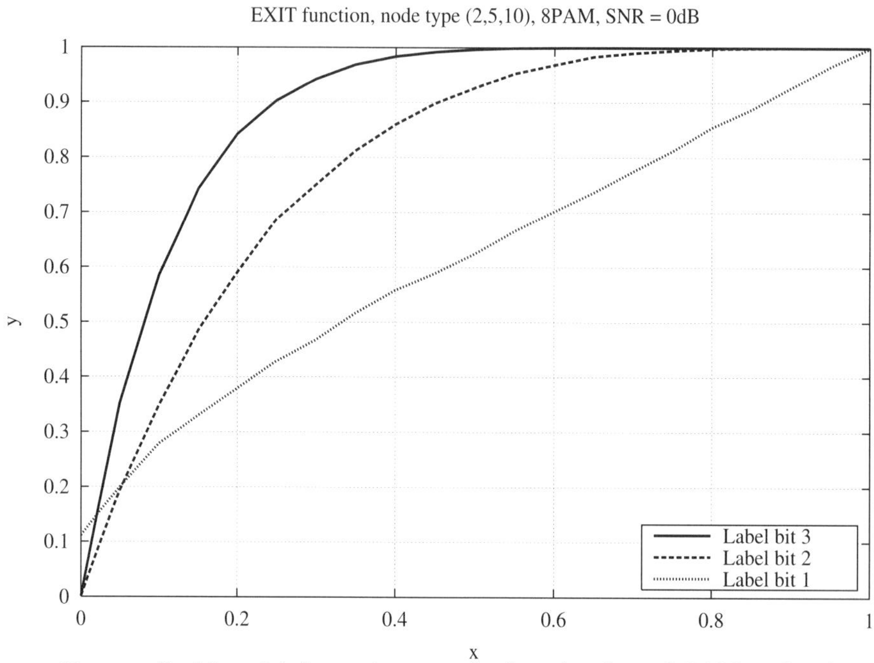

FIGURE 8. Mutual information transfer function for a 8-PAM node of type $(2,5,10)$ for SNR$=0$ dB.

Eventually, the one-dimensional DE approximation is given by the recursion

$$\mathsf{x}^{(\ell)} = \sum_{t,i} \lambda_{t,i}\Gamma_{t,i}\left(1 - \sum_j \rho_j J\left((j-1)J^{-1}\left(1-\mathsf{x}^{(\ell-1)}\right)\right)\right) \tag{4.9}$$

for $\ell = 1, 2, \ldots$, with initial condition $\mathsf{x}^{(0)} = 0$.

The recursion (4.9) has a unique stable fixed point in $\mathsf{x} = 1$ (corresponding to vanishing bit-error probability), if and only if

$$\mathsf{x} > \sum_{t,i} \lambda_{t,i}\Gamma_{t,i}\left(1 - \sum_j \rho_j J\left((j-1)J^{-1}(1-\mathsf{x})\right)\right), \quad \forall\, \mathsf{x} \in [0,1) \tag{4.10}$$

Hence, we can sample the above equation for x taking on values in a fine grid of points in the interval $[0,1)$ and for each point we obtain a linear constraint in the variables $\lambda_{t,1}$ (recall that $\lambda_{t,i} = \frac{d_{t,i}}{d_{t,1}}\lambda_{t,1}$, therefore the only independent variable of the optimization problem are $\{\lambda_{t,1}\}$). Since both the constraints and the objective function (4.3) are linear in the $\{\lambda_{t,1}\}$, the solution is readily obtained by linear programming.

4.3. Results. Fig. 9 shows the rate versus E_b/N_0 of some multilevel LDPC codes and LDPC-coded M-PAM codes obtained according to the methods described above, for the AWGN Dirty-Tape channel. In all cases, the block length (in PAM symbols) is 20000 and the marks correspond to bit-error probability $\leq 10^{-4}$ with a maximum of 100 decoder iterations.

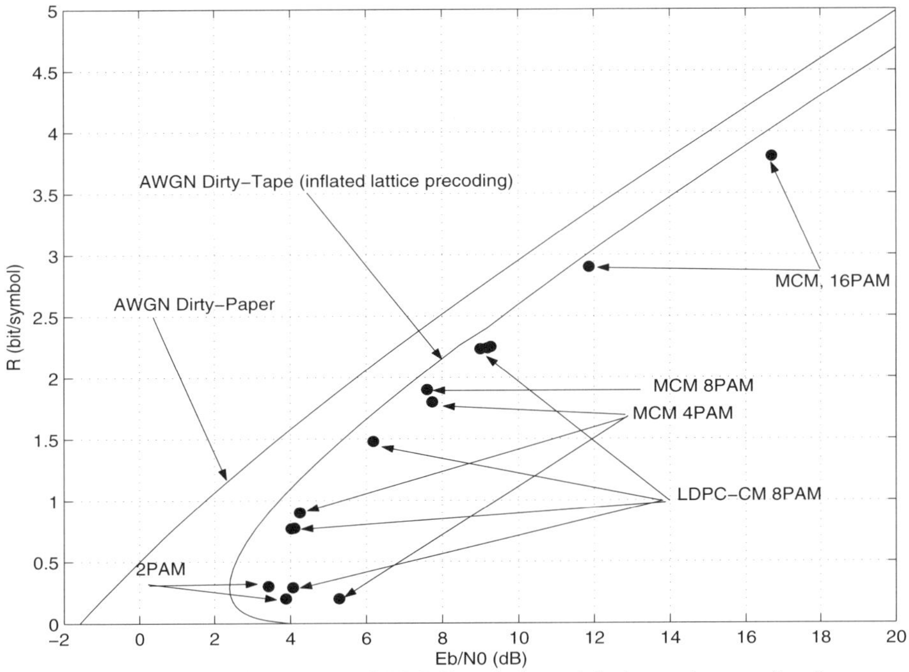

FIGURE 9. Performance of LDPC-coded modulation schemes for the dirty-tape channel with AWGN.

Marks labeled by "2PAM" correspond to standard binary LDPC, marks labeled by "MCM MPAM" correspond to scheme obtained by multilevel coded modulation and marks labeled by "LDPC-CM MPAM" correspond to direct optimization of LDPC-coded modulation.

We observe that the theoretical achievable rate $R^{\text{prec.}}$ given by (3.9) can be closely approached by our LDPC-coded PAM modulation, especially for high SNR. For low SNR, the best results are provided by standard binary LDPC codes ($M = 2$), while for high SNR the multilevel construction with one or two coding levels and the remaining levels left uncoded proves to be both efficient and simple in terms of complexity. There exists a region of intermediate SNR where it is indeed worthwhile to construct explicitly optimized LDPC-coded modulation, although better understanding and more refined optimization are called for, since the simple schemes that we experimented do not show a dramatic improvement with respect to multilevel codes. For direct optimization we restricted to codes over the 8-PAM alphabet, but the coding design ideas apply immediately to other cases.

It is interesting to notice that the minimum E_b/N_0 (corresponding to the inverse of the capacity per unit-cost) achieved by inflated lattice precoding with $k = 1$ is not obtained for vanishing SNR. This is due to the fact that the rate $R^{\text{prec.}}(\mathcal{E}/\sigma^2)$ (seen as a function of $\mathcal{E}/\sigma^2$) is not concave, therefore, in a region of low SNR a better rate is obtained by time-sharing between SNR$= 0$ and some positive SNR*. Fig. 10 shows $(E_b/N_0)^{\text{prec.}}$, defined

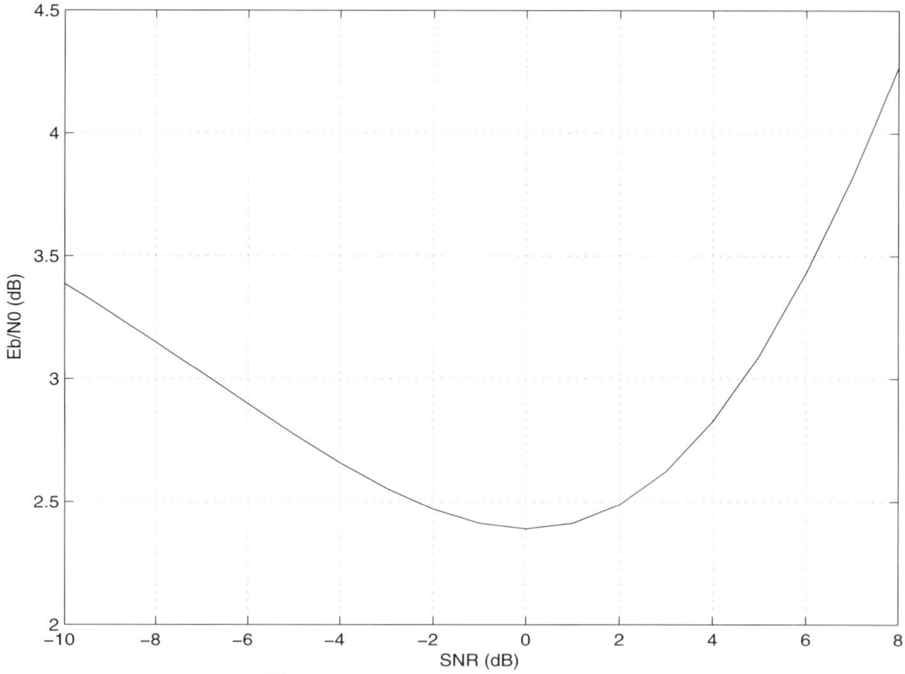

FIGURE 10. E_b/N_0 corresponding to the Dirty-Tape inflated lattice achievable rate vs. SNR.

implicitly by the equation

$$\left(\frac{E_b}{N_0}\right)^{\text{prec.}} R^{\text{prec.}}(\mathcal{E}/\sigma^2) = \frac{\mathcal{E}}{2\sigma^2} \tag{4.11}$$

versus SNR. From this figure we find $\text{SNR}^\star \approx 0$ dB.

5. Discussion

From the results of previous section we observe that inflated lattice precoding incurs a significant loss in achievable rates for low SNR. It is natural to ask whether this loss is due to the suboptimality of the inflated lattice strategy alphabet, or it is due to causal versus non-causal interference knowledge. Assuming i.i.d. interference with known distribution P_S, the capacity per unit-cost of the AWGN Dirty-Tape channel can be investigated by using the general formula of [**21**]. In particular, let $\mathcal{T}$ denote the set of all functions $t : \mathbb{R} \to \mathbb{R}$ mapping S into X, and let

$$P_{Y|T}(y|t) = \int P_{Y|X,S}(y|t(s), s) dP_S(s)$$

where $P_{Y|X,S}(y|x, s) = \frac{1}{\sqrt{2\pi}\sigma} e^{-\frac{(y-x-s)^2}{2\sigma^2}}$. Then,

$$U^{\text{non-causal}} = \sup_{t \in \mathcal{T}} \frac{D\left(P_{Y|T=t} \| P_{Y|T=\mathbf{0}}\right)}{E[t(S)^2]} \tag{5.1}$$

where, as before, $\mathbf{0}$ denotes the identically zero function. The supremization in (5.1) does not yield, in general, a simple solution. However, a lower-bound to $U^{\text{non-causal}}$ for some

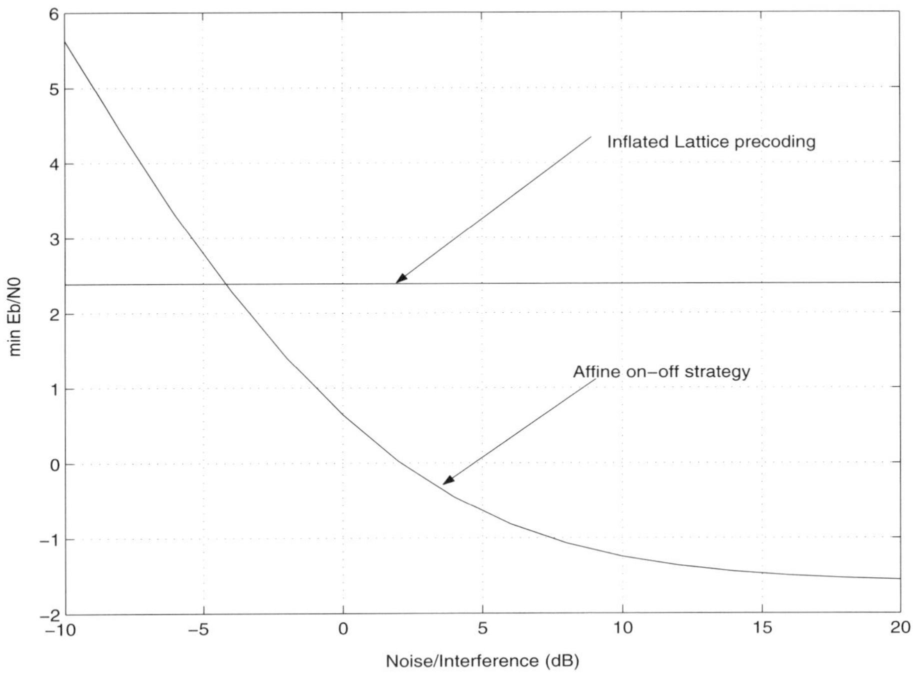

FIGURE 11. Minimum E_b/N_0 corresponding to the on-off affine strategy signaling and to inflated lattice precoding for the AWGN Dirty-Tape channel.

specific P_S can be found by restricting t to take on a particular form. For example, assume $S \sim \mathcal{N}(0, \sigma_S^2)$, define $\gamma = \sigma^2/\sigma_S^2$, and consider the set of affine functions

$$t(s) = as + b$$

We find [**31**]

$$U^{\text{non-causal}} \geq \frac{1}{2\sigma_S^2} \sup_{a \geq -1} \frac{-\frac{1-(1+a)^2}{1+\gamma} - \log\left(1 - \frac{1-(1+a)^2}{1+\gamma}\right)}{a^2} \tag{5.2}$$

where the RHS is the rate per unit-cost achieved by on-off affine-strategy signaling. The "on" input is $t^\star(s) = a^\star s$, with $a^\star$ given by the supremization in (5.2), and the "off" input is $\mathbf{0}$, where the duty-cycle is $\theta = \mathcal{E}/((a^\star)^2\sigma_S^2)$.

Fig. 11 shows the minimum E_b/N_0 corresponding to the above on-off signaling versus γ. This is compared with the minimum $(E_b/N_0)^{\text{prec.}}$ (which is independent of γ). As expected, for weak interference power (i.e., for large γ) on-off signaling approaches -1.59 dB, the minimum E_b/N_0 of the AWGN interference-free channel. However, on-off affine signaling is not robust to the interference power, since its minimum E_b/N_0 increases as γ decreases. On the contrary, the inflated lattice strategy yields minimum E_b/N_0 independent of the interference power. This example shows that for some P_S we can find coding strategies that outperform inflated lattice precoding in some range of non-asymptotically large SNR. Whether the convex upper envelope of the achievable rates $R^{\text{prec.}}$ is the worst-case Dirty-Tape capacity over all possible interference distributions P_S is still an open conjecture.

An important generalization of the coding schemes presented here considers the case of finite anticipation $k > 1$. Multilevel coding and direct optimization of LDPC-coded modulation can be applied to codes constructed over a k-dimensional constellation carved from a lattice Λ. Let $\mathbf{G} \in \mathbb{R}^{k\times k}$ be the generator matrix of Λ. Then, the code $\mathcal{C}$ will be constructed on the Voronoi constellation $\mathcal{A} = \mathbf{G}(M-\text{PAM})^k$, of M^k points in the fundamental Voronoi cell $\mathcal{V}$ of Λ (after appropriate scaling and translation). Depending on Λ, the modulo-Λ operations at the transmitter and receiver can be implemented either by some ML lattice decoder (e.g., based on a trellis representation of Λ) or by the general-purpose sphere decoder [**32, 33**]. As noticed in [**6**], the shaping gain plays a very important role also in the low-SNR region, and there is no hope of approaching Costa's Dirty-Paper limit for low SNR by using low-dimensional lattices (small k).

Finally, an interesting information-theoretic problem consists of studying the achievable rates with some restrictions of the code alphabet, e.g., the constraint that $\mathcal{C}$ must be a binary code.

References

[1] C. Shannon, "Channels with side information at the transmitter," *IBM J. Res. & Dev.*, pp. 289–293, 1958.

[2] A. Kusnetsov and B. Tsybakov, "Coding in a memory with defective cells," *Probl. Pered. Inform.*, vol. 10, no. 2, pp. 52–60, 1974.

[3] S. Gelfand and M. Pinsker, "Coding for channel with random parameters," *Problems of Control and Information Theory*, vol. 9, no. 1, pp. 19–31, January 1980.

[4] M. Costa, "Writing on dirty paper," *IEEE Trans. on Inform. Theory*, vol. 29, no. 3, pp. 439–441, May 1983.

[5] U. Erez, S. Shamai, and R. Zamir, "Capacity and lattice-strategies for cancelling known interference," submitted to IEEE Trans. on Inform. Theory. See also ISITA 2000, Honolulu, Hawaii (USA) November 2000.

[6] R. Zamir, S. Shamai, and U. Erez, "Nested linear/lattice codes for structured multiterminal binning," *IEEE Trans. on Inform. Theory*, vol. 48, no. 6, pp. 1250–1276, June 2002.

[7] G. Caire and S. Shamai, "On achievable rates in a multi-antenna broadcast downlink," in *38th Annual Allerton Conf. on Commun., Cont. and Comp.*, Monticello, IL, October 2000.

[8] G. Caire and S. Shamai, "On the achievable throughput of a multi-antenna Gaussian broadcast channel," submitted to IEEE Trans. on Inform. Theory, July 2001.

[9] P. Viswanath and D. Tse, "Sum capacity of the multiple antenna Gaussian broadcast channel and uplink-downlink duality," submitted to IEEE Trans. on Inform. Theory, August 2002.

[10] S. Vishwanath, N. Jindal, and A. Goldsmith, "Duality, achievable rates and sum capacity of Gaussian MIMO broadcast channels," submitted to IEEE Trans. on Inform. Theory, August 2002.

[11] M. Schubert and H. Boche, "Joint "Dirty-Paper" precoding and downlink beamforming," in *Proc. of Int. Symp. on Inform. Theory and Appl.*, Prague, September 2002.

[12] G. Kramer, S. Vishwanath, S. Shamai, and A. Goldsmith, "Information-theoretic issues concerning broadcasting," in *Proc. Workshop on Sig. proc. for Wireless Comm. (DIMACS 2002)*, Rutgers University, New Jersey, October 2002.

[13] G. Ginis and J. Cioffi, "A multi-user precoding scheme achieving crosstalk cancellation with application to DSL systems," in *34-th Asilomar Conference on Signals, Systems and Computers*, Pacific-Groove, CA, USA, November 2000.

[14] B. Chen and G. Wornell, "Quantization index modulation: a class of provably good methods for digital watermarking and information embedding," *IEEE Trans. on Inform. Theory*, vol. 47, no. 4, pp. 1423–1443, May 2001.

[15] J. Eggers, R. Bauml, R. Tzschoppe, and B. Girod, "Scalar Costa scheme for information embedding," to appear on IEEE Trans. on Sig. Proc., 2002.

[16] A. Cohen and A. Lapidoth, "The Gaussian watermarking game," *IEEE Trans. on Inform. Theory*, vol. 48, no. 6, pp. 1639–1667, June 2002.

[17] U. Erez and R. Zamir, "Lattice decoded nested codes achieve the Poltyrev exponent," in *Proc. IEEE Int. Symp. on Inform. Theory (ISIT 2002)*, Lausanne, Switzerland, June 2002.

[18] S. Pradhan, J. Chou and K. Ramchandran, "Duality between source coding and channel coding with side information," UCB/ERL Technical Memorandum No. M01/34, UC Berkeley, Dec. 2001.

[19] R. Gallager, *Information theory and reliable communication*, Wiley, New York, 1968.

[20] T. Cover and J. Thomas, *Elements of information theory*, Wiley, New York, 1991.

[21] S. Verdú, "On Channel Capacity per Unit Cost," *IEEE Trans. on Inform. Theory,* Vol. 36, No. 5, pp. 1019–1030, September 1990.

[22] R. Ahlswede, "Arbitrarily varying channels with states sequence known to the sender," *IEEE Trans. on Inform. Theory*, vol. 32, no. 5, pp. 621–629, September 1986.

[23] U. Wachsmann, R. Fisher, and J. Huber, "Multilevel codes: theoretical concepts and practical design rules," *IEEE Trans. on Inform. Theory*, vol. 45, no. 5, pp. 1361–1391, July 1999.

[24] R. Urbanke et al., "`http://lthcwww.epfl.ch/research/ldpcopt/`," 2002.

[25] G. Ungerboeck, "Channel coding with multilevel/phase signals," *IEEE Trans. on Inform. Theory*, vol. 28, no. 1, pp. 55–67, January 1982.

[26] T. Richardson and R. Urbanke, "The capacity of low-density parity-check codes under message-passing decoding," *IEEE Trans. on Inform. Theory*, vol. 47, no. 2, pp. 599–618, February 2001.

[27] S.-Y. Chung, T. Richardson, and R. Urbanke, "Analysis of sum-product decoding of low-density parity-check codes using a Gaussian approximation," *IEEE Trans. on Inform. Theory*, vol. 47, no. 2, pp. 657–670, February 2001.

[28] T. Richardson, A. Shokrollahi, and R. Urbanke, "Design of capacity-approaching irregular low-density parity-check codes," *IEEE Trans. on Inform. Theory*, vol. 47, no. 2, pp. 619–637, February 2001.

[29] S. ten Brink, "Convergence behavior of iteratively decoded parallel concatenated codes," *IEEE Trans. on Commun.*, vol. 49, no. 10, pp. 1727–1737, October 2001.

[30] A. Ashikhmin, G. Kramer, and S. ten Brink, "Extrinsic information transfer functions: A model and two properties," in *Proc. 36th Annual Conf. on Inform. Sc. and Syst. (CISS 2002)*, Princeton, New Jersey, March 2002.

[31] S. Shamai and A. Lapidoth, Private communication.

[32] E. Viterbo and J. Boutros, "A universal lattice code decoder for fading channels," *IEEE Trans. on Inform. Theory*, vol. 45, No. 7, pp. 1639–1642, July 1999.

[33] E. Agrell, T. Eriksson, A. Vardy and K. Zeger, "Closest point search in lattices," *IEEE Trans. on Inform. Theory*, vol. 48, No. 8, pp. 2201–2214, Aug. 2002.

INSTITUT EURECOM, DEPARTMENT OF MOBILE COMMUNICATIONS, 2229 ROUTE DES CRETES, 06094 SOPHIA-ANTIPOLIS, FRANCE

E-mail address: `giuseppe.caire@eurecom.fr`

TECHNION, ISRAEL INSTITUT OF TECHNOLOGY, DEPARTMENT OF ELECTRICAL ENGINEERING, 32000 HAIFA, ISRAEL

E-mail address: `sshlomo@ee.technion.ac.il`

DIMACS Series in Discrete Mathematics
and Theoretical Computer Science
Volume **62**, 2003

Dirty Paper Coding: Perturbing Off The Infinite Dimensional Lattice Limit

Gerard J. Foschini and Alvaro H. Diaz

ABSTRACT: Motivated by its potential role in reducing the need for orthogonal use of wireless communication resources in downlink signalling, we focus here on so-called Dirty Paper Coding (DPC). DPC can immunize a link against interference when the erstwhile interferer is launched from the same transmit site. This promotes reuse of space×spectral resources even at the same base station since would be interferers can be made phantoms (invisible interferencewise) to communicators on the same resource.
For best performance, DPC signals uses special high dimensional lattices. We derive a means of perturbing off of the ∞-dimensional (∞-D) limit to estimate Shannon capacities in very high dimensions. At the opposite extreme, we look at simple to use 1-D lattices. A wide range of SNRs are of interest. Use of higher dimensional lattices is important at low SNRs. For both high and low dimensions we emphasize the importance of DPC's "noise cooling" feature. At 0dB, where the Shannon capacity is 1 bps/Hz, a 1-D lattice requires an extra 4.5 dB. However, with proper cooling, it requires an additional 2.5dB. A 48-D lattice, with optimally cooled noise, is estimated to require only an extra 0.5 dB.
When the transmitter relies on receiver channel measurements, we find that the least number of reals that must be fedback.

1. Introduction And Motivation

We will discuss and analyze some basics aspects of "Dirty Paper Coding" (DPC), a means of communication whose history traces back several decades. DPC is a method of known interference cancellation. See [1-5] for some of the earlier references and see [6-8] for examples of recent related activity. As we will review in Section 3, DPC makes use of large dimensional lattices, signal space tessellations and modular number system-like properties for signalling. The analysis to support how and why DPC works on a link will be included in the sections that follow. While the paper focuses on DPC communication over a single, spatially one dimensional (1-D) link, for motivation, we next mention why there is interest in DPC in wireless networks.

A potential application of DPC is in wireless communication networks on the downlink (base to terminals), where multi-use of the same communication modes is made possible. By communication modes we mean {spatial modes as used in multielement array to multielement array communication}×{frequency subbands}. For example, the base can signal to say a user, B, over {space}×{frequency} modes that have been allocated to another user, called A, who is in the same cell. (Instead of frequency subbands we could use other means of organizing modes like CDMA codes or time slots in a TDMA system.) As expected, user B perceives user A as interference. However, user A does not perceive the presence of B. Without DPC, A's perception would have been analogous to B having written in ink over the same sheet of paper that he writes his message on. Instead, with DPC, A's message is not impaired by B's message at all - as if B's ink were made invisible to A. In theory, this holds no matter what B's power is.

The opportunity to operate on the downlink in this way with DPC balances what happens on the uplink with Multiuser Detection (MUD) [9]. To explain, consider what is possible on the uplink, again with A and B. Say a single base station is serving user A. Now user B is to be accommodated with multiple antenna propagation modes powered with full acccount of user A as interference. However, communications to A can be prepared as if B did not exist. That is acceptable since MUD can be employed to first detect B and then subtract B's signal from the total received signal, to make the reception of signal A free of B's interference for detection purposes. It might seem that there is an uplink-downlink asymmetry since this process does not (generally) work for the downlink[#], because it is not expected that the two downlink sites can detect signal B. By contrast, on the uplink, the base's job is to detect both signals. Using DPC in place of MUD completes the uplink-downlink symmetry. A phantom is a user such as B who communicates over a link (up or down) overlaying propagation modes already in use, yet maintaining invisibility interferencewise.

The standard means of DPC assumes that the transmitter and receiver know the channel. Then uncoupled spatial eigenmodes can be used for communication. It is within each of these eigenmodes that DPC takes place. It is assumed that each receiver knows the channel and the transmitter knows enough about the channel to appropriately prepare the transmitted signal for DPC on spatial modes. (Reference [11] reviews the basics of spatial eigenmode transmission when there are multiple antennas at the transmit and receive sites.)

A network exploiting MUD to create phantoms on the uplink, and DPC to create them on the downlink, might be called a PhantomNet. Since space×frequency water pouring is used to optimally allocate resources, we can say that PhantomNet is enabled by water, dirt and MUD. For references on network level work see [12-15].

The primary focus now will be on one spatial mode of one link in a downlink network such as we have just described. We want to quantify the capacities that are possible even down to lower SNRs as might arise in downlink network arrangements.

2. Assumptions And Outline Of Results

For simplicity, where possible, we will emphasize an idealized infomation theory view. Then, ideal transmission is assumed where the code book uses Gaussian waveforms

---

[#] See [10] for a framework with a single antenna base station accommodating users each with any number of antennas. Then MUD can be made to work even on the downlink so long as users are specially ordered.

composed so all signals can be detected error free (really a succession of progressively more advanced codebooks corresponding to a limiting behavior of error free transmission). In a multiple antenna downlink, to attain capacity, water pouring is done over space×frequency modes in the additive white gaussian noise context in which the user's signal preparation takes place. However, as already mentioned, for simplicity, in the following sections it suffices for us to focus on what happens in one eigenmode [11] of a base-to-user link. We take the frequency band to be so narrow that it is represented as frequency flat. As we will discuss, in order to induce the phantom property, finite dimensional lattice based communication is employed. The transmitted signals are then not gaussian, but, as is now well established, capacity is achieved and the gaussian signalling restored in the limit as the lattice dimension tends to infinity.

We review lattice communication in Section 3.0 Then, after some mathematical preliminaries in section 4.0, we proceed to the sphere channel which is the topic of Section 5.0. The "sphere channel" represents communication in the limit of wisely chosen high dimensional lattices. As the dimension of the (appropriately chosen) lattices approaches infinity, Shannon Capacity can be approached. We show how to perturb off of infinitely many dimensions to see what happens as lattice dimension increases in the neighborhood of infinity.

Even leaving DPC aside, a similar analysis is well established for the high SNR asymptote where a 1.53 dB separation exists between the Shannon limit for signalling with 1-D equally spaced lattice points and signalling using Gaussian waveforms corresponding to an infinite dimensional alphabet. Spherical alphabets of increasing dimension are used to first assess the possible value of signalling using lattices offering progressively more spherically shaped alphabets. The results for spheres motivate what has proved to be a successful examination of lattices for those that can come close to meeting the spherical ideal in each dimension. Certain dimensions fare better than others. The deficit relative to 1.53 dB is termed shaping loss. G. David Forney Jr. has pioneered a substantial body of work in this area, see, e.g., [16] and references there.

We here devise means of extending the spherical idealization to lower SNRs. As SNR is lowered we will see that a much greater gap than 1.53dB exists between the 1-D and ∞-D lattices. The perturbation analysis we devise provides the high SNR spherical alphabet asymptote in all dimensions. It provides the infinite dimensional asymptote at all SNRs and it also gives exact results in one dimension. A stochastic differential equation called the Bessel equation is used as a base for a perturbation analysis from infinitely many dimensions to large finite dimensions. Then, the asymptotic form of the Shannon capacity versus SNR in high dimensions can be estimated.

A tacit assumption here is that the channel is known to both the transmitter and the receiver. The knowledge is used to derive the spatially 1-D modes for DPC signalling. In section 7.0 we determine how many real numbers the receiver needs to send back to the transmitter to determine these modes. We do this for a channel with M transmitters and N receivers.

3. The Lattice Based Communication Link

We start with some mathematics needed to explain lattice based communication. Let V be a nonsingular $k \times k$ matrix operating on vectors in Euclidean k-space, E^k. The column vectors of V, $\{v_1, v_2 \ldots v_k\}$, are a basis for E^k. Let m be a k-D vector all of whose components are integers and denote the set of all such vectors by Z^k. Then the set of vectors $\{Vm | \, m \in Z^k\}$ defines a lattice denoted V_k. Taking E^k, Z^k and V_k to include the property of vector addition, they are all groups, with Z^k and V_k subgroups of E^k. Next we point to some quotient groups.

The lattice gives rise to a fundamental paralellepiped, P_k, by defining its vertex set as $\{x \in E^k \mid x=V\,\underline{m},\ \underline{m} = (\underline{m}_1,\underline{m}_2,\ldots\underline{m}_k)$ *where each* $\underline{m}_j = 0$ *or* $1\}$. In preparation for tessellation of E^k, we can avoid overlaping surrfaces by adding the requirement that the closure of all those $(k\text{-}1)\text{-}D$ faces that include the $k\text{-}D$ vector $V(1,1,\ldots1)'$ are not part of P_k. Then the set of translates of P_k by lattice vectors comprise a tesellation of E^k. This fundamental region repeats with multiple periodicities in E^k. Two points $(X_1, X_2 \in E^k\}$ are translations of each other by a lattice vector if: $X_1 = X_2 + v$ where $v = \Sigma\, m_j v_j$ where $m_j \in Z$. Then X_1 and X_2 which differ by a multiple of the lattice periodicity, are said to be equivalent modulo V_k. In such a case we write $X_1 \approx X_2$ *or* $[X_1 - X_2]mod(V_k) = 0$. Under vector addition $mod(V_k)$, P_k is a subgroup of E^k. In terms of quotient groups $P_k = E^k/V_k$.

P_k gives rise to another important tessellation of E^k, the Voronoi tessellation. A Voronoi region surrounding any specified lattice point l_i of V_k, is a special set of points which are one to one equivalent $mod(V_k)$ to the points in P_k, particularly, those points x in E^k which satisfy $\{\|x - l_i\| \le \|x - l_j\|,$ *for all* $l_j \in V_k\}$. Each point in the Voronoi region around l_i is closer to the "center of mass", l_i, than any of the points to which it is equivalent $mod(V_k)$. So, with energy measured by the squared distance to l_i, in the set of points that are in the same equivalence class under $mod(V_k)$, the Voronoi about l_i contains the minimal energy point. Later, when quantizing signals in E^k, we will use this means of quantization to the nearest lattice point.

We will be especially interested in measuring energy relative to the 0 vector. We denote the Voronoi region about the lattice vector 0 by V_{0k}. Like P_k, V_{0k} also represents the quotient group E^k/V_k. Indeed, V_{0k} and P_k are isomorphic groups under addition $mod(V_k)$. For references on lattices and Voronoi fundamental regions see [8,17].

FIGURE 1: Block diagram of wireless link using dirty paper

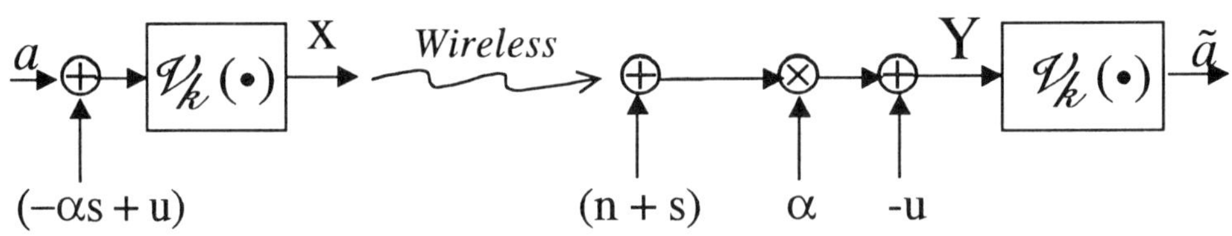

$$CHANNEL: \quad a \rightarrow \tilde{a} = a + (1\text{-}\alpha)\,\mathcal{U} + \alpha\,\mathcal{N}$$

Unif. Dist.↑ Normal↑(wrapped)

OVER LATTICE REGION V

AS $k \rightarrow \infty$, UNIFORM → WHITE GAUSSIAN & CHANNEL BECOMES
$a \rightarrow a$ + AWGN. OPTIMIZING α GIVES CAPACITY $\text{LOG}_2(1+\text{SNR})$.

We need more definitions in preparation for writing the equations for communication using a $k\text{-}D$ lattice, V_k,:

a: the encoded message to be transmitted. This k-D sequence of real numbers is in the Voronoi region V_{0k}.

u: one-time pad sequence of numbers. u is also a k-D vector. This pseudorandom vector, uniformly distributed over V_{0k}, is assumed known at both the transmitter and receiver.

s: arbitrary interfering sequence of k numbers known only to the transmitter *n:* k-D sequence of additive white gaussian noise

$\mathcal{V}_k(z)$: is the aformentioned modular function $z(mod\ V_k)$. $\mathcal{V}_k(z)$ takes values in V_{0k} and is described via vector quantization [17-21]: Specifically, the function $\mathcal{V}_k(z) = z - Q_k(z)$ where $Q_k(z)$ is the vector quantizer operating on z: which maps z to the closest lattice vector. Then $\mathcal{V}_k(z)$ is seen as the quantization error.

α: a parameter to be optimized over the interval $0 < \alpha < 1$ [4]. As we will see, this parameter serves to "cool off" the noise at the receiver.

See Figure 1 where the transmitted signal, X, is seen to be

$$X = a - \alpha s + u + \nu \tag{3.1}$$

where ν is the vector equal to the zero vector, modulo V_k, that is needed to reduce X to V_{0k}. We take the view that received signal already includes the amplification needed for the receiver to restore the level of the transmitted signal. So, the received signal is

$$X + n + s \tag{3.2}$$

where n includes the effect of additive white gaussian noise after amplification.

The receiver does the following: First multiply by α, subtract the one-time pad and pass the result through the k-D vector quantizer to get

$$Y = (X + n + s)\,\alpha - u + \nu' \tag{3.3}$$

where the ν' is equivalent to zero mod(V_{0k}). Next add $0 = X - X$ to Y to get

$$\begin{aligned} Y &= (X + n + s)\,\alpha - u + \nu' \\ &= a - (1-\alpha)[\,a - \alpha s + u + \nu] + \alpha n \\ &\quad -\alpha s + \alpha s + u - u + \nu'. \end{aligned} \tag{3.4}$$

The term $[a - \alpha s + u + \nu]$ has a signal independent uniform probability density over the lattice. So, the output of the mod(V_{0k}) operator expressed by the right hand side of the second equality in (3.4) has the form

$$\tilde{a} = a + (1-\alpha)\mathcal{U} + \alpha\mathcal{N} \tag{3.5}$$

where $\mathcal{U}$ is uniformly distributed over V_{0k} and $\mathcal{N}$ has an independent white gaussian distribution.

For an appropriate lattice sequence of increasing dimension, the limiting Voronoi region becomes spherical [22]. By a Theorem of Poincare [23], in the limit of large k, the k-D vector formed from the projections of $\mathcal{U}$ onto the k successive time coordinates, becomes an additive white gaussian noise vector of dimension k. The expression for the energy of a sphere of unit radius (including both signal and noise in our application) is k/(k+2). Per complex dimension the energy is 1/(k+2). This energy expression follows easily from the area formula for the n-D sphere as is reviewed in detail in Section 4.

From equation (3.5) the total noise energy is

$$[(1-\alpha)^2 + (\alpha^2/\rho)]\cdot(k/k+2). \tag{3.6}$$

To minimize the noise energy, take the derivative with respect to α and equate to zero to get the equation

$$-(1-\alpha) + \alpha/\rho = 0 \tag{3.7}$$

which we solve for α to get

$$\alpha_* = \rho/(\rho+1) \tag{3.8}$$

Substituting using (3.8), the minimum total noise energy becomes

$$[(1-\alpha_*)^2 + (\alpha_*^2/\rho)]\cdot(k/(k+2)) = \left[\frac{1}{(1+\rho)^2} + \frac{\rho^2}{(1+\rho)^2}\frac{1}{\rho}\right]\cdot(k/(k+2)) \tag{3.9}$$
$$= \frac{1+\rho}{(1+\rho)^2}\cdot\frac{k}{k+2} = \frac{1}{1+\rho}\cdot\frac{k}{k+2}.$$

In the last step, we take the ratio of the total average signal plus noise energy to the total noise energy, the k/(k+2) cancels to give a capacity, in units of bits per complex dimension, of

$$C = \log_2(1+\rho). \tag{3.10}$$

The important point is that the capacity is the same as if the interferer were not present. Moreover, the argument, and hence the final answer does not depend on the probability distribution of the interferer.

4. Some Mathematical Preliminaries

We require some capacity formulas as well as some basic formulas for higher dimensional spheres. We present these in this section.

4.1 Capacity Formulas. We will be interested here in a capacity formula when both the output and input space are the same set, $\mathcal{A}$. One class of examples has $\mathcal{A} = E^k/V_k$. Another is when $\mathcal{A}$ is the surface of the unit sphere in E^k with antipodes identified. We begin with the setup for E^k/V_k.

X, represents the input vector and the vector, Y, the output. We start with the mutual information formula [24]

$$I(Y,X) = \iint p_{YX}(y,x)\log_2\frac{p_{YX}(y,x)}{p_Y(y)p_X(x)}dxdy \tag{4.1}$$

where it its understood that the double integral is over $\mathcal{A}\times\mathcal{A}$. Reexpressing (4.1) in terms of the conditional density, $p_{Y|X}(y,x)$, we get

$$I(Y;X) = \iint p_X(x)p_{Y|X}(y,x)\log_2\frac{p_{Y|X}(y,x)}{p_X(x)}dxdy. \tag{4.2}$$

The channel in our discussion is homogeneous, meaning that for each point x in the input space the conditional density $p_{Y|X}(y,x) = p_{Y|X}(y-x,0)$. Consequently, equation (4.2) can be rewritten

$$
\begin{aligned}
I(Y,X) &= \iint p_X(x)p_{Y|X}(y-x,0)\log_2\frac{p_{Y|X}(y-x,0)}{p_X(x)}dxdy \\
&= \iint p_X(x)p_{Y|X}(y-x,0)\log_2 p_{Y|X}(y-x,0)dxdy \\
&- \iint p_X(x)p_{Y|X}(y-x,0)\log_2 p_X(x)dxdy.
\end{aligned} \tag{4.3}
$$

Consider integration over the y variable. Since integration is over the entire output group we see that the double integrals in equation (4.3) are invariant under the change of argument $y - x \to y$. Therefore, we see that equation (4.3) above then becomes

$$
\begin{aligned}
I(Y,X) &= \iint p_X(x)p_{Y|X}(y,0)\log_2 p_{Y|X}(y,0)dxdy \\
&- \iint p_X(x)p_{Y|X}(y,0)\log_2 p_X(x)dxdy.
\end{aligned} \tag{4.4}
$$

On the right hand side of (4.4) we integrate out the input density in the first term and the conditional output density in the second term to get

$$
I(Y,X) = \int p_{Y|X}(y,0)\log_2 p_{Y|X}(y,0)dy - \int p_X(x)\log_2 p_X(x)dx. \tag{4.5}
$$

As is well known, a simple variational argument shows that the constant density achieves maximum entropy when densities are compactly supported. It is easy to see then, that (5.5) can be rewritten as the following formula for capacity

$$
C = log_2 VOL - \int p_{Y|X}(y,0)\log_2 p_{Y|X}(y,0)dy \ \textit{bits/complex dim} \tag{4.6}
$$

where by *VOL* we mean volume of the input alphabet. The second negatively signed term in (4.6) is called the "noise" entropy. (We mention in passing that in the simple case when $p_{Y|X}(y,0)$ vanishes except on a set of volume *"vol"* where it is a constant (4.6) becomes simply $C = \text{Log}_2(\text{VOL/vol})$.

We can arrive at (4.6) under a weaker conditions on the channel which we call weak homogeniety. We forgo dealing with a group structure for this. Assume (4.2) in the form

$$
\begin{aligned}
I(Y;X) &= \\
&\iint p_X(x)p_{Y|X}(y,x)\log_2 p_{Y|X}(y,x)dxdy - \\
&\iint p_X(x)p_{Y|X}(y,x)\log_2 p_X(x)dxdy.
\end{aligned} \tag{4.7}
$$

which tacitly assumes a meaningful integration context as will be the case in our application in **5**. In formula (4.7), again consider integrating over the y variable. Of course, $\int p_{Y|X}(y,x)dy = 1$ uniformly in x. We call the channel *weakly homogeneous* if the integrals $\int p_{Y|X}(y,x)\log_2 p_{Y|X}(y,x)dy$ are likewise invariant with respect to x. Then once again (4.6) follows directly by essentially the same argument as before.

4.2 Formulas For A Unit Sphere. In **5** we will treat a channel for which the unit sphere will figure prominently. In preparation for this we record some basic formulas that we will need. We recall that the volume of a sphere of unit radius in n-D, V_1, is expressed by integrating over one radial coordinate and all angular coordinates is [25]

$$V_1 = \int_{\measuredangle} d\Omega \int_0^1 r^{n-1} dr = \pi^{n/2} / \Gamma\left(1+(n/2)\right). \tag{4.8}$$

Regarding the integral over the angular coordinates in (4.8), we conclude from (4.8) that

$$\int_{\measuredangle} d\Omega = n \cdot \pi^{n/2} / \Gamma\left(1+(n/2)\right) = nV_1. \tag{4.9}$$

Consider a probability density function, φ, that is spherically symmetric about the origin of a unit sphere. Using polar coordinates, r for radius and Ω for the vector of angular coordinates, we write the density function of the noise as $\varphi(r,\Omega) = \xi(r)\eta(\Omega) = \xi(r)\eta$ where the equality expresses the fact that the angular density is a constant. From the above equation, we see that for the integral to integrate to one, the constant density function, η, must equal $1/(nV_1)$. So the negative entropy of this density, after some elementary steps, becomes

$$\begin{aligned} &\int_{\measuredangle} d\Omega\, r^{n-1} dr\, \varphi \log_2 \varphi = \int_{\measuredangle} d\Omega \int_0^1 dr \xi(r)\eta(\Omega) \log_2[\xi(r)\eta(\Omega)] r^{n-1} = \\ &\int_{\measuredangle} d\Omega \int_0^1 dr \xi(r)\eta \log_2[\xi(r)] r^{n-1} + \int_{\measuredangle} d\Omega \int_0^1 dr \xi(r)\eta \log_2[\eta] r^{n-1} = \\ &\int_0^1 dr \xi(r) \log_2[\xi(r)] r^{n-1} + \int_0^1 dr \xi(r) \log_2[1/(nV_1)] r^{n-1} = \\ &\int_0^1 dr \xi(r) r^{n-1} \log_2[\xi(r) r^{n-1}] + \int_0^1 dr \xi(r) r^{n-1} \log_2[1/(r^{n-1} nV_1)]. \end{aligned} \tag{4.10}$$

Define h(r) to be the 1-D radial density function of the noise, namely,

$$h(r) \triangleq \xi(r) r^{n-1}. \tag{4.11}$$

The negative entropy term now becomes

$$\begin{aligned} &\int_0^1 dr\, h(r) \log_2 h(r) + \int_0^1 dr\, h(r) \log_2[1/(r^{n-1} nV_1)] = \\ &\int_0^1 dr\, h(r) \log_2 \frac{h(r)}{nr^{n-1}} + \int_0^1 dr\, h(r) \log_2[1/(V_1)] = \\ &\int_0^1 dr\, h(r) \log_2 \frac{h(r)}{nr^{n-1}} - \log_2 V_1. \end{aligned} \tag{4.12}$$

5. The Sphere Channel

In the limit of large lattice dimensions, the Voronoi region for the lattices of interest to us become more and more spherical [22]. Here we compute the capacity formula for what we will call the *sphere channel,* a channel which is defined in such a way that we can compute capacity versus SNR results for any finite dimension. This, despite the fact, that except for one dimension, translates of a sphere cannot perfectly pack Euclidean space in finite dimensions.

Parallelepiped fundamental regions tile all of space as do their Voronoi equivalents. A Voronoi region is also a polytope - a volume preserving rearrangement of a parallelepiped. Post-quantizer, that is after factoring the k-D lattice out of k-D Euclidean space, within either of these regions representing the quotient, E^k/V_k, one has spatial homogeneity and isotropy. At the same time, packing of k-D Euclidean space with the Voronoi polytopes about each lattice point allows known interference to be effectively cancelled. The sphere channel serves to give weak homogeneity and a spherical signalling alphabet in any dimension. Thus, the sphere channel provides a setting for perturbing off of infinitely many dimensions to enable the estimation of capacity versus SNR curves in the infinite dimensional asymptote. In the context of this perturbation setup we will compute capacities for the cases, 12 , 24 and 48.

5.1 First View of Sphere Channel. We take $\mathcal{S}_n$ to be a unit radius n-D sphere, with each antipodal surface pair identified as a single point. $\mathcal{S}_n$ will serve as our signalling alphabet as well as the set of possible received alphabet symbols. As a communication channel, each point in $\mathcal{S}_n$ is an equally likely alphabet point for the input. We begin by describing the noisy output on $\mathcal{S}_n$ for the case $\alpha = 1$. If we send the origin, O, we receive instead O* obtained as follows. A standard n-D Brownian motion is launched from O into $\mathcal{S}_n$. Let O*, be the random stopping point of the motion at a specified time T. O* is channel output. Since each point of the spherical boundary is identified with its antipode the radial coordinate of the Brownian motion reflects at r = 1. Only the radial coordinate matters for finding the output distribution since the geometry is angularly symmetric.

When a point, Q, in $\mathcal{S}_n$, other then the origin is used, we erect a copy of $\mathcal{S}_n$ denote $\mathcal{S}_{nQ}$, centered at Q and repeat just as with standard n-D Brownian motion launched from O except that the arena for the motion is $\mathcal{S}_{nQ}$. What happens if the stopping point Q′ lies outside of $\mathcal{S}_n$? Then we map to the output point in $\mathcal{S}_n$ that is the mirror image of Q* inside of $\mathcal{S}_n$, denote Q**. The mirror is the (n-1)-D hyperplane that is the perpendicular bisector of the line OQ. The arrow in Figure 2 illustrates the measure preserving mapping that preserves entropy of the distribution of Q* in $\mathcal{S}_{nQ}$ when mapping back to the sphere $\mathcal{S}_n$. (There is of arbitrariness in this mapping, eg, we could replace Q** by its reflection about the line containing OQ. This reflection is also in $\mathcal{S}_n$.)

Figure 2: The Sphere Channel Input And Output Space

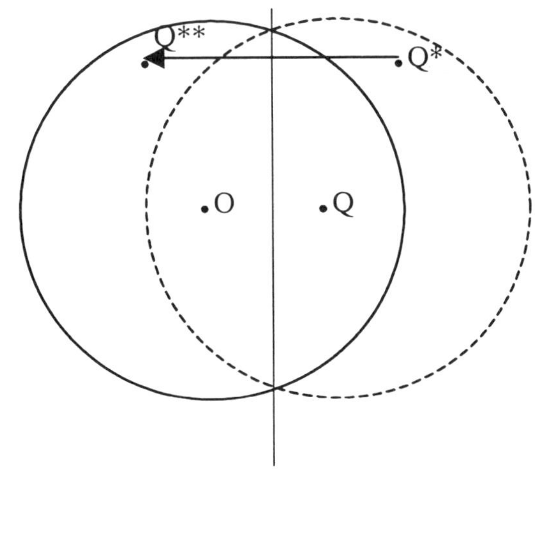

5.2 A Second Look At The Sphere Channel Using Views Of Projective Space. Topologically, latttices are homogeneous isotropic spaces. We know that there are certain, appropriately chosen, Voronoi of progressively higher dimension that become spherical in the infinite dimensional limit. We observed that the equivalent noise, really known interference plus additive noise, likewise tends toward spherically symmetry. For each finite dimension, through the use of the vector modular mapping, the Euclidean space in which a lattice is imbedded is mapped infinity-to-one onto a compact manifold.

In E^{n+1} identify all the lines parallel to each other as one point. The n-D projective space, P^n, is the collection of all such points. P^n is a topologically homogeneous isotropic compact manifold that cannot be realized in E^n [23]. However, it is realizable as the n-D spherical surface, S^n, in E^{n+1} if we identify each antipodal pair of S^n as a single point. The identification stems from the choice of "lines" rather than rays as the points in P^n. We will use of this realization of P^n to describe a signalling alphabet where each point in it is a center point. We use a continuum of orthogonal projections of P^n to create harmonized viewpoints from which each point of P^n is seen as occupying the center of a unit n-D sphere. Thereby, we avoid the perception that the alphabet has the origin distinct from the other points in it.

We will use z to denote the center of $\mathcal{S}_{n+1}$, an (n+1)-D unit sphere just as in **5.1**, but one dimension higher. Consider any point, ζ, outside of $\mathcal{S}_{n+1}$. Connect ζ to z with a straight line. The point (pair), Q, on the surface of $\mathcal{S}_{n+1}$ pierced by the directed line segment ζz is mapped to z. The point z is the center of the n-D unit sphere that is the equatorial hyperplanar section that is orthogonal to ζz and that is intersected, and thereby truncated, by the surface of $\mathcal{S}_{n+1}$. We take antipodal points of this intersection to be identified by the antipodal identification furnished by $\mathcal{S}_{n+1}$. The n-D sphere formed in this way is $\mathcal{S}_{nQ}$. In the direction ζ to z, the *entire* surface of $\mathcal{S}_{n+1}$ is orthogonally mapped 1:1 onto $\mathcal{S}_{nQ}$ with Q represented at the center by z.

For each choice of ζ outside of the n+1 sphere we get a 1:1, orthogonally projected, image of P^n, but, as ζ changes, the center point of the spherical section can change. Note that, with ζ situated outside of the unit sphere $\mathcal{S}_{n+1}$, anywhere along a particular line through z, from either side of the n+1 sphere we get the very same view of P^n. Each line through the center serves to map a new point of P^n to z. Considering all possible vantages, each point of P^n gets a chance to be mapped to the center. With these orthogonal projections we see that each point of P^n is viewed as a center of an n-D unit sphere with antipodes identified and no point of P^n is special. Any point in P^n can be used as the one from which we elect to measure the total energy of the transmit alphabet as the integral of the sum of the squares of the orthonormal coordinate values in the n-D spherical section of unit radius, with the special point z serving as the origin.

Next, in **5.3,** a specially launched, but otherwise standard Brownian motion, reflected at radius one, represents the noise dynamic. The motion is the standard one in the spherical equatorial sections, *not* the standard one on the curved surface of the n+1 sphere. The motion in each $\mathcal{S}_{nQ}$ can be lifted back to the surface that is P^n, but we will not make direct use of that.

5.3 The Stopped Bessel Process As The Noise. The radial coordinate of a standard n-D Brownian Motion launched from the origin is called a Bessel process [27] (because the solution for the probability density of the stopped motion, when there is no reflecting sphere, is expressed using a modified Bessel function). We use a Monte Carlo evaluation to find the probability density of the motion in presence of the reflecting spherical boundary. For 1-D spheres (line segments with endpoints identified = circles) the solution follows easily from the method of images. The stochastic differential equation for the radial distance, r(t), from the origin of an n-D standard Brownian motion is given in terms of a 1-D motion, b(t). The dynamics of this diffusion are easy to compute to get the following well known equation [27]:

$$dr(t) = \frac{n-1}{2r}dt + db(t). \tag{5.1}$$

In cases when $0 < \alpha < 1$, the motion to randomly map from input Q to the output is described exactly as for $\alpha = 1$, except that the launch point is taken to be a uniformly distributed point in an n-D sphere of radius $(1 - \alpha)$ centered at Q. Upon launch, the radial distance evolves in the unit sphere as described by (5.1), but with $b(t)$ replaced by $\alpha b(t)$. This is the noise cooling. We can write $r_{n,\alpha}(t)$ for $r(t)$ to emphasize the dependence on n and α. So, corresponding to the equation above, we have

$$dr_{\alpha,n}(t) = \frac{n-1}{2r}dt + \alpha \cdot db(t) \tag{5.2}$$

We have completed the definition of the sphere channel.

5.4 Capacity Formula. Basic formulas are needed to get the signal SNR, ρ, for the Sphere Channel. We need the area for the n-D sphere, $\mathcal{S}$, of radius a [25]

$$Area(\mathcal{S}(a)) = n\pi^{n/2}a^{n-1} / \Gamma((n/2)+1). \tag{5.3}$$

Integrating we get that the volume of a sphere of radius a is:

$$Vol(\mathcal{S}(a)) = \pi^{n/2}a^{n} / \Gamma((n/2)+1). \tag{5.4}$$

The average energy of a signal selected from a uniform distribution over the unit sphere follows by a simple integral of the area formula

$$\mathcal{E}(\mathcal{S}) \triangleq [Vol(\mathcal{S}(a))]^{-1} \int_0^1 a^2 \cdot Area(\mathcal{S}(a))\,da = n/(n+2). \tag{5.5}$$

Consequently, per real dimension, we get an energy of 1/(n+2) and per complex dimension an energy of 2/(n+2). For a standard Brownian Motion the noise energy per real dimension is $\sigma^2 t$ and for a standard Brownian motion $\sigma^2 = 1$. So the SNR, ρ, is

$$\rho = 1/[(n+2)t]. \tag{5.6}$$

Figure 3

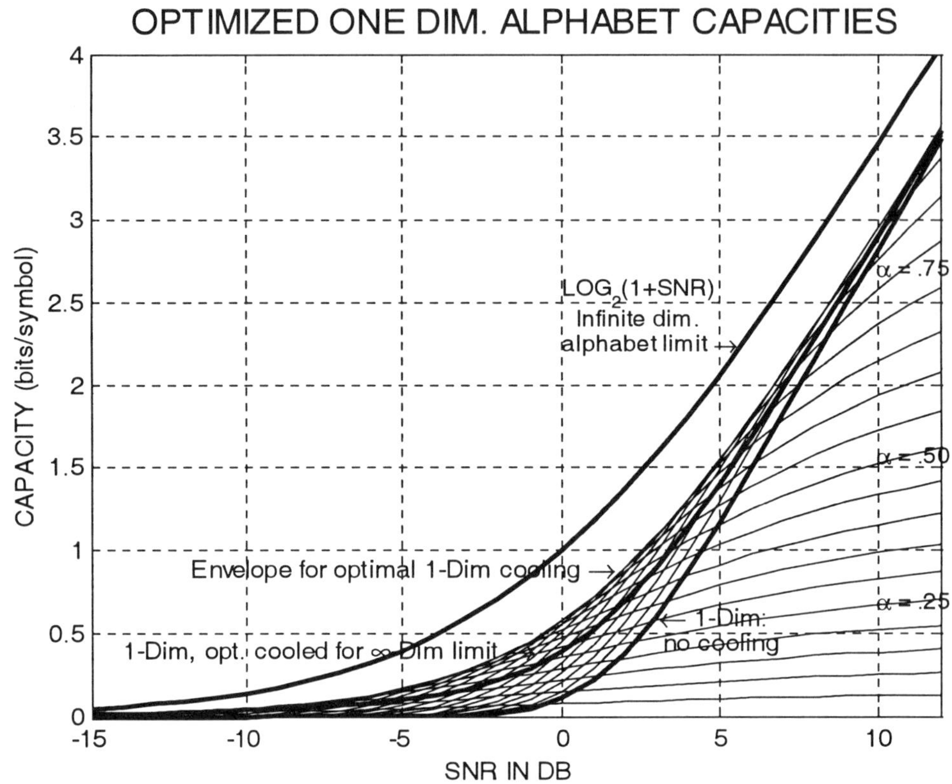

Using $p_{n,\alpha,\rho}(r)$, the probability density on the interval [0,1] of the standard motion stopped at time $t = T_{n,\rho} = 1/[(n+2)\rho]$, we can express the capacity for the sphere channel using equation (4.12). We get that the Sphere Channel capacity, per complex symbol by the following formula

$$C_{n,\alpha}(\rho) = \frac{2}{n}\int_0^1 \log_2(p_{n,\alpha,\rho}(r)/nr^{n-1})dr \, bits/symbol. \tag{5.7}$$

5.5 Numerical Results Next we discuss obtaining the density function $p_{n,\alpha,\rho}(r)$. For fixed α we can run a Monte Carlo simulation with thousands of random n-D radial paths $r_{n,\alpha}(t)$ and record the evolved radial coordinate at a sequence of times corresponding to $\rho_{dB} = 10\log_{10}\rho$ in increments of a dB or so.

Figure 4

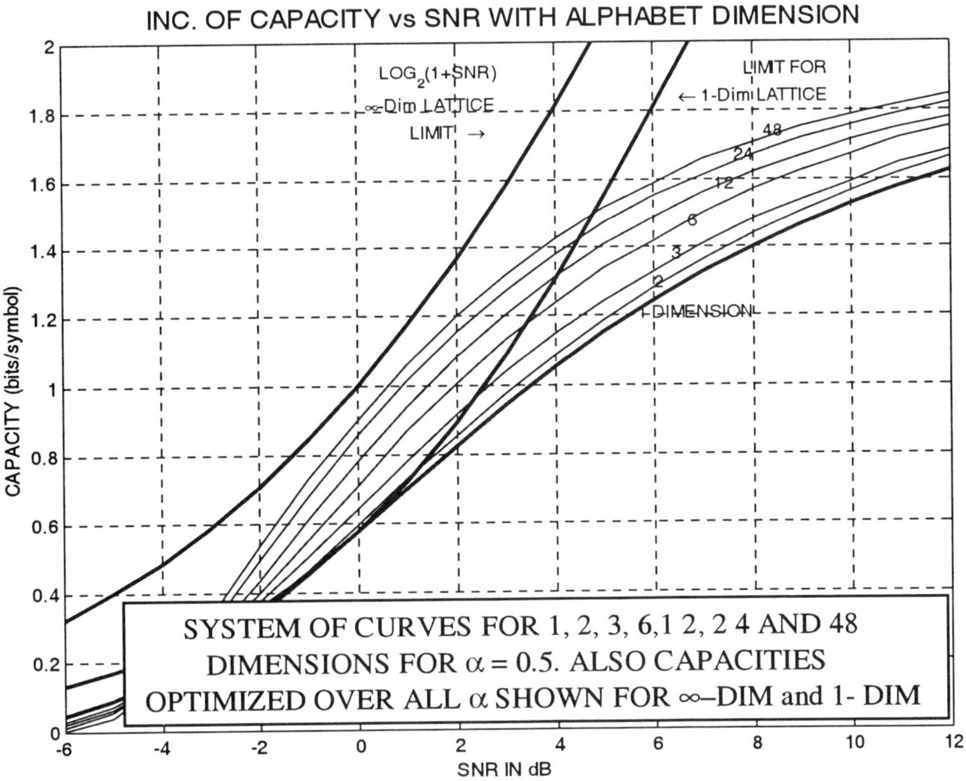

This was done for the 1-D case with results shown in Figure 2. In this case the 1-D lattice is a 1-D sphere, namely, the perimeter of a circle (alternately a line segment with its endpoints identified as one point). In this 1-D case the numerical solutions can be checked with the method of images. The curve marked 1-D no cooling show the 1-D case with $\alpha = 1$. The infinite dimensional Shannon capacity is depicted by the leftmost thick line graph. The middle thick line curve again shows the 1-D case. However, it is optimized for the infinite dimensional limit with α given by equation (3.8). This curve shows some substantial improvement. Figure 2 also shows the optimized cooling for the 1-D case. The thin line curves show the graphs for $\alpha = .05, .10, .15, \ldots 1.00$. The upper envelope of these graphs gives the capacity for 1-D. Considerably more improvement is shown for the lower SNR values.We stress that at low SNRs there is a significant growing gap between the infinite dimensional Shannon capacity and even the optimally cooled 1-D alphabet capacity characteristic.

We can also explore cooled capacity versus SNR characteristics for a succession of dimensions of interest by using the asymptotic form given by the sphere channel. See Figure 4 where this was done for n = 12, 24 and 48 for the value $\alpha = .5$. See Figure 5 where the process was repeated for 20 values of α from zero to one. The upper envelope gives the curves in Figure 5

Figure 5

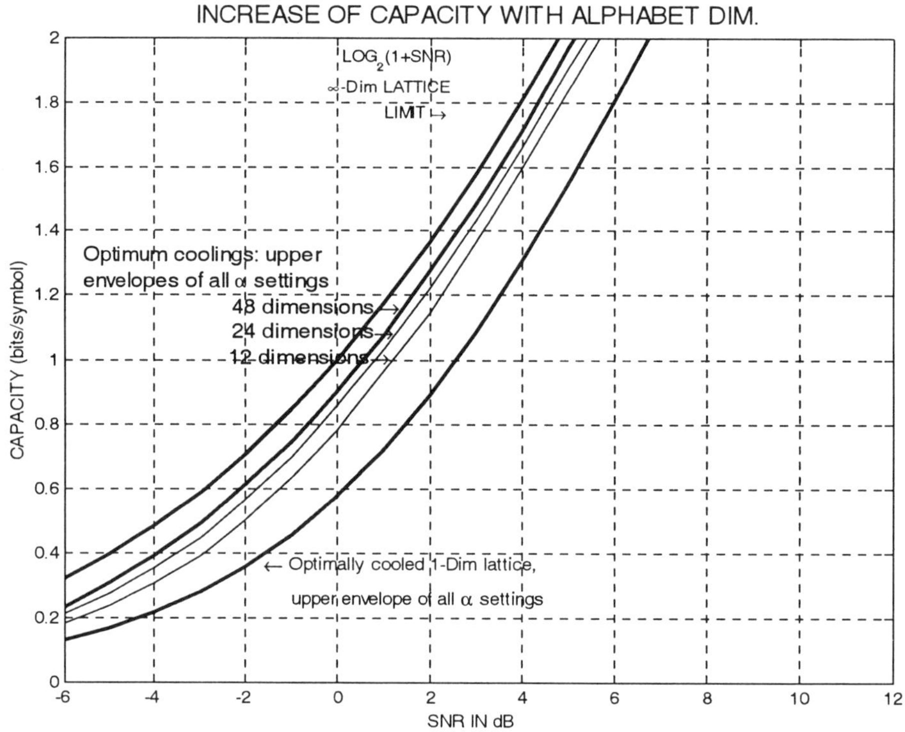

6. Feeding Back To The Transmitter

Sections 3-5 assumed a context of a single spatial mode for transmission. Here we derive how many real numbers a receiver, who knows the channel, must transmit back to the transmitter so that the transmitter can signal over the proper spatial modes. Within each powered eignmode DPC as described in **3** takes place. See [11] for a discussion of channel eigenmodes for a system with M transmit antennas and N receive antennas.

The setup is that the transmit site knows the variance of the additive white gaussian receiver noise. It does not know the matrix channel characteristic, H, but it does not need to know this characteristic. It needs to know the system of complex weights to drive the optimal set of virtual antennas formed from the actual antennas. Furthermore, the transmit site needs to know the rates to send on each of these virtual transmit antennas. We determine next the amount of information the receiver needs to convey back to the transmitter, measured not in bits, but more crudely measured in terms of how many real numbers need to be fed back.

Let $\{u_j: j = 1,2, \ldots M\}$ denote the canonical M-D spatial basis vectors at the transmitter. By this we mean that u_j corresponds to a unit (voltage) signal on antenna j and the zero signal on the other M-1 antennas. The vector u_j is the probe that is received as the N-D vector h_j, the jth column of the H matrix. Once the M antennas are probed, each sufficiently often so that receiver noise is not an issue, the receiver has learned the entire N×M H matrix. Then the receiver can compute the M spatial M-D eigenvectors of

$H^{\dagger}H$, denoted by $\{v_j: j = 1,2, \ldots M\}$. The receiver will inform the transmitter of these eigenvectors, at least as many of them as is needed, as indicated by spatial water pouring. When normalized to unity, these eigenvectors are the complex weights that transmitter uses to form the virtual antennas.

FIGURE 6: Feedback for up to 16 transmit antennas

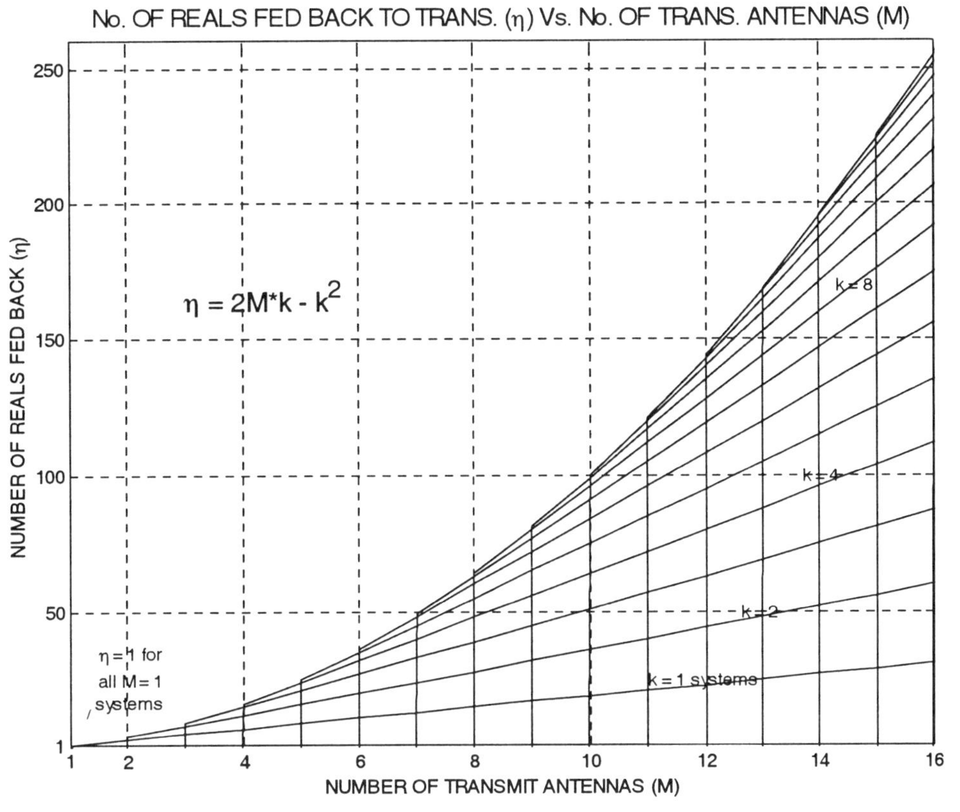

By the absolute continuity of the probability measure of the channel matrix outcomes we can take the first component of each eigenvector to be a nonnegative real. The receiver can leave it to the transmitter to normalize the k eigenvectors to one, so instead the receiver normalizes them to a norm equal to the total received power level under the hypothetical circumstance that unit power is launched on this eigenvector. That way, the transmitter, upon receipt of v_j, knows how much throughput to assign to the jth eigenmode.

Say k eigenvectors are needed at the transmitter, clearly, $k \leq \min(M,N)$. We assess what the receiver should send back to the transmitter, assuming that the most, namely, $k = \min(M,N)$ are needed. The trick to keeping the number of reals fed back to a minimum, is for the transmitter and receiver to have agreed beforhand on a basis for the transmitter to begin to use to interpret the eigenvectors that are fed back. The transmitter will iteratively transition this basis from all the u_js to the v_js. This enables an economy in the quantity of reals fed back. As each v_j is received by the transmit site it displaces u_j as a basis vector for expressing the next eigenvector. To see the reason for this, note, e.g., that with a basis $\{v_1, v_2, v_3, u_4, u_5, \dots u_M\}$ at the transmitter, when the receiver wants to convey v_4 it can express it in terms of $u_4, u_5, \dots u_M$. because v_4 is orthogonal to v_1, v_2, v_3. So v_4 is conveyed with 2M – 7 real numbers, instead of 2M – 1 real numbers as was needed for conveying v_1. Consequently we see that to convey all k vectors the total of real numbers to be fed back is

$$(2M - 1) + (2M - 3) + \dots + (2M - (2(k-1)-1)) = k\times(2M - 1) - 2(k-1)k/2 = 2Mk - k^2. \quad (6.1)$$

See Figure 6 for a plot of the system of linear functions of M: $\eta = 2Mk - k^2$, where M is the ordinate and k is a parameter.

When spatial water pouring gives that only $\kappa < k$ eigenmodes are to be driven, then

$$\underline{\eta} = 2M\kappa - \kappa^2 \quad (6.2)$$

real numbers is all that the receiver need send back to the transmitter.

7. Acknowledgement

It is a pleasure to thank J. Salz, Wireless Laboratory Consultant, at Bell Labs, Crawford Hill, for his invaluable and generous contributions.

References

[1] K. Marton, *A Coding Theorem For The Descrete Memory Broadcast Channel*, IEEE Transactions on Information Theory, Volume 25, No. 3, May 1979, pp 306-311
[2] S. I. Gelfand and M. S. Pinsker, *Coding for Channel with Random Parameters*, Problems of Control and Information Theory, Volume 9, No. 1, January 1980, pp 19-31
[3] A. El Gamal and C. Heegard, *On the Capacity of Computer Memories With Defects,* IEEE Transactions on Information Theory, Volume 29, September 1983, pp 731–739
[4] Max H. N. Costa, *Writing On Dirty Paper,* IEEE Trans. on Infor. Theory, Vol. 29, No. 3, 5/1983, pp 439-441
[5] R. Ahlswende, *Arbitrarily Varying Channels With State Sequence Known to the Sender*, IEEE Transactions on Information Theory, Volume 32, No. 5, September 1986, pp 621–629
[6] R. Zamir, S. Shamai and U.Erez, *Nested Linear/Lattice Codes for Structured Multiterminal Binning,* submitted to IEEE Trans. on Information Theory
[7] R. Zamir and M. Feder, *On Lattice Quantization Noise,* IEEE Trans. on Information Theory, Vol. 42, No. 4, 7/96, pg 1152-1159
[8] G. Poltyrev, *On Coding Without Restriction for the AWGN Channel,* IEEE Transactions on Information Theory, Volume 40, No. 2, March 1994, pg 409-417
[9] S. Verdu, *Multiuser Detection*, Cambridge University Press, London, 1998
[10] D. Tse, *Optimum Power Allocation Over Parallel Gaussian Broadcast Channels,* submitted to IEEE Transactions on Information Theory
[11] D. Chizhik, G. J. Foschini, and R. A. Valenzuela, *Capacities of Multielement Transmit and Receive Antennas: Correll. and Keyholes,* IEE Elec. Lett., 22/6/00, Vol. 36(13), pp. 1099-1110
[12] S. Shamai and B. M. Zaidel, *Enhancing the Cellular Downlink Capacity Via Co-processing at the Transmitting End*, VTC 2001, Rhodes Greece, pp 1745-1749
[13] G. Caire amd S. Shamai, *On the Multiple Antenna Broadcast Channel*, 35th Asilomar Conference, Pacific Grove, California, USA, October/November 2001
[14] H. C. Huang, S. Venkatesan and H Viswanathan, *Downlink Capacity Evaluation of Cellular Networks with Known Interference Cancelling,* To be pub. IEEE JSAC special issue on MIMO

[15] S. Jafar, G. J. Foschini and A. Goldsmith, *Optimal Multiuser, Multicellular, Multiple Antenna Systems,* VTC, September 2002
[16] G. D. Forney, *Trellis Shaping*, IEEE Trans. on Infor. Th., Vol. 38, No. 2, 3/1992, pg 281-909
[17] J. H. Conway and N. J. A. Sloane, Sphere Packing Lattices and Groups, Springer-Verlag, NY, 2nd Ed. 1992
[18] J. H. Conway and N. J. A. Sloane, *Voronoi Regions of Lattices, Second Moments of Polytopes and Quantization,* IEEE Trans. on Information Theory, Vol. 28, No. 2, March, 1982, pg 211-226
[19] G. D. Forney, *a Bounded Distance Decoding Algorithm for the Leech Lattice, with Generalizations,* IEEE Trans. on Information Theory, Vol. 35, No. 4, 7/8 1989, pg 906-909
[20] J. H. Conway and N. J. A. Sloane, *A Fast Encoding Method for Lattice Codes and Quantizers,* IEEE Trans. on Information Theory, Vol. 29, No. 6, November, 1983, pg 820-824
[21] J. H. Conway and N. J. A. Sloane, *Soft Dec. Techiniques for Codes and Lattices, Including the Golay Code and the Leech Lattice*, IEEE Trans. on Info. Theory, Vol. 32, (1986) pg 41-50
[22] C. A. Rogers, *Lattice Coverings of Space,* Mathematika, No. 6, 1959, pg 33-39
[23] P. Diaconis and D Freedman, *A Dozen De-Finetti Style Results in Search of A Theory*, Annals Insitute Henri Poincare, Prob. Stat.,Volume 23, suppl. 2, pp 397-423, 1987
[24] R. G. Gallager, *Info. Theory and Reliable Infor.*, New York, John Wiley and Sons, 1968
[25] H. Flanders, *Differential Forms*, Academic Press, New York, 1963
[26] D. W. Kahn, *Topology, An Intro. to Point Set and Algebraic Areas*, Dover, NY, 1995, pg 112
[27] K. Ito and H. P. McKean, *Diff. Proc. and Their Sample Paths,* Springer-Verlag, NY, 1965

LUCENT TECHNOLOGIES, BELL LABS, CRAWFORD HILL LAB, HOLMDEL, NJ
E-dresses: gjf@.lucent.com and ahd@lucent.com

Part III

Signal Processing for Multiantenna Channels

DIMACS Series in Discrete Mathematics
and Theoretical Computer Science
Volume **62**, 2003

Maximum-Likelihood Decoding and Integer Least-Squares: The Expected Complexity

Babak Hassibi and Haris Vikalo

Abstract. The problem of finding the least-squares solution to a system of linear equations where the unknown vector is comprised of integers, but the matrix coefficient and given vector are comprised of real numbers, arises in many applications: communications, cryptography, GPS, to name a few. The problem is equivalent to finding the closest lattice point to a given point and is known to be NP-hard. In communications applications, however, the given vector is not arbitrary, but rather is an unknown lattice point that has been perturbed by an additive noise vector whose statistical properties are known. Therefore in this paper, rather than dwell on the worst-case complexity of the integer-least-squares problem, we study its expected complexity, averaged over the noise and over the lattice. For the "sphere decoding" algorithm of Fincke and Pohst we find a closed-form expression for the expected complexity and show that, for a wide range of noise variances and dimensions, the expected complexity is polynomial, in fact often roughly cubic. Since many communications systems operate at noise levels for which the expected complexity turns out to be polynomial, this suggests that maximum-likelihood decoding, which was hitherto thought to be computationally intractable, can in fact be implemented in real-time—a result with many practical implications.

1. Introduction and Problem Statement

In this paper we shall be concerned with the following so-called *integer least-squares problem*

$$\min_{s\in\mathcal{Z}^m} \|x - Hs\|_2 , \tag{1.1}$$

where $x \in \mathcal{R}^n$, $H \in \mathcal{R}^{n\times m}$, and $\mathcal{Z}^m$ denotes the m-dimensional integer lattice, i.e., s is an m-dimensional vector with integer entries. Often, the search space is a (finite) subset of the infinite lattice, $\mathcal{D} \subset \mathcal{Z}^m$, in which case we have

$$\min_{s\in\mathcal{D}\subset\mathcal{Z}^m} \|x - Hs\|_2 . \tag{1.2}$$

The integer least-squares problem has a simple geometric interpretation. As the entries of s run over the integers, s spans the "rectangular" m-dimensional lattice, $\mathcal{Z}^m$. However, for any given *lattice-generating matrix* H, the n-dimensional vector Hs spans a "skewed" lattice. (When $n > m$, this skewed lattice lives in an

Key words and phrases. Integer least-squares problem, sphere decoding, wireless communications, multiple-antenna systems, lattice problems, NP hard, expected complexity, polynomial-time complexity.

m-dimensional subspace of $\mathcal{R}^n$.) Therefore, given the skewed lattice Hs, and given a vector $x \in \mathcal{R}^n$, the integer least-squares problem is to find the "closest" lattice point (in a Euclidean sense) to x—see Figure 1.

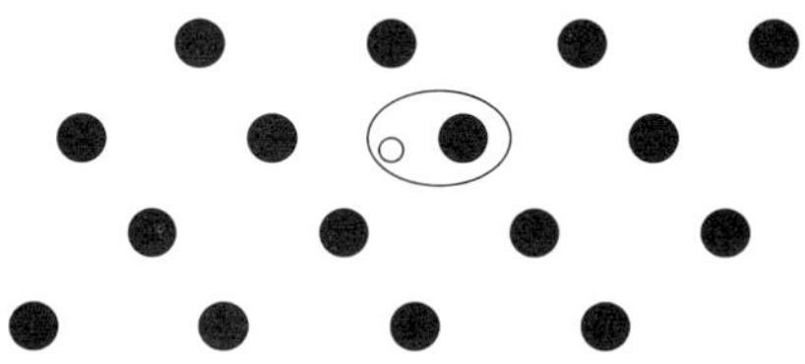

FIGURE 1. *Geometrical interpretation of the integer least-squares problem.*

Compared to the standard least-squares problem where the unknown vector s is an arbitrary vector in $\mathcal{R}^m$, and the solution is obtained via a simple pseudo-inverse, it is much more difficult to find the solution to (1.1) or (1.2). The main reason is that the search space is discrete (whether it is finite or infinite). In fact, it is well known that problems (1.1) and (1.2) are, for a general H, NP hard, both in a worst-case sense [**1**], as well as in an average sense [**2**].

Integer least-squares problems appear in a host of applications. In communications, when the channel is linear and the noise i.i.d. Gaussian, maximum-likelihood decoding leads to a least-squares cost. When the transmitted symbols are from a finite set, this can be often cast as an integer least-squares problem. Applications include lattice codes [**3, 4**], CDMA systems [**5, 6**], multi-antenna systems [**7, 8, 9**], etc. In all these applications, the unknown vector s represents the transmitted signal, the matrix H represents the channel, and the vector x represents the received signal. For example, in the multi-antenna context of V-BLAST [**7**] where we have M transmit and N receive antennas, H is the $(m = 2M) \times (n = 2N)$ real channel matrix, and for linear space-time codes (such as those in [**9**]) it is the equivalent channel matrix. The integer least-squares problem also arises when we have a frequency selective FIR channel $H(z) = h_0 + h_1 z^{-1} + \ldots h_L z^{-L}$, in which case the channel matrix takes the form of a Toeplitz matrix

$$H = \begin{bmatrix} h_0 & & & \\ h_1 & h_0 & & \\ \vdots & \ddots & \ddots & \\ h_L & \ddots & \ddots & h_0 \\ & h_L & \ddots & h_1 \\ & & \ddots & \vdots \\ & & & h_L \end{bmatrix}. \tag{1.3}$$

Other applications include global positioning systems (GPS) [**10**] and cryptography. In fact, there is a whole family of public-key cryptosystems based on the NP-hardness of the integer least-squares problem [**11, 12, 13**].

The remainder of this paper is organized as follows. In section 2 we give an overview of some heuristic and exact methods to solve the integer least-squares problem. We show that the exact methods can offer substantial gains over the heuristic ones. However, they generally require an exponential worst-case complexity, whereas the heuristic methods require only cubic, $O(m^3)$, computations.

Section 3 introduces the sphere decoding algorithm of Fincke and Pohst and argues that, if the received point is arbitrary, then the expected complexity of the algorithm is exponential. Section 4 introduces a random model for the integer least-squares problem, where the received point is assumed to be a lattice point perturbed by an additive Gaussian noise vector with known statistical properties. It then proceeds to compute the expected complexity of the sphere decoding algorithm, averaged over both the noise and the lattice, as a function of the noise variance, or SNR. This is done both for the infinite lattice, as well as some finite lattices encountered in communications problems. Simulations are presented in section 5, where it is shown that over a wide range of rates, SNRs and dimensions (in fact, those that are typically encountered in communications problems), the expected complexity of the sphere decoding algorithm is polynomial, often cubic. Section 6 mentions some generalizations of the Fincke-Pohst algorithm and section 7 provides the conclusion. The appendices give some mathematical background for the problems encountered in this paper.

2. Overview of Methods

Since the integer least-squares problem arises in many applications and finding the exact solution is, in general, NP hard, all practical systems employ some approximations, heuristics or combinations thereof. In communications applications, these approximations can be broadly categorized into three classes.

1. Solve the unconstrained least-squares problem to obtain $\hat{s} = H^{\dagger}x$, where $H^{\dagger}$ denotes the pseudo-inverse of H. Since the entries of $\hat{s}$ will not necessarily be integers, round them off to the closest integer (a process referred to as slicing) to obtain

$$\hat{s}_B = \left[H^{\dagger}x\right]_{\mathcal{Z}}. \tag{2.1}$$

   The above $\hat{s}_B$ is often called the Babai estimate [**1**]. In communications parlance, this procedure is referred to as *zero-forcing equalization.*
2. *Nulling and cancelling.* In this method, the Babai estimate is used for only one of the entries of s, say the first entry s_1. s_1 is then assumed to be known and its effect is cancelled out to obtain a reduced-order integer least-squares problem with $m-1$ unknowns. The process is then repeated to find s_2, etc. In communications parlance this is known as *decision-feedback equalization.*
3. *Nulling and cancelling with optimal ordering.* Nulling and cancelling can suffer from "error-propagation": if s_1 is estimated incorrectly it can have an adverse effect on the estimation of the remaining unknowns s_2, s_3, etc. To minimize the effect of error propagation, it is advantageous to perform nulling and cancelling from the "strongest" to the "weakest" signal. This is the method proposed for V-BLAST [**7**]—see also [**14**].

The above heuristic methods all require $O(m^3)$ computations, essentially because they all first solve the unconstrained least-squares problem.

2.1. Lattice Reduction. The aforementioned heuristic methods are exact only if the columns of H are orthogonal. In this case H can be diagonalized by a unitary transformation on the left, and so slicing the unconstrained least-squares solution yields the exact solution.

In practice, however, the columns of H are rarely orthogonal. Orthogonalizing the columns of H via a QR decomposition, or otherwise, generally destroys the lattice structure. (The reason being that, if s has integer entries, Rs need not have integer entries.) One method that attempts to alleviate this is *lattice reduction*. In these methods, one attempts to find an invertible $M \times M$ matrix T, such that T and T^{-1} have integer entries (thereby preserving the lattice structure), and such that the matrix $G = HT$ is as "orthogonal as possible". Having found such a T, rather than solve (1.1), one can solve the integer least-squares problem

$$\min_{t \in \mathcal{Z}^m} \|x - Gt\|_2 , \tag{2.2}$$

using the earlier mentioned heuristics and set $s = T^{-1}t$. Of course, lattice reduction is itself NP-hard. A common heuristic is the LLL (Lenstra, Lenstra and Lovasz [**15**]) algorithm which, permitting a gross oversimplification, can be regarded as Gram-Schmidt over integers.

While lattice reduction may lead to some improvement in the solution of (1.1), the integer least-squares problem over the infinite lattice, it is not useful for (1.2), which is over a subset of the lattice. The reason is that the lattice transforming matrix T often destroys the properties of the subset $\mathcal{D} \subset \mathcal{Z}^m$. Since in communications applications, we are always interested in a *finite* subset of the integer lattice, we shall therefore not consider lattice reduction methods in this paper.

2.2. Exact Methods. With the abundance of heuristic methods, it is natural to ask what their performance is, and how close they come to the optimal solution? In [**8**] this question is studied in the context of V-BLAST where it is shown that the exact solution significantly outperforms even the best heuristics. We also give an example here in the context of space-time codes from [**9**], which is shown in Figure 2. The example is a rate $R = 16$ space-time code for a system with $M = 8$ transmit and $N = 4$ receive antennas,. The resulting integer least-squares problem corresponds to dimension $m = 64$ and the entries of s each take on 4 integer values, say $\{-3, -1, 1, 3\}$. Therefore the number of lattice points in $\mathcal{D}$ is $4^{64} = 2^{128} \approx 3.4 \times 10^{38}$. As can be seen from Figure 2, the BER performance of the exact is integer least-squares solution is far superior to that of the best heuristic, which in this case is nulling and cancelling with optimal ordering.[1]

The above discussion shows that there is merit in studying exact solutions. The most obvious one is to form a search over the entire lattice which, although theoretically feasible for finite lattices, invariably requires an exponential search. There do, however, exist exact methods that are a bit more sophisticated than the above full search. These include Kannan's algorithm [**16**] (which searches only over restricted parallelograms), the KZ algorithm [**17**] (based on the Korkin-Zolotarev reduced basis [**18**]) and the sphere decoding algorithm of Fincke and Pohst [**19**]. Since the work of Fincke and Pohst the sphere decoding algorithm has been rediscovered in several contexts (see, e.g., [**10**] in the context of GPS systems) and is the algorithm we will be considering in this paper.

3. Sphere Decoding

The basic premise in sphere decoding is rather simple: we attempt to search over only lattice points that lie in a certain hypersphere of radius r around the given vector x, thereby reducing the search space and hence the required computations (see Figure 3). Clearly, the closest lattice point inside the hypersphere will also be

[1] Of course, at this point it may appear surprising that one can even generate Figure 2, since it requires finding the exact solution among a set of size 10^{38}—more on this later.

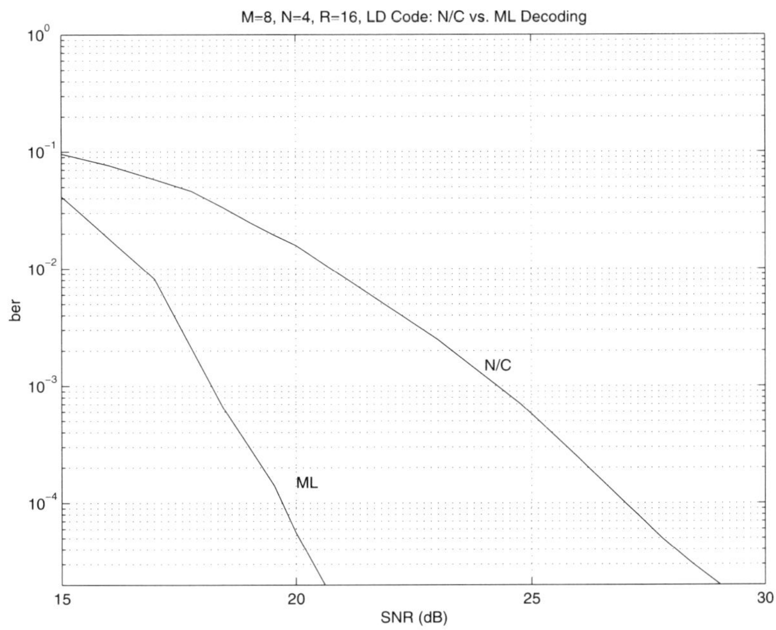

FIGURE 2. *Bit error performance of a rate* 16 *linear space-time code, corresponding to* $m = 64$. *Exact ML solution vs. nulling/cancelling with optimal ordering. (No. of lattice points*$= 2^{128} \approx 3.4 \times 10^{38}$*).*

the closest lattice point for the whole lattice. However, close scrutiny of this basic idea leads to two key questions.

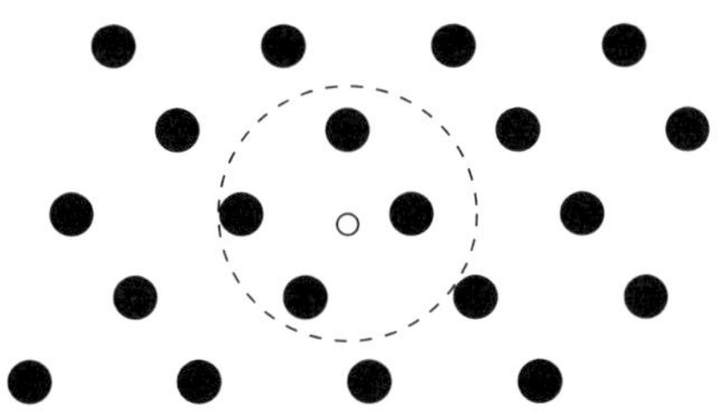

FIGURE 3. *Idea behind the sphere decoder*

1. *How to choose* r*?* Clearly, if r is too large, we obtain too many points and the search remains exponential in size, whereas if r is too small, we obtain no points inside the hypersphere.

   A natural candidate for r is the *covering radius* of the lattice, defined to be the smallest radius of spheres centered at the lattice points that cover the entire space. This is clearly the smallest radius that guarantees the existence of a point inside the hypersphere for any vector x. The problem with this choice of r is that determining the covering radius for a given lattice is itself NP hard [**20**].

Another choice is to use r as the distance between the Babai estimate and the vector x, i.e., $r = \|x - H\hat{s}_B\|$, since this radius guarantees the existence of at least one lattice point (here the Babai estimate) inside the hypersphere. However, it is again not clear in general whether this choice of radius leads to too many lattice point lying inside the hypersphere.

2. *How can we tell which lattice points are inside the hypersphere?* If this requires testing the distance of each lattice point from x (to determine whether it is less than r), then there is no point in sphere decoding as we will still need an exhaustive search.

Sphere decoding does not really address the first question. However, it does propose an efficient way to answer the second, and more pressing, one. The basic observation is the following. Although it is difficult to determine the lattice points inside a general m-dimensional hypersphere, it is trivial to do so in the (one-dimensional) case of $m = 1$. The reason is that a one-dimensional hypersphere is simply an interval and so the desired lattice points will be the integer values that lie in this interval. We can use this observation to go from dimension k to $k+1$. Suppose we have determined all k-dimensional lattice points that lie in a hypersphere of radius r. Then for any such k-dimensional point, the set of admissible values of the $k+1$-th dimensional coordinate that lie in the higher dimensional sphere of the *same* radius r forms an interval.

The above means that we can determine all lattice points in a hypersphere of dimension m and radius r by successively determining all lattice points in hyperspheres of lower dimensions $1, 2, \dots, m$ and the same radius r. Such an algorithm for determining the lattice points in an m-dimensional hypersphere essentially constructs a tree where the branches in the k-th level of the tree correspond to the lattice points inside the hypersphere of radius r and dimension k—see Figure 4. Moreover, the complexity of such an algorithm will depend on the *size* of the tree, i.e., on the number of lattice points visited by the algorithm in different dimensions.

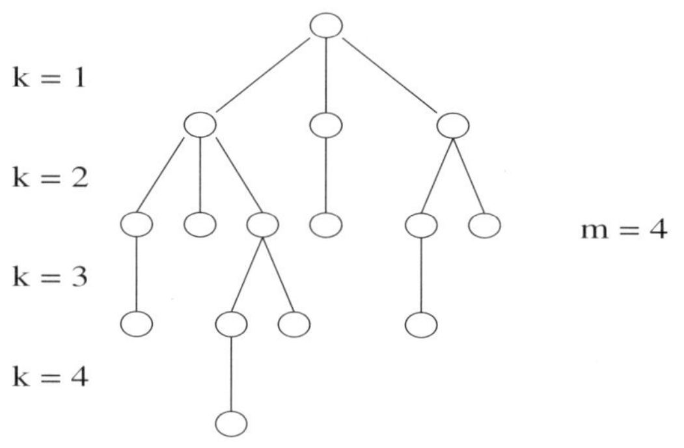

FIGURE 4. *Sample tree generated to determine lattice points in a 4-dimensional hypersphere.*

With this brief discussion we can now be more specific about the problem at hand. To this end, we shall assume that $n \geq m$, i.e., that there are at least as many equations as unknowns in $x \approx Hs$ (the case $n < m$ is considered further below). Note that the lattice point Hs lies in a hypersphere of radius r if, and only if,

$$r^2 \geq \|x - Hs\|^2 . \tag{3.1}$$

In order to break the problem into the subproblems described above, it is useful to consider the QR factorization of the matrix H

$$H = Q\begin{bmatrix} R \\ 0_{(n-m)\times m} \end{bmatrix}, \tag{3.2}$$

where R is an $m \times m$ upper triangular matrix and $Q = \begin{bmatrix} Q_1 & Q_2 \end{bmatrix}$ is an $n \times n$ orthogonal matrix. The matrices Q_1 and Q_2 represent the first m and last $n - m$ orthonormal columns of Q, respectively. The condition (3.1) can therefore be written as

$$\begin{aligned} r^2 &\geq \left\| x - \begin{bmatrix} Q_1 & Q_2 \end{bmatrix} \begin{bmatrix} R \\ 0 \end{bmatrix} s \right\|^2 = \left\| \begin{bmatrix} Q_1^* \\ Q_2^* \end{bmatrix} x - \begin{bmatrix} R \\ 0 \end{bmatrix} s \right\|^2 \\ &= \|Q_1^* x - Rs\|^2 + \|Q_2^* x\|^2 . \end{aligned}$$

Or in other words,

$$r^2 - \|Q_2^* x\|^2 \geq \|Q_1^* x - Rs\|^2 . \tag{3.3}$$

Defining $y = Q_1^* x$ and $r'^2 = r^2 - \|Q_2^* x\|^2$ allows us to rewrite this as

$$r'^2 \geq \sum_{i=1}^{m} \left(y_i - \sum_{j=i}^{m} r_{ij} s_j \right)^2 . \tag{3.4}$$

Here is where the upper triangular property of R comes in handy. The RHS of the above inequality can be expanded as

$$r'^2 \geq (y_m - r_{mm} s_m)^2 + (y_{m-1} - r_{m-1,m} s_m - r_{m-1,m-1} s_{m-1})^2 + \ldots \tag{3.5}$$

where the first term depends only on s_m, the second term on $\{s_m, s_{m-1}\}$ and so on. Therefore a necessary condition for Hs to lie inside the hypersphere is that $r'^2 \geq (y_m - r_{mm} s_m)^2$. This condition is equivalent to s_m belonging to the interval

$$\left\lceil \frac{-r' + y_m}{r_{mm}} \right\rceil \leq s_m \leq \left\lfloor \frac{r' + y_m}{r_{mm}} \right\rfloor . \tag{3.6}$$

Of course, (3.6) is by no means sufficient. For every s_m satisfying (3.6), defining $r'^2_{m-1} = r'^2 - (y_m - r_{mm} s_m)^2$ and $y_{m-1|m} = y_{m-1} - r_{m-1,m} s_m$, a stronger necessary condition can be found by looking at the first two terms in (3.5), which leads to s_{m-1} belonging to the interval

$$\left\lceil \frac{-r'_{m-1} + y_{m-1|m}}{r_{m-1,m-1}} \right\rceil \leq s_{m-1} \leq \left\lfloor \frac{r'_{m-1} + y_{m-1|m}}{r_{m-1,m-1}} \right\rfloor . \tag{3.7}$$

One can continue in a similar fashion for s_{m-2}, and so on until s_1, thereby obtaining all lattice points belonging to (3.1).

3.1. The Sphere Decoding Algorithm. We can now formalize the algorithm.

Input: $Q = \begin{bmatrix} Q_1 & Q_2 \end{bmatrix}$, R, x, $y = Q_1^* x$, r.

1. Set $k = m$, $r'^2_m = r^2 - \|Q_2^* x\|^2$, $y_{m|m+1} = y_m$
2. (Bounds for s_k) Set $UB(s_k) = \lfloor \frac{r'_k + y_{k|k+1}}{r_{k-1,k-1}} \rfloor$, $s_k = \lceil \frac{-r'_k + y_{k|k+1}}{r_{k-1,k-1}} \rceil - 1$
3. (Increase s_k) $s_k = s_k + 1$. If $s_k \leq UB(s_k)$ go to 5, else to 4.
4. (Increase k) $k = k + 1$; if $k = m + 1$ terminate algorithm, else go to 3.

5. (Decrease k) If $k = 1$ go to 6. Else $k = k - 1$, $y_{k|k-1} = y_k + \sum_{j=k+1}^{m} r_{kj} s_j$, $r_k^{'2} = r_{k+1}^{'2} - (y_{k+1} - r_{k+1,k+1} s_{k+1})^2$, and go to 2.
6. Solution found. Save s and go to 3.

We should mention that the original paper of Fincke and Pohst [**19**] used slightly different notation to the one we have used. For completeness, we shall include it here. The paper [**19**] makes use of the unconstrained least-squares solution $\hat{s} = H^{\dagger} x = R^{-1} Q_1^* x$. In this case, it follows that $\|Q_2^* x\|^2 = \|x\|^2 - \|H\hat{s}\|^2$ and so inequality (3.3) becomes

$$r^2 - \|x\|^2 + \|H\hat{s}\|^2 \geq \|R(s - \hat{s}\|^2. \tag{3.8}$$

The expansion (3.5) becomes

$$r^{'2} \geq r_{m-1,m-1}^2 \left(s_{m-1} - \hat{s}_{m-1} + \frac{r_{m-1,m}}{r_{m-1,m-1}} (s_m - \hat{s}_m) \right)^2 + \dots \tag{3.9}$$

and the intervals (3.6) and (3.7)

$$\left\lceil \hat{s}_m - \frac{r^{'}}{r_{mm}} \right\rceil \leq s_m \leq \left\lfloor \hat{s}_m + \frac{r^{'}}{r_{mm}} \right\rfloor \tag{3.10}$$

and

$$\left\lceil \hat{s}_{m-1|m} - \frac{r^{'}_{m-1}}{r_{m-1,m-1}} \right\rceil \leq s_{m-1} \leq \left\lfloor \hat{s}_{m-1|m} + \frac{r^{'}_{m-1}}{r_{m-1,m-1}} \right\rfloor \tag{3.11}$$

respectively, where we have defined $\hat{s}_{m-1|m} = \hat{s}_{m-1} + \frac{r_{m-1,m}}{r_{m-1,m-1}}(s_m - \hat{s}_m)$. We can now alternatively write the algorithm as

Input: R, x, $\hat{s}$, r.

1a. Set $k = m$, $r_m^{'2} = r^2 - \|x\|^2 + \|H\hat{s}\|^2$, $\hat{s}_{m|m+1} = \hat{s}_m$
2a. (Bounds for s_k) Set $z = \frac{r^{'}_k}{r_{kk}}$, $UB(s_k) = \lfloor z + \hat{s}_{k|k+1} \rfloor$, $s_k = \lceil -z + \hat{s}_{k|k+1} \rceil - 1$
3a. (Increase s_k) $s_k = s_k + 1$. If $s_k \leq UB(s_k)$ go to 5a, else to 4a.
4a. (Increase k) $k = k + 1$; if $k = m + 1$ terminate algorithm, else go to 3a.
5a. (Decrease k) If $k = 1$ go to 6a. Else $k = k-1$, $\hat{s}_{k|k-1} = \hat{s}_k + \sum_{j=k+1}^{m} \frac{r_{kj}}{r_{kk}}(s_j - \hat{s}_j)$, $r_k^{'2} = r_{k+1}^{'2} - r_{k+1,k+1}^2 (s_{k+1} - \hat{s}_{k+1|k+2})^2$, and go to 2.
6a. Solution found. Save s and go to 3a.

3.2. A First Look at Complexity. The paper [**19**] gives a complexity analysis of the above algorithm. The main result is that the number of arithmetic operations of the aforementioned algorithms (excluding Steps 1, 2, 3) is at most

$$\frac{1}{6}(2m^3 + 3m^2 - 5m) + \frac{1}{2}\left((2\lfloor \sqrt{r^2 d} \rfloor + 1) \binom{\lfloor 4r^2 d \rfloor + m - 1}{\lfloor 4r^2 d \rfloor} + 1 \right), \tag{3.12}$$

where $d = \max(r_{11}^2, \dots, r_{mm}^2)$. In practice d grows proportional to n (r_{11}^2, for example, is simply the squared norm of the first column of H, which has n entries) and r^2 grows proportional to m (for more on this see below) and so the upper bound on the number of computations in (3.12) can be quite large. Our experience with numerical implementations of the algorithm shows that the bound is quite loose. Moreover, although it does depend on the lattice-generating matrix H (through the quantity d), it offers little insight into the complexity of the algorithm. We will therefore not further consider it.

In this paper we propose to study the complexity of the sphere decoding algorithm using the geometric interpretation we have developed so far. As mentioned earlier, the complexity of the sphere decoding algorithm depends on the size of the generated tree in Fig. 4, which is equal to the sum of the number of lattice points in hyperspheres of radius r and dimensions $k = 1, \ldots, m$. The size of this tree depends on the matrix H, as well as on the vector x. Therefore, unlike the complexity of solving the unconstrained least-squares problem which only depends on m and n and *not* on the specific H and x the complexity of the sphere decoding algorithm is *data-dependent*.

3.2.1. *Expected Complexity.* Of course since the integer least-squares problem is NP hard, the worst-case complexity of sphere decoding is exponential. However, if we assume that the matrix H and vector x are generated randomly (according to some known distributions), then the complexity of the algorithm will itself be a random variable. In this case, it is meaningful to study the expected (or average) complexity of sphere decoding, and perhaps even some of its higher order moments.[2]

In what follows we will give a rough argument for the expected complexity of sphere decoding, although it is not too difficult to make it rigorous. (For a rigorous treatment, albeit using a different approach, see [**2**].) For an arbitrary point x, and an arbitrary lattice H, it is not too difficult to show that the expected number of lattice points inside the k-dimensional sphere of radius r is proportional to its volume,

$$\frac{\pi^{k/2}}{\Gamma(k/2+1)} r^k.$$

Therefore the expected total number of points is

$$\sum_{k=1}^{m} \frac{\pi^{k/2}}{\Gamma(k/2+1)} r^k > \sum_{k=1}^{\frac{m}{2}} \frac{\pi^k}{\Gamma(k+1)} r^{2k} \approx e^{\pi r^2}, \text{ for large } m.$$

To have a nonvanishing probability of finding a point in the m-dimensional sphere, its volume must be

$$\frac{\pi^{m/2}}{(m/2)!} r^m = O(1).$$

But from Stirling's formula this implies that $r^2 = O(m)$ and that the expected complexity of the algorithm is exponential, $e^{O(m)}$.

4. A Random Model

Although not unexpected, the above is a discouraging result. In communications applications, however, the vector x is not arbitrary, but rather is a lattice point perturbed by additive noise with known statistical properties. Thus, we will assume

$$x = Hs + v, \tag{4.1}$$

where the entries of v are independent $N(0, \sigma^2)$ random variables.

[2] In passing, we should mention that there is recent interest in studying the expected, rather than worst-case, complexity of various algorithms in the computer science literature. The reader may wish to refer to the survey paper [**21**] and the references therein, as well as the influential papers [**22, 23**]. In these works, a uniform distribution on the underlying problem is often (artificially) assumed and complexity issues such as NP-completeness, etc., are re-visited. However, as we shall see below, our problem allows for a very natural stochastic model.

4.1. Choice of the Radius. The first by-product of this assumption is a method to determine the desired radius r. Note that $\frac{1}{2\sigma^2} \cdot \|v\|^2 = \frac{1}{2\sigma^2} \cdot \|x - Hs\|^2$ is a Ξ^2 random variable with $n/2$ degrees of freedom. Thus we may choose the radius to be a scaled variance of the noise,

$$r^2 = \alpha n \sigma^2,$$

in such a way that with a high probability p_{fp} we find a lattice point inside the sphere,

$$\int_0^{\alpha n/2} \frac{\lambda^{n/2-1}}{\Gamma(n/2)} e^{-\lambda} d\lambda = p_{\text{fp}},$$

where p_{fp} is set to a value close to 1, say, $p_{\text{fp}} = 0.99$. [If the point is not found, we can increase the probability p_{fp}, adjust the radius, and search again.]

The important point is that the radius r is chosen based on the statistics of the noise, and *not* based on the lattice H. Making the choice based on H quickly leads us to NP hard problems (such as determining the covering radius). Moreover, choosing the radius based on the noise has a beneficial effect on the computational complexity.

4.2. Implications for Complexity. Clearly, when $\sigma^2 = 0$, i.e., when there is no noise, the exact solution can be found in $O(m^3)$ time. (The pseudo-inverse does the trick). On the other hand, when $\sigma^2 \to \infty$, the received vector x becomes arbitrary, for which we argued in section 3.2 that the expected complexity is exponential. What we are interested in is what happens at intermediate noise levels. In other words, how do we transition from cubic-time to exponential complexity?

In our analysis we shall compute the expected complexity averaged over both the noise v, as well as over the lattice-generating matrix H. Thus, we need a random model for H and will assume that it is comprised of independent $N(0,1)$ entries. This assumption is made for two reasons:

1. It makes the problem analytically tractable.
2. It is also a very reasonable assumption for large, unstructured, matrices H. (There exist many results in random matrix theory, such as Wigner's semicircle law, mixing conditions, etc. that are not very sensitive to Gaussian assumptions–see e.g., [**24**].)

Of course, if H possesses special structure, such as the Toeplitz structure (1.3), then this is not a reasonable assumption and the structure must be explicitly taken into account. However, this merits a separate analysis and is beyond the scope of the current paper.

Now, as argued in the previous section, the complexity of sphere decoding algorithm is proportional to the number of nodes visited on the tree in Figure 4 and, consequently, to the number of points visited in the spheres of radius r and dimensions $k = 1, 2, \ldots, m$. Hence the expected complexity is proportional to the number of points in such spheres that the algorithm visits on average. Thus the expected complexity of sphere decoding algorithm is given by

$$C(m, \sigma^2) = \sum_{k=1}^{m} \underbrace{(\text{expected \# of points in } k\text{-sphere of radius } r)}_{\triangleq E_p(k, r^2 = \alpha m \sigma^2)} \cdot \underbrace{(\text{flops/point})}_{2k+17}. \tag{4.2}$$

The summation in (4.2) goes over the dimensions $k = 1$ through $k = m$. The coefficient $2k + 17$ is the number of elementary operations (additions, subtractions, and multiplications) that the Fincke-Pohst algorithm performs per each visited point in dimension k.

We need to compute $E_p(k, r^2)$, the expected number of points inside the k-dimensional hypersphere of radius r. Let us first begin with the highest dimension, i.e., $k = m$.

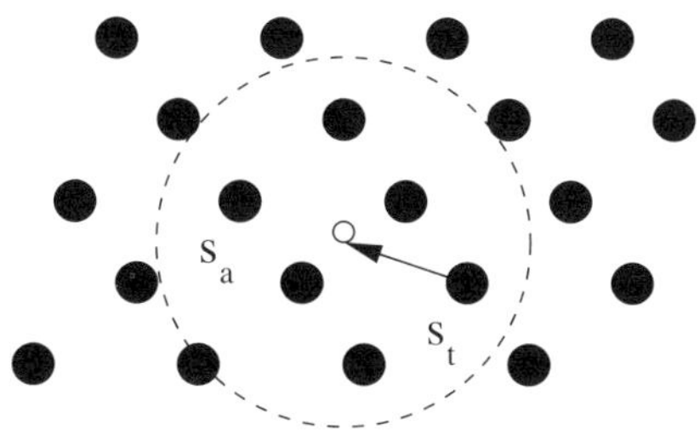

FIGURE 5. *s_t transmitted and x received. We are interested whether an arbitrary point s_a lies in a hypersphere of radius r centered around x.*

4.2.1. $k = m$. If the lattice point s_t was transmitted and the vector $x = Hs_t + v$ received, we are interested in the number of arbitrary lattice points s_a such that

$$\|x - Hs_a\|^2 \leq r^2.$$

(See Figure 5.) But, since $x = Hs_t + v$, this is just

$$\|v + H(s_t - s_a)\|^2 \leq r^2. \tag{4.3}$$

Now the vector $w = v + H(s_t - s_a)$ is clearly a zero-mean Gaussian random vector, since its entries are the sums of zero-mean Gaussian random variables. Now the covariance matrix has (i, j) entry

$$\begin{aligned} Ew_iw_j &= E\left(v_i + \sum_{k=1}^{m} h_{ik}(s_{t,k} - s_{a,k})\right)\left(v_j + \sum_{l=1}^{m} h_{jl}(s_{t,l} - s_{a,l})\right) \\ &= \sigma^2\delta_{ij} + \sum_{k=1}^{m}\sum_{l=1}^{m} \delta_{ij}\delta_{kl}(s_{t,k} - s_{a,k})(s_{t,l} - s_{a,l}) \\ &= \delta_{ij}(\sigma^2 + \|s_t - s_a\|^2), \text{ where } \|s\|^2 = \textstyle\sum_{k=1}^{m} s_k^2. \end{aligned}$$

Thus, $w = v + H(s_t - s_a)$ is an n-dimensional vector of zero-mean iid Gaussian random variables with variance $\sigma^2 + \|s_t - s_a\|^2$. This implies that $\frac{\|w\|^2}{2(\sigma^2 + \|s_t - s_a\|^2)} = \frac{\|v + H(s_t - s_a)\|^2}{2(\sigma^2 + \|s_t - s_a\|^2)}$ is a χ^2 random variable with $n/2$ degrees of freedom. Thus, the probability that the lattice point s_a lies in a hypersphere of radius r around x is

$$\gamma\left(\frac{r^2}{2(\sigma^2 + \|s_a - s_t\|^2)}, \frac{n}{2}\right) = \int_0^{\frac{r^2}{2(\sigma^2 + \|s_a - s_t\|^2)}} \frac{\lambda^{n/2-1}}{\Gamma(n/2)} e^{-\lambda} d\lambda. \tag{4.4}$$

Now that we have computed this probability, the expected number of points in the m-dimensional hypersphere can be evaluated. However, before doing so, let us turn to the $k < m$ case.

4.2.2. $k < m$. Referring back to (3.5), we are interested in all k-dimensional lattice points s such that

$$r'^2 \geq \sum_{i=m-k+1}^{m} \left(y_i - \sum_{j=i}^{m} r_{ij} s_j \right)^2 . \tag{4.5}$$

To better understand this set, let us again consider the QR decomposition of (3.2) to write

$$\begin{aligned} \|x - Hs_a\|^2 = \|v + H(s_t - s_a)\|^2 &= \left\| v + Q \begin{bmatrix} R \\ 0 \end{bmatrix} (s_t - s_a) \right\|^2 \\ &= \left\| Q^* v + \begin{bmatrix} R \\ 0 \end{bmatrix} (s_t - s_a) \right\|^2 . \end{aligned}$$

Now if we partition the upper triangular matrix R and the vector $u \stackrel{\Delta}{=} Q^* v$ as

$$R = \begin{bmatrix} R_{m-k,m-k} & R_{m-k,k} \\ 0 & R_{k,k} \end{bmatrix} \text{ and } u = \begin{bmatrix} u^{m-k} \\ u^k \\ u^{n-m} \end{bmatrix} \tag{4.6}$$

where the block matrices $R_{m-k,m-k}$, $R_{m-k,k}$ and $R_{k,k}$ are $(m-k) \times (m-k)$, $(m-k) \times k$ and $k \times k$, respectively, and the vectors u^{m-k}, u^k and u^{n-m} are $m-k$, k and $n-m$ dimensional, respectively, then we can write

$$\begin{aligned} \|x - Hs_a\|^2 &= \|u^{m-k} + R_{m-k,m-k}(s_t^{m-k} - s_a^{m-k}) + R_{m-k,k}(s_t^k - s_a^k)\|^2 + \\ &\quad \|u^k + R_{k,k}(s_t^k - s_a^k)\|^2 + \|u^{n-m}\|^2 . \end{aligned}$$

It is now straightforward to see that $r'^2 = r^2 - \|u^{n-m}\|^2$ and that $\|u^k + R_{k,k}(s_t^k - s_a^k)\|^2$ is simply the last k terms in the sum (3.4). Thus, we may rewrite the inequality (4.5) as

$$r^2 \geq \|u^k + R_{k,k}(s_t^k - s_a^k)\|^2 + \|u^{n-m}\|^2 = \left\| \begin{bmatrix} u^k \\ u^{n-m} \end{bmatrix} + \begin{bmatrix} R_{k,k} \\ 0 \end{bmatrix} (s_t^k - s_a^k) \right\|^2 . \tag{4.7}$$

Thus, to compute the expected number of k-dimensional lattice points that satisfy (4.5), we need to determine the probability distribution of the RHS of (4.7). For this we need the following result.

LEMMA 4.1. *Let H be an $n \times m$ (with $n \geq m$) random matrix with iid columns such that each column has a distribution that is rotationally-invariant from the left. In other words, for any $n \times n$ unitary matrix Θ, the distribution $\underline{h}_i$, the i-the column of H, satisfies*

$$p_{\underline{\mathbf{h}}}(\underline{h}_i) = p_{\underline{\mathbf{h}}}(\Theta \underline{h}_i).$$

Consider now the QR decomposition

$$H = Q \begin{bmatrix} R \\ 0 \end{bmatrix},$$

with Q $n \times n$ and unitary, and R $m \times m$ and upper triangular with non-negative diagonal entries. Then Q and R are independent random matrices, where

1. *Q has an isotropic distribution, i.e., one that is invariant under pre-multiplication by any $n \times n$ unitary matrix:*

$$p_{\mathbf{Q}}(Q) = p_{\mathbf{Q}}(\Theta Q), \ \forall \Theta\Theta^* = \Theta^*\Theta = I.$$

2. *Consider the partitioning of R according to (4.6) and further partition H as*

$$H = \begin{bmatrix} H_{m-k,m-k} & H_{m-k,k} \\ H_{n-m+k,m-k} & H_{n-m+k,k} \end{bmatrix},$$

where the subscripts indicate the dimensions of the sub-matrices. Then $R_{k,k}$ has the same distribution as the R obtained from the QR decomposition of the $(n-m+k) \times k$ matrix $H_{n-m+k,k}$.

Proof: See Appendix A.

□

Remarks:

1. What is interesting about the above Lemma is that even though the $(n-m+k) \times k$ submatrix $\begin{bmatrix} R_{k,k} \\ 0 \end{bmatrix}$ is *not* the R of the QR decomposition of the $(n-m+k) \times k$ submatrix $H_{n-m+k,k}$ it has the *same* distribution.
2. Lemma 4.1 clearly holds for an H with iid zero-mean unit-variance Gaussian entries. In this case, one can be explicit about the distribution of R: the entries are all independent with the i-th diagonal having a χ^2 distribution with $(n-i+1)/2$ degrees of freedom and the strictly upper triangular entries having iid zero-mean unit-variance Gaussian distributions.

Let us now apply Lemma 4.1 to the problem at hand. First, since v has iid zero-mean σ^2-variance Gaussian entries and Q is unitary, the same is true of $u = Q^* v$ and also of the sub-vectors u^{m-k}, u^k and u^{n-m}. Moreover, since Q is indepedent of R, the same is true of u. Returning to the inequality (4.7) let us multiply the vector inside the norm by an isotropically-random unitary matrix Θ. Since this does not change norms, we have

$$r^2 \geq \left\| \Theta \begin{bmatrix} u^k \\ u^{n-m} \end{bmatrix} + \Theta \begin{bmatrix} R_{k,k} \\ 0 \end{bmatrix} (s_t^k - s_a^k) \right\|^2.$$

Now clearly, the $n-m+k$-dimensional vector $\bar{v} = \Theta \begin{bmatrix} u^k \\ u^{n-m} \end{bmatrix}$ has iid zero-mean σ^2-variance Gaussian entries. Also, from Lemma 4.1 part 2, the $(n-m+k) \times k$ matrix $\bar{H} = \Theta \begin{bmatrix} R_{k,k} \\ 0 \end{bmatrix}$ has iid zero-mean unit-variance Gaussian entries. Thus, we may write (4.7) as

$$r^2 \geq \|\bar{v} + \bar{H}(s_t^k - s_a^k)\|^2, \tag{4.8}$$

which is precisely (4.3), except that the dimensions have changed from n and m to $n-m+k$ and k. Thus, using the same argument as presented after (4.3), we conclude that the probability that the k-dimensional lattice point s_a^k lies in a hypersphere of radius r around x is

(4.9)

$$\gamma\left(\frac{r^2}{2(\sigma^2 + \|s_a^k - s_t^k\|^2)}, \frac{n-m+k}{2}\right) = \int_0^{\frac{r^2}{2(\sigma^2 + \|s_a^k - s_t^k\|^2)}} \frac{\lambda^{(n-m+k)/2-1}}{\Gamma((n-m+k)/2)} e^{-\lambda} d\lambda.$$

Given this probability and the one in (4.4), one could in principle proceed by finding the argument of the gamma function in (4.4) and (4.9) for each pair of points (s_a, s_t), and sum their contributions; however, even for a finite lattice this would clearly be a computationally formidable task (and not doable at all in the infinite lattice case). Therefore, we shall find it useful to enumerate the lattice, i.e., count the number of points with the same argument of the gamma function in (4.4) and (4.9). Enumeration of infinite and finite lattices is treated separately.

4.3. The Infinite Lattice Case. The above probability (4.9) depends only on the Euclidean distance between the points s_t^k and s_a^k, that is, on $\|s_a - s_t\|^2 . \|s_a^k - s_t^k\|^2 = \|s\|^2$. Now, since in an infinite integer lattice $s_a - s_t = s$ is just another lattice point, we conclude that the probability in (4.9) depends only on the squared norm of an arbitrary lattice point in the k-dimensional lattice. It is thus straightforward to see that the expected number of of lattice points inside a k-dimensional hypersphere of radius r is given by:

(4.10)
$$E_p(k, r^2) = \sum_{l=0}^{\infty} \gamma(\frac{r^2}{2(\sigma^2 + n)}, \frac{n - m + k}{2}) \cdot (\# \text{ of } k\text{-dim lattice points s.t. } \|s^k\|^2 = l).$$

Since $\|s\|^2 = s_1^2 + \ldots + s_k^2$, we basically need to figure out how many ways a non-negative integer l can be represented as the sum of k squared integers. This is a classic problem in number theory and the solution is denoted by $r_k(l)$ [**25**]. There exist a plethora of results on how to compute $r_k(l)$. We only mention one here due to Euler: $r_k(l)$ is given by the coefficient of x^n in the expansion

$$\left(1 + 2\sum_{m=1}^{\infty} x^{m^2}\right)^k = 1 + \sum_{l=1}^{\infty} r_k(l) x^l. \tag{4.11}$$

(For more on the problem of representing integers as the sum of squares see Appendix B.)

The above arguments lead to the following result.

THEOREM 4.2 (Expected complexity of sphere decoding over infinite lattice). *Consider the model*

$$x = Hs + v,$$

where $v \in \mathcal{R}^{n \times 1}$ is comprised of i.i.d. $\mathcal{N}(0, \sigma^2)$ entries, $H \in \mathcal{R}^{n \times m}$ is comprised of i.i.d. $\mathcal{N}(0, 1)$ entries, and $s \in \mathcal{Z}^m$ is an m-dimensional vector whose entries are integer numbers. Then the expected complexity of the sphere decoding algorithm of section 3.1 with a search radius r for solving the integer least-squares problem,

$$\min_{s \in \mathcal{Z}^m} \|x - Hs\|^2,$$

is given by

$$C(m, \sigma^2, r) = \sum_{k=1}^{m} (2k + 17) \sum_{l=0}^{\infty} \gamma\left(\frac{r^2}{2(\sigma^2 + l)}, \frac{n - m + k}{2}\right) r_k(l). \tag{4.12}$$

Proof: Follows from the earlier discussions.

□

We should remark that, for any given search radius r, there always exists a probability that no lattice point is found. Therefore, to obtain the optimal solution, it is necessary to increase the search radius. One plausible way of doing this is to start with a radius for which the probability of finding a point is $1 - \epsilon$, then if no point is found to increase the search radius to a value such that the probability of finding no point is $1 - \epsilon^2$, and so on. For such a strategy, we have the following result.

COROLLARY 1 (Expected complexity for finding the optimal solution). *Consider the setting of Theorem 4.2. Given any $0 < \epsilon \ll 1$, consider a strategy where we first choose a radius such that we find a lattice point with probability $1 - \epsilon$, and then increase it to a probability of $1 - \epsilon^2$, and so on, if no point is found. Then the expected complexity of the sphere decoding algorithm to find the optimal solution is given by*

(4.13)
$$C(m, \sigma^2, \epsilon) = \sum_{i=1}^{\infty} (1-\epsilon)\epsilon^{i-1} \sum_{k=1}^{m} (2k+17) \sum_{l=0}^{\infty} \gamma\left(\frac{\alpha_i m \sigma^2}{2(\sigma^2 + l)}, \frac{n-m+k}{2}\right) r_k(l),$$

where α_i is chosen such that

$$\gamma\left(\frac{\alpha_i n}{2}, \frac{n}{2}\right) = 1 - \epsilon^i,\ i = 1, 2, \ldots \tag{4.14}$$

4.3.1. *The Complex Case.* In many applications, one is confronted with a complex version of the integer least-squares problem. In this case, we may assume that the model is

$$x = Hs + v, \tag{4.15}$$

where now $v \in \mathcal{C}^{n\times 1}$ is comprised of i.i.d. $\mathcal{C}N(0, \sigma^2)$ (circularly-symmetric complex normal) entries, $H \in \mathcal{C}^{n\times m}$ is comprised of i.i.d. $\mathcal{C}N(0, 1)$ entries, and $s \in \mathcal{C}Z^m$ is an m-dimensional complex vector whose entries have real and imaginary parts that are integers. As before, we are interested in the problem:

$$\min_{s \in \mathcal{C}Z^m} \|x - Hs\|^2. \tag{4.16}$$

The sphere decoding algorithm of section 3.1 can again be applied, provided we use the complex QR decomposition, and replace with real operations with appropriate complex ones. In this case, it can be shown (we are omitting the details for brevity and because they closely parallel the real case) that (4.12) is now replaced by

$$C(m, \sigma^2, r) = \sum_{k=1}^{m} (8k+17) \sum_{l=0}^{\infty} \gamma\left(\frac{r^2}{\sigma^2 + l}, n-m+k\right) r_{2k}(l). \tag{4.17}$$

Moreover, (4.13) is replaced by

$$C(m, \sigma^2, \epsilon) = \sum_{i=1}^{\infty} (1-\epsilon)\epsilon^{i-1} \sum_{k=1}^{m} (8k+17) \sum_{l=0}^{\infty} \gamma\left(\frac{\alpha_i m \sigma^2}{(\sigma^2 + l}, n-m+k\right) r_{2k}(l), \tag{4.18}$$

where α_i is chosen such that

$$\gamma(\alpha_i n, n) = 1 - \epsilon^i,\ i = 1, 2, \ldots \tag{4.19}$$

4.3.2. *Simulation Results.* As a measure of complexity, instead of the complexity itself, it is often useful to look at the *complexity exponent*, defined as

$$e_c = \frac{\log C(m, \sigma^2)}{\log m}. \tag{4.20}$$

When plotted, e_c is more visually appealing since complexity exponent approaches a constant if the expected complexity is polynomial, and grows like $\frac{m}{\log m}$ if $C(m, \sigma^2)$ is exponential.

The complexity exponent is plotted as a function of m for different values of σ^2 in Figure 6. As can be seen from the figure, for small enough noise the expected complexity is polynomial, indicated by the constant e_c over a wide range of m. On the other hand, for large noise e_c clearly exhibits the $\frac{m}{\log m}$ behavior and the computational complexity of the algorithm is exponential. We thus see the transition from polynomial-time to exponential complexity that we were seeking.

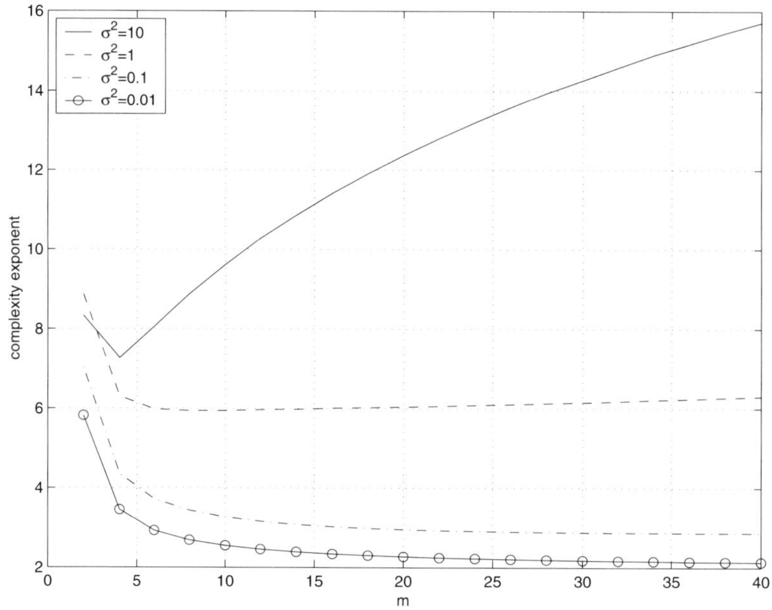

FIGURE 6. *The complexity exponent as a function of* m *for* $\sigma^2 = 0.01, 0.1, 1, 10$ *with* $\epsilon = .1$ *chosen for the sphere decoder.*

4.4. Finite Lattice Case. In communications problems, rather than being unbounded integers, the entries in the unknown m-dimensional vector s are often points that belong to an L-PAM constellations,

$$\mathcal{D}_L = \left\{ -\frac{L-1}{2}, -\frac{L-3}{2}, \ldots, \frac{L-3}{2}, \frac{L-1}{2} \right\}. \tag{4.21}$$

In fact, L is often taken as power of 2. We say that the point s then belongs to the lattice $\mathcal{D}_L^m$,

$$\mathcal{D}_L^m = \underbrace{\mathcal{D}_L \times \mathcal{D}_L \times \cdots \times \mathcal{D}_L}_{m\text{-times}},$$

where $\times$-operation denotes the Cartesian product.

Furthermore, in this case, rather than the noise variance σ^2, one is interested in the signal-to-noise ratio ρ,

$$\rho = \frac{m(L^2-1)}{12\sigma^2}. \tag{4.22}$$

The probability expression (4.9) for finding an arbitrary lattice point s_a^k inside a sphere around the given point x when the point lattice s_t^k was transmitted , holds for the finite lattice case as well. However, counting lattice points which have the same argument of the gamma incomplete function in (4.9) is not as easy. The reason is that unlike in the infinite lattice case, the difference between two lattice points, $s_a^k - s_t^k$, is not necessarily another lattice point. Thus, the lattice enumeration that we used in the previous section needs to be performed over pairs of points, (s_a^k, s_t^k).

More formally, the number of subset lattice points in the k-dimensional sphere is given by

$$\frac{1}{L^k} \sum_l \sum_{s_t^k, s_a^k \in \mathcal{D}_L^k, \|s_t^k - s_a^k\|^2 = l} \gamma\left(\frac{\alpha m \sigma^2}{2(\sigma^2+n)}, \frac{n-m+k}{2}\right),$$

and enumerating the set

$$\left\{(s_t^k, s_a^k) \mid s_t^k, s_a^k \in \mathcal{D}_L^k, \|s_t^k - s_a^k\|^2 = l\right\},$$

appears to be complicated.

For this, we propose a modification of Euler's generating function technique. In particular, for various finite lattices we will define generating polynomials that, when combined appropriately, perform the counting operations for us.

Let us do the case study for various values of L:

1. $\mathcal{D}_2^k$: The constellation $\mathcal{D}_2^k$ consists of the corners of a k-dim hypercube, as illustrated in Figure 7. Due to symmetry, all points in the hypercube are

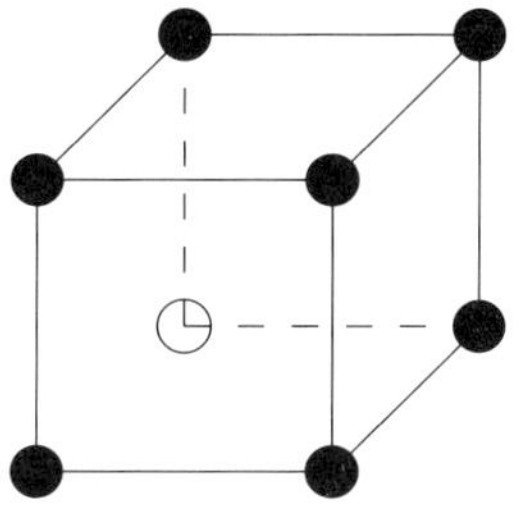

FIGURE 7. *Counting for* $\mathcal{D}_2^k$

essentially equivalent. Therefore, without loss of generality, we can assume that the "lower-left-corner" point $s_t^k = \left[-\frac{1}{2}, -\frac{1}{2}, \ldots, -\frac{1}{2}\right]^T$ has been transmitted. Then, depending on whether the corresponding entry of s_t^k is $-\frac{1}{2}$ or $\frac{1}{2}$, the vector $s_a^k - s_t^k$ is comprised of zero and one entries. The number of such vectors whose squared norm is l is clearly given by the number of vectors that have l entries at one, which is $\binom{k}{l}$. This gives the number of points in $\mathcal{D}_2^k$ at distance l from s_t^k.

2. $\mathcal{D}_4^k$: In this case, all points in the constellation are *not* the same. For each entry of s_t^k, we can distinguish between the "corner" and the "center" points, as illustrated in Figure 8. Extending Euler's idea, for the corner points we identify the generating polynomial

$$\theta_0(x) = 1 + x + x^4 + x^9, \tag{4.23}$$

and for the center points the polynomial

$$\theta_1(x) = 1 + 2x + x^4. \tag{4.24}$$

Essentially, the powers in the polynomials $\theta_0(x)$ and $\theta_1(x)$ contain information about possible squared distances between an arbitrary point s_a^k and the

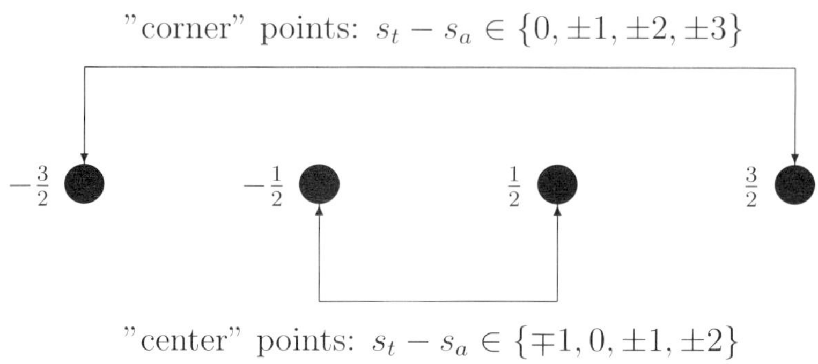

FIGURE 8. *Counting for* $\mathcal{D}_4^k$

transmitted point s_t^k. For instance, if an entry in the transmitted vector s_t^k, say $s_{t,1}$, is a corner point, then $s_{a,1} - s_{t,1} \in \{0, \pm 1, \pm 2, \pm 3\}$, depending on $s_{a,1} \in \mathcal{D}_4$. Thus the squared norm of their difference $|s_{a,1} - s_{t,1}|^2$ can be either 0, 1, 4, or 9, as described by the powers of $\theta_0(x)$. On the other hand, if $s_{t,1}$, is a center point, then $s_{a,1} - s_{t,1} \in \{0, \pm 1, \mp 1, \pm 2\}$ (which explains coefficient 2 in front of term x in $\theta_1(x)$). Now, if among the k entries of s_t^k, we choose a corner point j times, the number of ways $\|s_t^k - s_a^k\|^2$ can add up to l is given by the coefficient of x^l in the polynomial

$$\begin{pmatrix} k \\ j \end{pmatrix} \phi_0^j(x)\phi_1^{k-j}(x). \tag{4.25}$$

3. $\mathcal{D}_8^k$: Note that

$$\mathcal{D}_8^k = \{-\frac{7}{2}, -\frac{5}{2}, -\frac{3}{2}, -\frac{1}{2}, \frac{1}{2}, \frac{3}{2}, \frac{5}{2}, \frac{7}{2}\}^k.$$

Let us define the following subsets of $\mathcal{D}_8$:

$$\begin{array}{ll} \mathcal{S}_1 = \{-\frac{7}{2}, \frac{7}{2}\} & \mathcal{S}_2 = \{-\frac{5}{2}, \frac{5}{2}\} \\ \mathcal{S}_3 = \{-\frac{3}{2}, \frac{3}{2}\} & \mathcal{S}_4 = \{-\frac{1}{2}, \frac{1}{2}\} \end{array}. \tag{4.26}$$

Similar to the $L = 4$ case, we can identify the following polynomials for counting $s_a - s_t$ in $\mathcal{D}_8^k$ lattice:

$$\begin{aligned} \psi_0(x) &= 1 + x + x^4 + x^9 + x^{16} + x^{25} + x^{36} + x^{49}, \\ \psi_1(x) &= 1 + 2x + x^4 + x^9 + x^{16} + x^{25} + x^{36}, \\ \psi_2(x) &= 1 + 2x + 2x^4 + x^9 + x^{16} + x^{25}, \\ \psi_3(x) &= 1 + 2x + 2x^4 + 2x^9 + x^{16}. \end{aligned} \tag{4.27}$$

Therefore, if among k entries of s_t^k, we choose j_i points from $\mathcal{S}_i$, $i \in \{1, 2, 3, 4\}$, then the number of ways $\|s_t^k - s_a^k\|^2$ can add up to l is given by the coefficient of x^l in the polynomial

$$\begin{pmatrix} k \\ j_1, j_2, j_3, j_4 \end{pmatrix} \psi_0^{j_1}(x)\psi_1^{j_2}(x)\psi_0^{j_3}(x)\psi_1^{j_4}(x), \tag{4.28}$$

where $j_1 + j_2 + j_3 + j_4 = k$ and $\begin{pmatrix} k \\ j_1, j_2, j_3, j_4 \end{pmatrix} = \frac{k!}{j_1! j_2! j_3! j_4!}$.

4. Counting for $\mathcal{D}_{16}^k$ and higher order lattices is done similarly.

We can now summarize the above results for the expected computational complexity of the Fincke-Pohst algorithm for finite lattices in the following theorem.

THEOREM 4.3. **[Expected complexity of the sphere decoding over a finite lattice]** *Consider the model*

$$x = Hs + v,$$

where $v \in \mathcal{R}^{n\times 1}$ is comprised of i.i.d. $\mathcal{N}(0,1)$ entries, $H \in \mathcal{R}^{n\times m}$ is comprised of i.i.d. $\mathcal{N}(0,\rho/m)$ entries, and $s \in \mathcal{D}_L^m$ is an m-dimensional vector whose entries are elements of an L-PAM constellation. Then the expected complexity of the sphere decoding algorithm of section (3.1) with a search radius of r for solving the integer least-squares problem

$$\min_{s\in\mathcal{D}_L^m} \|x - Hs\|^2,$$

1. *For a 2-PAM constellation is*

$$C(m,\rho,r) = \sum_{k=1}^{m}(2k+17)\sum_{l=0}^{k}\binom{k}{l}\gamma\left(\frac{\alpha m}{2(1+\frac{12\rho l}{m(L^2-1)})}, \frac{n-m+k}{2}\right). \tag{4.29}$$

2. *For a 4-PAM constellation is*

(4.30)
$$C(m,\rho,r) = \sum_{k=1}^{m}(2k+17)\sum_{q}\frac{1}{2^k}\sum_{l=0}^{k}\binom{k}{l}g_{kl}(q)\gamma\left(\frac{\alpha m}{2(1+\frac{12\rho q}{m(L^2-1)})}, \frac{n-m+k}{2}\right),$$

where $g_{kl}(q)$ is the coefficient of x^q in the polynomial

$$(1+x+x^4+x^9)^l(1+2x+x^4)^{k-l}.$$

3. *For a 8-PAM constellation is*

(4.31)
$$C(m,\rho,r) = \sum_{k=1}^{m}(2k+17)\sum_{q}\frac{1}{4^k}\sum_{l=0}^{k}g_{kj_1j_2j_3j_4}(q)\gamma\left(\frac{\alpha m}{2(1+\frac{12\rho q}{m(L^2-1)})}, \frac{n-m+k}{2}\right),$$

where $g_{kj_1j_2j_3j_4}(q)$ is the coefficient of x^q in the polynomial (4.28).

4. *Similar expressions can be obtained for 16-PAM, etc., constellations.*

Proof: Follows from the above discussions.

□

We remark that to obtain the optimal solution to the integer least-squares problem we will occasionally need to increase the search radius r, and so we can obtain a result similar to that of Corollary 1, which we omit for brevity.

4.4.1. *The Complex Case.* When confronted with a complex integer least-squares problem, results similar to Theorem 4.3 hold. For complex 2-PAM (i.e., 4-QAM) constellations we have

$$C(m,\rho,r) = \sum_{k=1}^{m}(8k+17)\sum_{l=0}^{2k}\binom{2k}{l}\gamma\left(\frac{\alpha m}{1+\frac{12\rho l}{m(L^2-1)}}, n-m+k\right), \tag{4.32}$$

for complex 4-PAM (i.e., 16-QAM),

(4.33)
$$C(m,\rho,r) = \sum_{k=1}^{m}(8k+17)\sum_{q}\frac{1}{2^{2k}}\sum_{l=0}^{2k}\binom{2k}{l}g_{2kl}(q)\gamma\left(\frac{\alpha m}{1+\frac{12\rho q}{m(L^2-1)}}, n-m+k\right),$$

and for complex 8-PAM (i.e., 64-QAM) we have

(4.34)

$$C(m,\rho,r) = \sum_{k=1}^{m}(8k+17)\sum_{q}\frac{1}{4^{2k}}\sum_{l=0}^{2k} g_{2kj_1j_2j_3j_4}(q)\gamma\left(\frac{\alpha m}{1+\frac{12\rho q}{m(L^2-1)}}, n-m+k\right).$$

5. Simulation Results

We shall illustrate the complexity calculations with a communications example. Figure 9 shows the multiple antenna system with M-transmit and N-receive antennas.

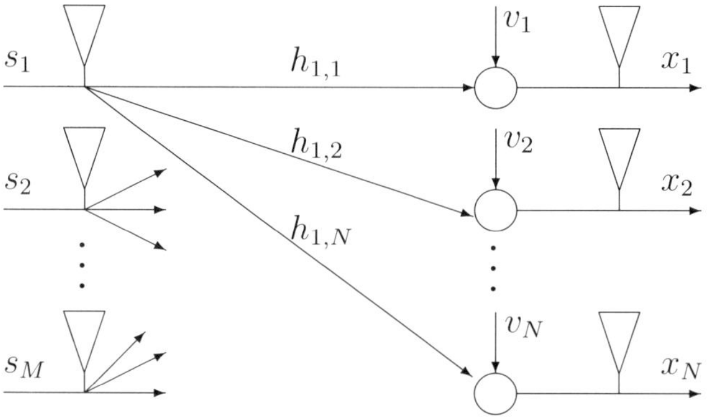

FIGURE 9. *Multiple antenna system*

The received signal $\mathbf{x}$ is related to the transmitted symbol $\mathbf{s}$ via

$$\mathbf{x} = \sqrt{\frac{\rho}{M}}\mathbf{H}\mathbf{s} + \mathbf{v}, \tag{5.1}$$

where $\mathbf{H} \in \mathcal{C}^{N\times M}$ is the known channel matrix, and $\mathbf{v} \in \mathcal{C}^{N\times 1}$ is the additive noise vector, both comprised of independent, identically distributed complex-Gaussian entries $\mathcal{C}(0,1)$. Furthermore, entries in the symbol vector $\mathbf{s}$ are chosen from a complex-valued QAM constellation. If we assume that the entries of $\mathbf{s}$ and $\mathbf{H}$ have, on the average, unit variance, then ρ is the expected received signal-to-noise ratio (SNR) at each receive antenna. To find a real-valued equivalent to the model (9), we let $m = 2M$, $n = 2N$, and let s denote an m-dimensional real vector obtained from the M-dimensional complex vector $\mathbf{s}$,

$$s = \left[\mathcal{R}(\mathbf{s})^T \quad \mathcal{I}(\mathbf{s})^T\right]^T.$$

Furthermore, let $x \in \mathcal{R}^{n\times 1}$ denotes

$$x = \left[\mathcal{R}(\mathbf{x})^T \quad \mathcal{I}(\mathbf{x})^T\right]^T,$$

let $v \in \mathcal{R}^{n\times 1}$ denotes

$$v = \left[\mathcal{R}(\mathbf{v})^T \quad \mathcal{I}(\mathbf{v})^T\right]^T,$$

and let $H \in \mathcal{R}^{n\times m}$ be given by

$$H = \sqrt{\frac{\rho}{M}}\begin{bmatrix} \mathcal{R}(\mathbf{H}) & \mathcal{I}(\mathbf{H}) \\ -\mathcal{I}(\mathbf{H}) & \mathcal{R}(\mathbf{H}) \end{bmatrix}.$$

Then the real-valued equivalent of (9) is given by

$$x = Hs + v.$$

We consider the expected complexity of sphere decoding algorithm for signal detection in the system shown in Figure 9 for various QAM modulation schemes. The expected complexity $C(\rho, m)$ is a function of both the symbol vector size m and the SNR ρ.[3] We shall consider "snapshots" in each dimension, i.e., we keep either m or ρ variable fixed and plot the complexity as a function of the other variable.

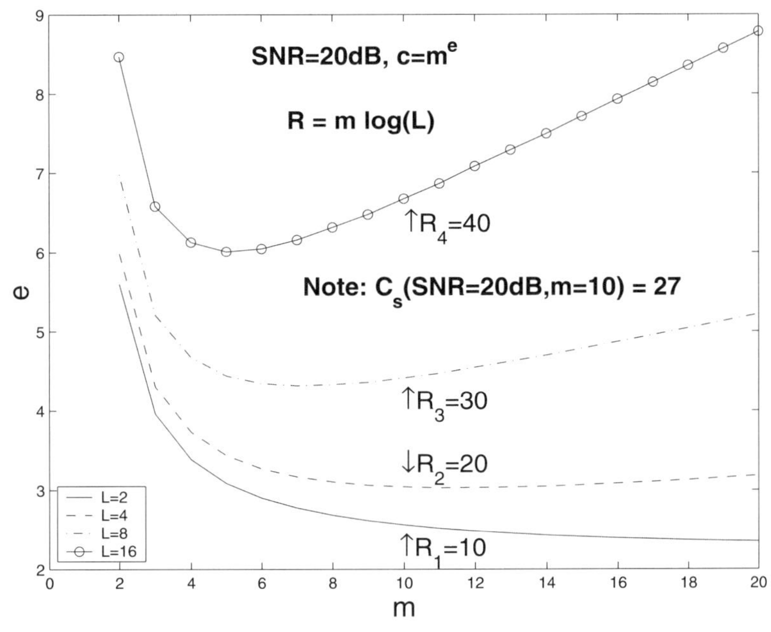

FIGURE 10. *The complexity exponent as a function of* m *for* $\rho = 20db$ *and* $L = 2, 4, 8, 16$.

Figure 10 shows the complexity exponent as a function of m for a fixed SNR $\rho = 20$db and L-PAM constellations with $L = 2, 4, 8, 16$. For low rates (i.e., small constellations) the expected complexity is polynomial, whereas for high rates (i.e., large constellations) it is exponential. Simulation results suggest that the complexity is polynomial as long as the rate is sufficiently, but not necessarily all that much, below the Shannon capacity corresponding to the SNR. Since this is the regime at which most communication systems operate, it suggests that ML decoding can be feasible. For instance, the complexity exponents curves in Figure 10 that correspond to $L = 8$ and $L = 16$ modulation schemes appear to be in the exponential regime. However, as is illustrated in Figure 10 for $m = 10$, the data rates corresponding to the points on those two curves are larger than the corresponding ergodic capacity,

$$C_s = E\left[\log\det\left(I_m + \frac{\rho}{M} HH^T\right)\right].$$

[3]In all the simulations presented, the complexities are for $\epsilon = .1$. In other words, our initial radius is determined so that we find a lattice point with probability .9. If no lattice point is found, we increase the radius so that this probability increases to .99, and so on.

For instance, when $m = 10$ (and SNR $20dB$), ergodic capacity is $C_{\text{erg}} = 27$. For the same system parameters, only the rates provided by the modulation schemes corresponding to $L = 2$ and $L = 4$ ($R = 10$ and $R = 20$, respectively) can be supported by the channel. The other two modulation schemes cannot be employed (we assume uncoded transmission). Note that expected complexity exponent in the data transmission regime that is supportable by the channel complexity is roughly cubic – which, in fact, is the complexity of the heuristic techniques.

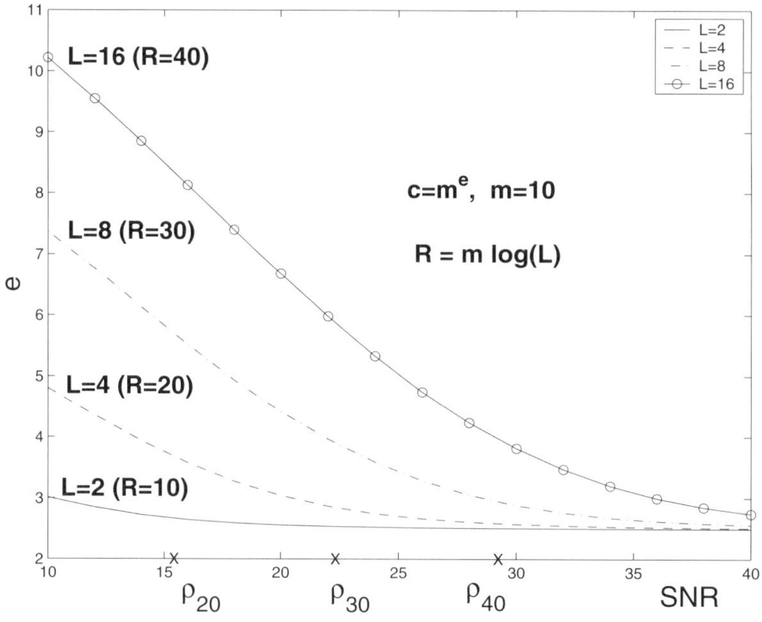

FIGURE 11. *The complexity exponent as a function of* ρ *for* $m = 10$ *and* $L = 2, 4, 8, 16$.

Figure 11 shows the complexity as a function of SNR for a fixed $m = 10$ (i.e., $M = N = 5$ transmit and receive antennas) and L-PAM constellations with $L = 2, 4, 8, 16$. A particular modulation scheme can be used only in the range of SNRs that supports transmission at the rate corresponding to that modulation scheme. We note that in such a range, the complexity exponent is roughly cubic. For instance, although the complexity for $L = 16$ appears to be high over a wide range of SNR, it is only for $\rho > \rho_{40}$ that this modulation scheme can be employed (ρ_{40} is the SNR for which the capacity $C_{\text{erg}} = 40 = R(L = 16)$). The complexity exponent at ρ_{40} and $L = 16$ is $c_e \approx 4$. The other SNRs marked on Figure 11, ρ_{30}, and ρ_{20}, have similar meanings (only for $L = 8$ and $L = 4$, respectively).

Figures 10-11 show the expected complexity, that is, the first-order statistics. In Figure 12, the empirical distribution of the complexity exponent is shown for $M = N = 5$ transmit and receive antennas, 16-QAM modulation scheme, and for 4 different SNR values. From Figure 11, we see that the lowest SNR in Figure 12 (16dB) roughly corresponds to the minimum SNR required for transmission on the particular system with the modulation scheme of choice. The outer dashed lines in each graph of Figure 12 correspond to three standard deviations of the

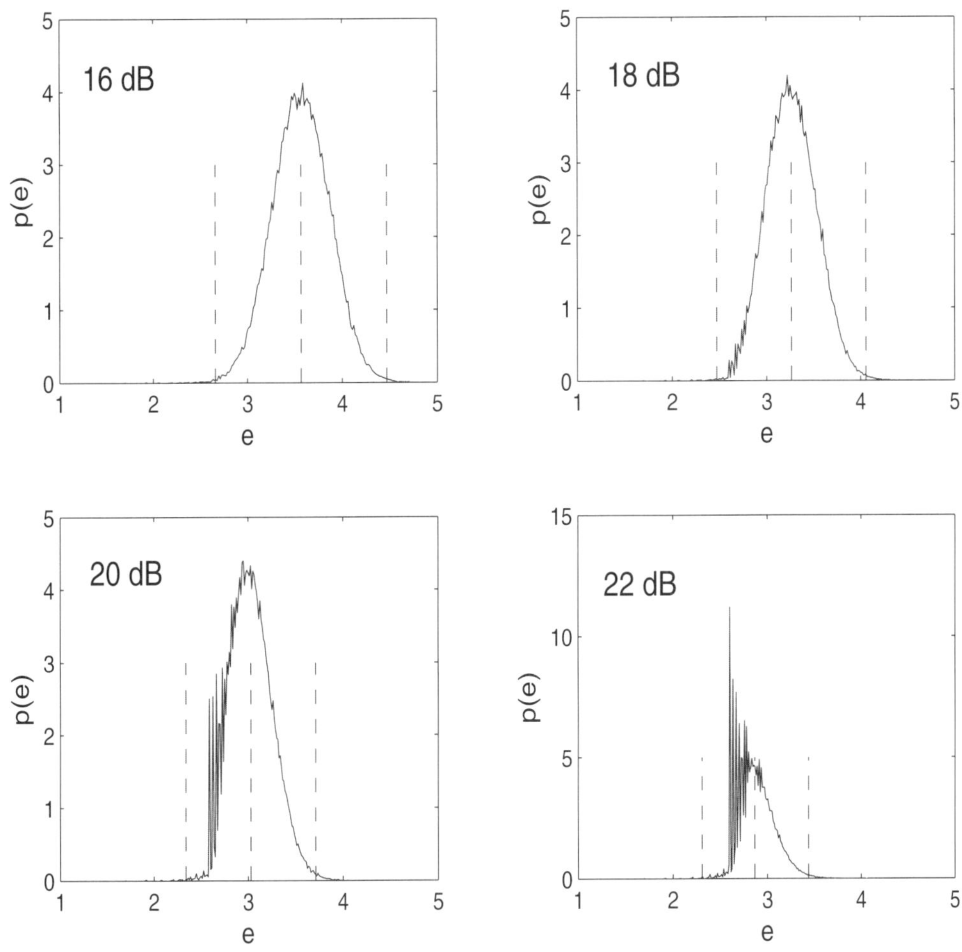

FIGURE 12. *The complexity exponent distribution for* $M = N = 5$, $L = 4$, *and* $SNR = 16, 18, 20, 22dB$.

corresponding distribution. The middle dashed line denotes the mean, i.e., the expected complexity that we previously obtained analytically. We can make the following observations in relation to the distributions as a function of the SNR:

- The expected complexity decreases, which was already anticipated from the results illustrated in Figure 11.
- The variance of the complexity decreases, as illustrated with tightening of the standard deviation.
- As the SNR increases, some "point-mass" like segments occur in the distribution. This is expected: for large SNRs, the radius of the sphere will be small and only a small (discrete) number of lattice points are found inside.

For comparison, exhaustive search in $M = N = 5$, 16-QAM system requires examining $k = 4^{10} \approx 10^6$ points, which is of the $O(m^{\log_{10} 10^6}) = O(m^6)$ order.

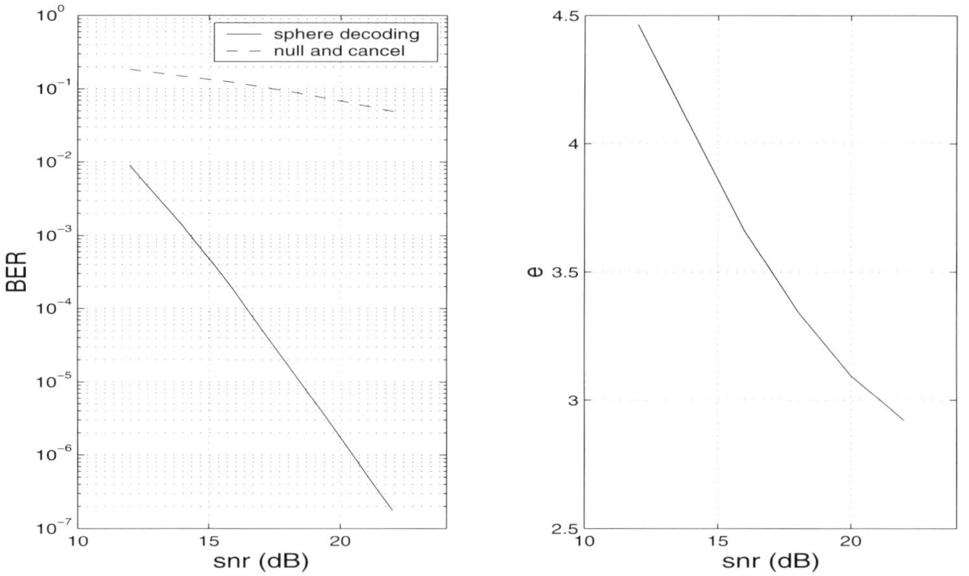

FIGURE 13. *Sphere decoder vs. nulling and cancelling,* $M = N = 5$, $L = 4$.

Figure 13 shows the improvement in performance of sphere decoding over nulling and cancelling for a multi-antenna system employing $M = N = 5$ transmit and receive antennas and 16-QAM modulation scheme. The complexity of ML decoding via sphere decoding here is comparable to that of nulling and cancelling, whereas the performance improvement is significant. The range of signal-to-noise ratios in Figure 13 is typical for indoor applications ([**7**]).

Figure 14 compares the performance of sphere decoding and nulling and cancelling for a multi-antenna system with $M = N = 2$ antennas, employing 16-QAM. The expected complexity exponent is also shown in Figure 14. In the SNR range of interest, the expected complexity of sphere decoding is again comparable to that of nulling and cancelling.

Comparing Figure 13 and Figure 14 we note that the performance gap between ML detection (provided by sphere decoding) and nulling and cancelling increases as the system employs more antennas. The analytical expression for the average probability of error appears difficult to derive. As an alternative, a pairwise error probability, as in [**26**], may be considered instead. The pairwise error probability that the vector $s^{(j)}$ was detected while $s^{(i)}$ was transmitted can be upper bounded at high SNRs as [**26**]

$$P(s^{(i)} \to s^{(j)}) \leq \frac{1}{\left(\frac{\rho}{4M}\right)^N \|d_{i,j}\|_F^2}, \tag{5.2}$$

where $d_{i,j} = s^{(i)} - s^{(j)}$. The bound (5.2) indicates that BER decreases exponentially with *receive diversity* – the number of receive antennas. ML decoder, as evident from Figure 13, fully exploits the receive diversity – the slope of the BER curve implies improvement by $N = 4$ orders of magnitude per SNR decade. On the other hand, nulling and cancelling (assuming no error propagation) converts the

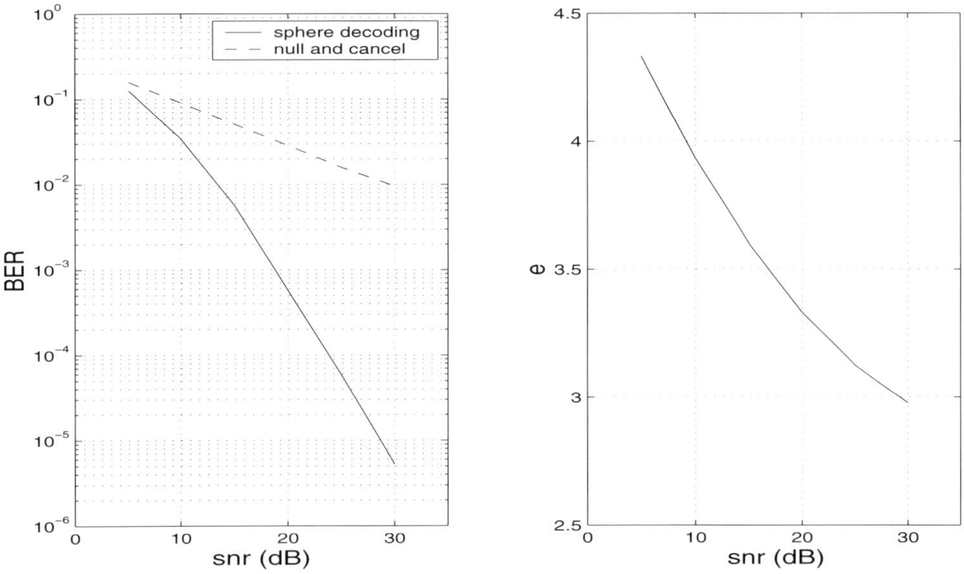

FIGURE 14. *Sphere decoder vs. nulling and cancelling,* $M = N = 2$, $L = 4$.

channel into a set of parallel channels with increasing diversity [**27**]. However, due to the error propagation, the performance is dominated by the first stream decoded by the receiver. Thus to improve the performance of nulling and cancelling, the decoding is often ordered according to the signal-to-interference-and-noise ratio of the incoming data streams.

Finally, Figure 15 illustrates the symbol error rate performance comparison of the sphere decoding and nulling and cancelling for the system employing $M = 8$ transmit and $N = 12$ receive antennas and 16-QAM modulation, the system specifications of V-BLAST [**7**]. The corresponding expected complexity is sub-cubic over the entire SNR range of interest.

6. Some Remarks

The expected complexity that we discussed in this paper accounts for finding all the lattice points in the sphere. The point among those found ones that is closest to x is the solution to the maximum-likelihood problem. There are some variations on the basic sphere decoding algorithm which we briefly mention here.

- *Sphere decoding with radius update.*
  Every time a point s_i in the sphere is found, we set the new radius of the sphere $r^2 = \|x - Hs_i\|^2$ and restart the algorithm. The radius update may be particularly useful at lower SNRs, where the number of points in the initial sphere is relatively large. However, it may not be benificial at high SNR, since restarting the sphere decoder may be costly. In any event, computing the expected complexity for this algorithm appears to be complicated, since it requires the calculation of the distribution of the radii that are updated.
- *Schnorr-Euchner version of the sphere decoding.*

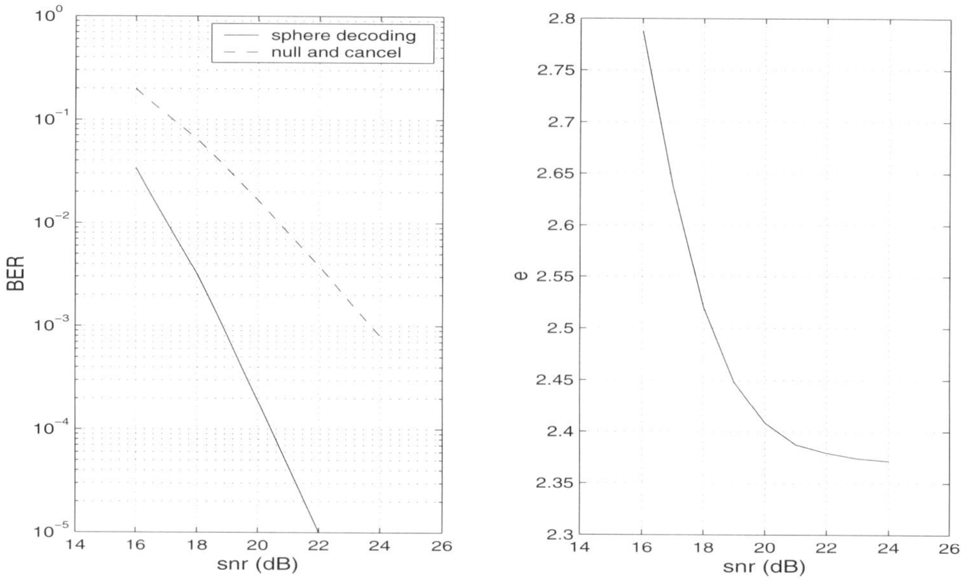

FIGURE 15. *Sphere decoder vs. nulling and cancelling,* $M = 8$, $N = 12$, *16-QAM.*

This strategy was proposed in [**28**]. The likelihood that the point will be found early is maximized if the search at each dimension k is performed in the order

$$[\hat{s}_d], [\hat{s}_d] - 1, [\hat{s}_d] + 1, [\hat{s}_d] - 2, \ldots$$

and if radius update (as described above) is used. The expected complexity of the Schnorr-Euchner version of the sphere decoding algorithm is no greater than the expected complexity of the basic algorithm that we derived in this chapter. However, computing the expected complexity of this algorithm appears to be formidable.

7. Conclusion

In many communication problems, maximum-likelihood detection reduces to solving an integer least-squares problem. In such applications ML detection is rarely performed, on the grounds that it requires exponential complexity and is therefore computationally intractable. In this paper we obtained a closed-form expression for the expected complexity of sphere decoding in terms of the noise variance, the dimension of the lattice, and (for subsets of lattices) the constellation. It turns out that over a wide range of noise variances and dimensions the expected complexity is often cubic or sub-cubic. Since many communications systems operate at noise levels for which this is the case, this suggests that maximum-likelihood decoding, which was hitherto thought to be computationally intractable, can in fact be implemented with complexity similar to heuristic methods, but with siginificant performance gains—a result with many practical implications.

Appendix A. Proof of Lemma 4.1

Let us start with part 1. Since H is rotationally-invariant from the left, the matrix ΘH has the same distribution as H, for any unitary matrix Θ. Since this distribution is independent of Θ, we conclude that the same is true for any random unitary Θ that is independent of H (since the distribution of Θ is readily integrated out as $p(\Theta H)$ does not depend on Θ). In other words, for any random unitary matrix Θ, the matrix ΘH has the same distribution as H. This is true, in particular, for an isotropically random random unitary Θ.[4] Now, for such a Θ, we have

$$\Theta H = \Theta Q R,$$

from which, due to the uniqueness of the QR decomposition when R has positive diagonal entries (see, e.g., [**31, 32**]), we conclude that ΘQ is the Q and R remains the R in the QR decompositon of ΘH. Now, since Θ is isotropically random, ΘQ is also isotropically random and, moreover, it is independent of Q. Therefore ΘQ must be independent of R, as well. Since ΘH and H have the same distribution, the Q's in their QR decompositions must have the same distribution, from which we conclude that Q must be an isotropically random unitary matrix, independent of R.

This concludes the proof of part 1.

We remark that the prrof of part 1 only required that H be rotationally-invariant. We did not require the independence of the columns of H. This independence is required for the proof of part 2, to which we now turn our attention.

Consider the partitioning

$$H = \begin{bmatrix} H_{m-k,m-k} & H_{m-k,k} \\ H_{n-m+k,m-k} & H_{n-m+k,k} \end{bmatrix},$$

where the subscripts indicate the dimensions of the sub-matrices. Now consider the QR decomposition of of the leading $m-k$ columns of H, i.e.,

$$\begin{bmatrix} H_{m-k,m-k} \\ H_{n-m+k,m-k} \end{bmatrix} = Q_1 \begin{bmatrix} R_{m-k,m-k} \\ 0 \end{bmatrix},$$

where Q_1 is unitary and $R_{m-k,m-k}$ is upper triangular with non-negative diagonal entries. Now since $R^*_{m-k,m-k}R_{m-k,m-k} = H^*_{m-k,m-k}H_{m-k,m-k} + H^*_{m-k,k}H_{m-k,k}$, which is the leading $(m-k)\times(m-k)$ submatrix of H^*H, we conclude that, as the notation suggests, $R_{m-k,m-k}$ is indeed the leading $(m-k)\times(m-k)$ submatrix of R in the partitioning (4.6).

Applying the unitary matrix Q_1^* to the full H (and not just its leading $m-k$ columns) we have

$$Q_1^* H = \begin{bmatrix} R_{m-k,m-k} & \bar{H}_{m-k,k} \\ 0 & \bar{H}_{n-m+k,k} \end{bmatrix}. \tag{A.1}$$

Now, since Q_1 depends only on the first $m-k$ columns of H and these are independent of the remaining k columns, by the rotational-invariance of the columns of H, we conclude that $\bar{H}_{n-m+k,k}$ has the *same* distribution as $H_{n-m+k,k}$.[5] Now

[4]For some of the properties of isotropically random unitary matrices the reader may refer to [**29, 30**].

[5]This is also true of $\bar{H}_{m-k,k}$ and $H_{m-k,k}$, though we shall not need this fact.

if we consider the QR decomposition of $\bar{H}_{k,k}$:

$$\bar{H}_{n-m+k,k} = Q_2 \begin{bmatrix} R_{k,k} \\ 0 \end{bmatrix},$$

combining this with (A.1), we have

$$Q_1^* H = \begin{bmatrix} I & 0 \\ 0 & Q_2 \end{bmatrix} \begin{bmatrix} R_{m-k,m-k} & \bar{H}_{m-k,k} \\ 0 & R_{k,k} \\ 0 & 0 \end{bmatrix},$$

and so

$$H = Q_1 \begin{bmatrix} I & 0 \\ 0 & Q_2 \end{bmatrix} \begin{bmatrix} R_{m-k,m-k} & \bar{H}_{m-k,k} \\ 0 & R_{k,k} \\ 0 & 0 \end{bmatrix}. \tag{A.2}$$

Since $Q_1 \begin{bmatrix} I & 0 \\ 0 & Q_2 \end{bmatrix}$ is unitary and since the diagonal entries of $R_{m-k,m-k}$ and $R_{k,k}$ are non-negative, we conclude that this is indeed the QR decomposition of H (which justifies our use of the notation $R_{k,k}$ for the R in the QR of $\bar{H}_{k,k}$).[6] Since $\bar{H}_{n-m+k,k}$ and $H_{n-m+k,k}$ have the same distribution, we conclude that $R_{k,k}$ has the *same* distribution as the R obtained from the QR decomposition of $H_{n-m+k,k}$.

This concludes the proof of part 2.

Appendix B. Representing Integers as a Sum of Squares

The problem of representing a positive integer as the sum of squares has a long history in mathematics and number theory. The problem of determining the number of ways that a non-negative integer l can be represented as k squares was first posed by Waring in 1770 and is denoted by $r_k(l)$. [7] The first known result in this direction is due to Diophantus of Alexandria (325-409 A.D.) who showed that no integer of the form $4m+3$ can be represented as the sum of two squares. In other words, $r_2(4m+3) = 0$. In 1632, Girard conjectured that l is the sum of two squares if the prime divisors of l of the form $4m+3$ occur in l in an even power. (For example, $l = 3^2 \cdot 5 = 3^2 + 6^2$, while $l = 3^3 \cdot 5$ cannot be represented as the sum of two squares.) Euler proved this conjecture in 1749. However, he did not give an explicit formula for $r_2(l)$. This was done by Legendre in 1798 and Gauss in 1801, who showed that

$$r_2(l) = 4(d_1(l) - d_3(l)), \tag{B.1}$$

where $d_1(l)$ and $d_3(l)$ are the number of divisors of l congruent to 1 and 3 mod 4, respectively.

In 1770, Lagrange proved his famous Four Squares Theorem, which states that every positive integer can be represented as the sum of four squares. This essentially establishes that $r_4(l) > 0$ for all positive integers l; however, Lagrange did not give an explicit formula for $r_4(l)$.

---

[6] Thus, for example, $\bar{H}_{m-k,k} = R_{m-k,k}$.

[7] In fact, Waring considered the much more general problem of determining the number of ways am integer can be represented as the sum of k integers raised to the power q. In this sense, the number of ways an integer can be represented as the sum of k squares is essentially the $q = 2$ Waring problem.

In terms of computing the value of $r_k(l)$, the first result is due to Euler who introduced (what is now known as) the Jacobi theta function

$$\theta(x) = \sum_{m=-\infty}^{\infty} x^{m^2} = 1 + 2\sum_{m=1}^{\infty} x^{m^2} \tag{B.2}$$

and established the following.

THEOREM B.1. *Let $\theta(x)$ be given by (B.2). Then*

$$\theta^k(x) = 1 + \sum_{l=1}^{\infty} r_k(l)x^l. \tag{B.3}$$

In other words, the number of ways a non-negative integer l can be represented as the sum of k squares is given by the coefficient of x^l in the expansion of $\theta^k(x)$.

This can be illustrated as follows: $\theta^k(x)$ is clearly a series in which each term has an exponent that is obtained as the sum of k squares; since the summation in (B.2) goes over all integers, the coefficients in front of each term in the series $\theta^k(x)$ must be equal to the number of ways that the exponent in that same term can be represented as a sum of two squares.

Using the connection between the above theta function and elliptic functions, Jacobi in 1829 obtained closed-form expressions for $r_k(l)$ when $k = 2, 4, 6, 8$ (see [**25**], chapter 9). His formula for $k = 4$ immediately yields Lagrange's Four Squares Theorem. Solutions for $k = 10$ and $k = 12$ were found by Liouville and Eisenstein. Later, Ramanujan, Hardy, and Littlewood obtained formulas for even $k \leq 24$. For odd k, the only results are due to Dirichlet, who found $r_3(l)$, and Eisenstein, Smith, and Minkowski, who found $r_5(l)$ and $r_7(l)$.

For a long time, these were the only known explicit formulas for $r_k(l)$. Indeed, results by Glaisher, and by R. Rankin (1965), using the theory of modular forms, discouraged many researchers from obtaining further closed-form expressions. The subject was therefore presumed to be "dead" until very recently. In 1994, as a consequence of their study of certain affine super-algebras, V. Kac and M. Wakimoto conjectured formulas for $\theta^k(x)$ when $k = 4m^2$ and $k = 4m(m+1)$ [**33**]. In 1996, these conjectures were proved by S. Milne using Jacobi's elliptic functions, Hankel determinants and continued fractions [**34**]. For an expository review of this, and subsequent results, the interested reader is referred to [**35**].

This exhausts known closed-form solutions for $r_k(l)$. There exist many asymptotic results (in both k and l)—see e.g., [**36**], chapter 5. In anycase, for any given k and l, the value of $r_k(l)$ can be numerically computed using Euler's formula (B.3). Moreover, $r_{k,l}$ is also a built-in function in *Mathematica*, `SumOfSquaresR[k,l]` [**37**].

References

[1] M. Grotschel, L. Lovasz, and A. Schriver, *Geometric Algorithms and Combinatorial Optimization*. Springer Verlag, 2nd ed., 1993.

[2] M. Ajtai, "The shortest vector problem in L_2 is NP-hard for randomized reductions," *Proceedings of the 30th Annual ACM Symposium on Theory of Computing*, pp. 10–19, 1998.

[3] A. Banihashemi and A. Khandani, "On the complexity of decoding lattices using the Korkin-Zolotarev reduced basis," *IEEE Transactions on Information Theory*, vol. 44, no. 2, pp. 162–171, 1998.

[4] E. Agrell, A. Vardy, and K. Zeger, "Closest point search in lattices," *submitted to IEEE Transactions on Information Theory*, 2002.

[5] C. Brutel and J. Boutros, "Euclidean space lattice decoding for joint detection in CDMA systems," *Proc. of the 1999 IEEE Info. Thy. and Comm. Workshop*, p. 129, 1999.

[6] E. Viterbo and J. Boutros, "A universal lattice decoder for fading channels," *IEEE Transacstions on Information Theory*, vol. 45, pp. 1639–1642, July 2000.

[7] G. J. Foschini, "Layered space-time architecture for wireless communication in a fading environment when using multi-element antennas," *Bell Labs. Tech. J.*, vol. 1, no. 2, pp. 41–59, 1996.

[8] M. O. Damen, A. Chkeif, and J.-C. Belfiore, "Lattice code decoder for space-time codes," *IEEE Comm. Let.*, pp. 161–163, May 2000.

[9] B. Hassibi and B. Hochwald, "High-rate codes that are linear in space and time," *IEEE Trans. Info. Theory*, vol. 48, pp. 1804–1824, July 2002.

[10] A. Hassibi and S. Boyd, "Integer parameter estimation in linear models with applications to GPS," *IEEE Transactions on Signal Processing*, vol. 46, pp. 2938–52, November 1998.

[11] M. Ajtai, "Generating hard instances of lattice problems," *Proceedings of the 28th Annual ACM Symposium on Theory of Computing*, pp. 99–108, 1996.

[12] O. Goldreich, S. Goldwasser, and S. Halevi, "Public-key cryptosystems from lattice reduction problems," *Advances in Cryptology - CRYPTO97. 17th Ann. Int. Crypto. Conf.*, pp. 112–31, 1997.

[13] R. Fischlin and J. Seifert, "Tensor-based trapdoors for cvp and their application to public key cryptography," *Cryptography and Coding. 7th IMA International Conference*, pp. 244–57, 1999.

[14] B. Hassibi, "An efficient square-root algorithm for BLAST," *submitted to IEEE Trans. Sig. Proc.*, 2000. Download available at `http://mars.bell-labs.com`.

[15] A. K. Lenstra, H. W. Lenstra, and L. Lovász, "Factoring polynomials with rational coefficients," *Math. Ann.*, pp. 515–534, 1982.

[16] R. Kannan, "Improved algorithms on integer programming and related lattice problems," *Proc. 15th Annu. ACM Symp. on Theory of Computing*, pp. 193–206, 1983.

[17] J. Lagarias, H. Lenstra, and C. Schnorr, "Korkin-Zolotarev bases and successive minima of a lattice and its reciprocal," *Combinatorica*, vol. 10, pp. 333–348, 1990.

[18] A. Korkin and G. Zolotaerv, "Sur les formes quadratiques," *Math. Ann.*, vol. 6, pp. 366–389, 1873.

[19] U. Fincke and M. Pohst, "Improved methods for calculating vectors of short length in a lattice, including a complexity analysis," *Mathematics of Computation*, vol. 44, pp. 463–471, April 1985.

[20] J. Conway and N. Sloane, *Sphere Packings, Lattices and Graphs*. Springer-Verlag, 1993.

[21] J. Wang, "Average-case computational complexity theory," *Complexity Theory Retrospective II*, pp. 295–328, 1997.

[22] L. Levin, "Average case complete problems," *SIAM Journal on Computing*, vol. 15, pp. 285–86, 1986.

[23] Y. Gurevich, "Average case completeness," *Journal of Computer and System Sciences*, vol. 42, no. 3, pp. 346–398, 1991.

[24] M. L. Mehta, *Random Matrices*. Academic Press, 2nd ed., 1991.

[25] G. Hardy, *Ramanujan: Twelve Lectures*. Chelsea Publishing, 1940.

[26] V. Tarokh, N. Seshadri, and A. R. Calderbank, "Space-time codes for high data rate wireless communication: Performance criterion and code construction," *IEEE Trans. Info. Theory*, vol. 44, pp. 744–765, 1998.

[27] A. Paulraj, R. Nabar, and D. Gore, *Introduction to Space-time Wireless Communications*. Cambridge University Press, to appear, 2003.

[28] C. P. Schnorr and M. Euchner, "Lattice basis reduction: improved practical algorithms and solving subset sum problems," *Mathematical Programming*, vol. 66, pp. 181–191, 1994.

[29] B. M. Hochwald and T. L. Marzetta, "Unitary space-time modulation for multiple-antenna communication in Rayleigh flat-fading," *IEEE Trans. Info. Theory*, vol. 46, pp. 543–564, Mar. 2000.

[30] B. Hassibi and T. L. Marzetta, "Block-fading channels and isotropically-random unitary inputs: The received signal density in closed-form," *IEEE Trans. on Info. Thy.*, vol. 48, pp. 1473–84, June 2002.

[31] R. Horn and C. Johnson, *Topics in Matrix Analysis*. Cambridge: University Press, 1991.

[32] G. Golub and C. V. Loan, *Matrix Computations*. Baltimore: Johns Hopkins University Press, 2nd ed., 1989.

[33] V. Kac and M. Wakimoto, "Integrable highest weight modules over affine super-algebras and Appell's function," *Comm. Math. Phys.*, vol. 215, no. 3, pp. 631–682, 2001.

[34] S. Milne, "Infinite families of exact sums of squares formulas, Jacobi elliptic functions, continued fractions, and Schur functions," *Ramanujan*, vol. 6, no. 1, pp. 7–149, 2002.

[35] I. Peterson, "Surprisingly square," *Science News*, vol. 159, June 2001.

[36] M. Knopp, *Modular Functions in Analytic Number Theory*. Chicago, IL: Markham Publishing Company, 1970.

[37] S. Wolfram, *The Mathematica Book*. Cambridge University Press, 4th ed., 1999.

Department of Electrical Engineering, California Institute of Technology, Pasadena, CA 91125

E-mail address: hassibi@caltech.edu

Department of Electrical Engineering, California Institute of Technology, Pasadena, CA 91125

E-mail address: hvikalo@caltech.edu

DIMACS Series in Discrete Mathematics
and Theoretical Computer Science
Volume **62**, 2003

Slowing the time-fluctuating MIMO channel

Dmitry Chizhik

ABSTRACT It has been reported [1], [2] that the number of transmitters that can be used beneficially in a MIMO system is limited by the coherence time of the channel due to difficulties with channel estimation at the receiver. Furthermore, rapid channel fluctuations degrade the performance of feedback schemes that would otherwise increase the information throughput that may be achieved. These impediments either reduce achievable capacity or impose an effective "speed limit" on the mobile, above which the effective throughput is reduced. In this paper the signal incident at the mobile is represented as a sum of plane waves. The channel transfer matrix is found to factor into time-dependent and time-independent parts. A processing method is proposed whereby the received signals are first pre-processed so as to produce signals that fluctuate on a much slower time scale. The pre-processing consists of beam forming, followed by Doppler compensation for signals received on each beam. Both operations are non-singular and do not alter the capacity of the MIMO channel. Beam forming effectively partitions the angular spectrum at the mobile, with each partition suffering a smaller Doppler spread, resulting in a "slowed down" channel fluctuation due to mobile motion. The coherence time of the pre-processed channel is found to increase by a factor on the order of the number of mobile receive antennas. In the limit of an infinite number of mobile receive antennas, the MIMO channel is shown to become static. These results remove constraints imposed by training requirements on the number of transmit antennas and increase allowed vehicle speeds. The "slowed down" channel characteristics may also be fed back to the transmitter, allowing an increase in information throughput.

I. Introduction

Rapid temporal fluctuations of channel characteristics present fundamental as well as practical difficulties and limitations on reliable communications. In mobile communications, the primary source of rapid channel fluctuations is the motion of the mobile, presumed to be in a vehicle [5]. For coherent reception, the receiver needs an estimate of the channel, usually obtained by transmitting a known training (pilot) signal. To reduce the impact of noise and interference on channel estimation, successive channel estimates are usually time-averaged. Time-averaging, however, introduces distortion in time-varying channels, as the averaging limits tracking of rapidly varying channels. Duration of the training interval is thus limited by the coherence time of the channel.

keywords: MIMO, channel estimation, beam forming, Doppler spread

In systems using MIMO communications, the channel must be learned from multiple transmitters, exacerbating the difficulties described above. It has been reported [1], [2] that the number of transmitters that may be used in MIMO is limited by the coherence time of the channel. In previous investigations, block-fading channels have been considered and the pilot symbols were time-multiplexed with data symbols. It was pointed out that the number of pilot symbols must not be smaller than the number of transmit antennas. As the number of transmit antennas increases, the required number of pilot symbols would exceed the coherence time of the channel before the data symbols are sent. Furthermore, rapid channel fluctuations degrade the performance of feedback schemes that would otherwise increase the information throughput that may be achieved. In the work by Q. Sun, et. al. [4], channel estimation in continuously faded MIMO channels was analyzed. It was found that the accuracy of channel estimation may be improved through linear interpolation of successive channel estimates. Nevertheless, the accuracy was observed to degrade at higher vehicle speeds. Beside the difficulties in accurate channel estimation, rapid temporal fluctuations of the channel require the receiver to adapt to the channel at a rate comparable with the rate of channel fluctuation. Computation of the receiver settings (e.g. space or space-time equalizer settings) demands computational power that increases with the rate of channel fluctuation and the number of transmit antennas.

The above difficulties persist in the case of systems where the pilot signals are sent on a separate parallel channel, e.g. a pilot code in CDMA [10]. As coherent MIMO communication requires that the channel be learned from every transmitter, the (finite) pilot signal power is split between different transmitters, making an estimate from each more prone to noise and interference. Here, again, duration of the available training period is limited by the coherence time of the channel.

Reduction of Doppler spread for a SISO channel through a combination of beam forming, followed by Doppler compensation has been proposed by Norklit and Vaughan [16]. There multiple antennas at the mobile have been used to form beams, with beam signals combined following Doppler compensation. The result was a reduction in fading rate. However, the multiple receive antennas were not used in [16] to combine the received signals coherently, thus producing no gain in received power or diversity. The technique proposed in [16] is generalized here for MIMO communications, allowing the full benefit of such systems [17], [18].

In this work the time–varying channel transfer matrix $\mathbf{H}(t)$ is presented in terms of plane waves incident at the mobile. Pre-processing of the signal vector at the mobile is proposed, consisting of beam forming and Doppler compensation for signals received on each beam. Both operations are non-singular and do not alter the underlying capacity of the MIMO channel. Reduction in the angular spread achieved by beam forming leads to a reduction of the Doppler spread. This is shown to effectively "slow down" the channel fluctuation due to mobile motion. The coherence time of the pre-processed channel is found to increase by a factor on the order of the number of mobile receive antennas. In the limit of an infinite number of mobile receive antennas, the MIMO channel is shown to become static. The pre-processed signals may be treated as signals arriving on virtual antennas, and may be further processed in an arbitrary way. This may be capitalized upon both to improve the performance of channel estimation and tracking, as well as to allow feedback to transmitter to increase the throughput.

It is further proposed that the same technique may be used to reduce the rate of channel fluctuation for the mobile-to-base (uplink) communication. To that end, transmitted signals from the mobile are sent along the beams, as opposed to individual antennas. Each signal is pre-compensated for Doppler prior to beam forming.

The paper is organized as follows: Section II describes the modeling of time-varying **H** matrices, Section III describes the receiver pre-processing in the case of an infinite number of mobile antennas. Receiver pre-processing in the case of a finite number of mobile receive antennas is described in Section IV, and the resulting increase in coherence time is quantified in Section V. Constraints on optimal number of base transmit antennas are discussed in Section VI, and Section VII outlines the conclusions.

II. Representation of time-varying H matrices

The environment being considered is that of a stationary base station, with the mobile moving with constant velocity through clutter. The scatterers are assumed stationary, with the understanding that the dominant source of rapid temporal variation is due to the mobile motion. A casual observation may be made that in terrestrial propagation environments, the most rapidly moving common scatterers are other vehicles, whose physical size is but a small fraction of that of buildings, terrain features, and trees slowly swaying in the wind. When the mobile is moving, however, waves scattered from stationary scatterers are no longer perceived as stationary. A vivid illustration of this may be found in Section 1.2.2 of Jakes [5], where the Doppler spectrum is seen to broaden considerably as the mobile increases its speed.

Here we will consider only fields in a single particular polarization. The field incident at the mobile may then be regarded as scalar. In the absence of motion, the field E for a particular polarization incident at the mobile antenna located at **r** may be represented as a superposition of plane waves:

$$E(\mathbf{r}) = \int A(\mathbf{k}) e^{i\mathbf{k}\bullet\mathbf{r}} d\mathbf{k} \tag{1}$$

where A(**k**) are complex amplitudes of plane waves arriving from the direction corresponding to the wave vector direction **k**. For a mobile moving with a constant velocity **v**, the location **r** becomes a function of time:

$$\mathbf{r}(t) = \mathbf{r}_0 + \mathbf{v}t \tag{2}$$

and the incident field becomes

$$E(\mathbf{r}(t)) = \int d\mathbf{k} A(\mathbf{k}) e^{i\mathbf{k}\bullet\mathbf{r}_0} e^{i\mathbf{k}\bullet\mathbf{v}t} \tag{3}$$

where $\omega_D = \mathbf{k} \bullet \mathbf{v}$ is the Doppler angular frequency. In the case of fields perceived by stationary terminals any complete orthogonal basis may be used to expand the spatial field in (1), and the choice of the plane wave expansion in (1) is arbitrary and not particularly better than, say, spherical harmonics. In the case of receivers undergoing uniform linear motion (2), however, the plane wave expansion is a natural expansion, as plane waves are eigenfunctions of induced temporal variations, with plane waves remaining uncoupled and each plane wave suffering multiplication by the Doppler factor $e^{i\mathbf{k}\bullet\mathbf{v}t}$, an eigenvalue of the temporal evolution.

The above expression may be generalized to the multiple antenna case where the channel is described by a channel matrix **H**, with an element h_{nm} describing the channel from the transmitter m to receiver n. Although the results below may be generalized to a frequency dependent case, only frequency independent channels will be treated here. The element h_{nm} may be recognized as the Green's function (i.e. point source response) $G(\mathbf{r}_n, \mathbf{r}'_m)$ of the channel, sampled spatially at the receiver location $\mathbf{r}_n$ and the transmitter location $\mathbf{r}'_m$. Like the electric field in (3), the Green's function may be represented formally as a sum of plane waves arriving at a receiver through a Fourier transform with respect to the receiver coordinates:

$$G(\mathbf{r}_n(t), \mathbf{r}'_m) = \int d\mathbf{k} A(\mathbf{k}, \mathbf{r}_m{}') e^{i\mathbf{k}\bullet\mathbf{r}_n(0)} e^{i\mathbf{k}\bullet\mathbf{v}t} \tag{4}$$

where the receive array is assumed to be moving uniformly, as in (2). The entire narrowband time-dependent channel **H** matrix of size $N_R \times N_T$. may then be written as

$$\mathbf{H}(t) = \begin{pmatrix} \int_{-\infty}^{\infty} d\mathbf{k} A(\mathbf{k}, \mathbf{r}'_1) e^{i\mathbf{k}\bullet\mathbf{r}_1} e^{i\mathbf{k}\bullet\mathbf{v}t} & \int_{-\infty}^{\infty} d\mathbf{k} A(\mathbf{k}, \mathbf{r}'_2) e^{i\mathbf{k}\bullet\mathbf{r}_1} e^{i\mathbf{k}\bullet\mathbf{v}t} & \bullet \\ \int_{-\infty}^{\infty} d\mathbf{k} A(\mathbf{k}, \mathbf{r}'_1) e^{i\mathbf{k}\bullet\mathbf{r}_2} e^{i\mathbf{k}\bullet\mathbf{v}t} & \int_{-\infty}^{\infty} d\mathbf{k} A(\mathbf{k}, \mathbf{r}'_2) e^{i\mathbf{k}\bullet\mathbf{r}_2} e^{i\mathbf{k}\bullet\mathbf{v}t} & \bullet \\ \bullet & \bullet & \bullet \end{pmatrix} \tag{5}$$

The angular spectrum coefficients $A(\mathbf{k}, \mathbf{r}'_m)$ describe the complex amplitudes of a plane waves arriving at the mobile receiver from direction of the wave vector $\boldsymbol{k}$ due to a source at $\mathbf{r}'_m$. Here N_R and N_T are the number of receive and transmit antennas, respectively. Discretizing the wave vector space **k** into a discrete set of N_L plane waves with a step $\Delta\mathbf{k}$, replacing the above integrals by corresponding sums, and defining $A_m(\mathbf{k}) = A(\mathbf{k}, \mathbf{r}'_m)$, the **H** matrix may be written as:

$$\mathbf{H}(t) = \frac{1}{\sqrt{N_L}} \begin{pmatrix} \sum_l A_1(\mathbf{k}_l) e^{i\mathbf{k}_l\bullet\mathbf{r}_1} e^{i\mathbf{k}_l\bullet\mathbf{v}t} & \sum_l A_2(\mathbf{k}_l) e^{i\mathbf{k}_l\bullet\mathbf{r}_1} e^{i\mathbf{k}_l\bullet\mathbf{v}t} & \bullet \\ \sum_l A_1(\mathbf{k}_l) e^{i\mathbf{k}_l\bullet\mathbf{r}_2} e^{i\mathbf{k}_l\bullet\mathbf{v}t} & \sum_l A_2(\mathbf{k}_l) e^{i\mathbf{k}_l\bullet\mathbf{r}_2} e^{i\mathbf{k}_l\bullet\mathbf{v}t} & \bullet \\ \bullet & \bullet & \bullet \end{pmatrix} \tag{6}$$

which may be factored as:

$$\mathbf{H}(t) = \frac{1}{\sqrt{N_L}} \underbrace{\begin{pmatrix} e^{i\mathbf{k}_1\bullet\mathbf{r}_1} & e^{i\mathbf{k}_2\bullet\mathbf{r}_1} & \bullet \\ e^{i\mathbf{k}_1\bullet\mathbf{r}_2} & e^{i\mathbf{k}_2\bullet\mathbf{r}_2} & \bullet \\ \bullet & \bullet & \bullet \end{pmatrix}}_{\mathbf{B}} \underbrace{\begin{pmatrix} e^{i\mathbf{k}_1\bullet\mathbf{v}t} & 0 & \bullet \\ 0 & e^{i\mathbf{k}_2\bullet\mathbf{v}t} & \bullet \\ \bullet & \bullet & \bullet \end{pmatrix}}_{\mathbf{D}} \underbrace{\begin{pmatrix} A_1(\mathbf{k}_1) & A_2(\mathbf{k}_1) & \bullet \\ A_1(\mathbf{k}_2) & A_2(\mathbf{k}_2) & \bullet \\ \bullet & \bullet & \bullet \end{pmatrix}}_{\mathbf{H_k}} \tag{7}$$

Or in short hand notation

$$\mathbf{H}(t) = \mathbf{BDH}_k \tag{8}$$

where $\mathbf{H}_k$ is the scattering matrix of the channel, each element of which relates the signal emitted from a particular base antenna to a plane wave incident on the receive antenna array. The diagonal Doppler shift matrix **D** represents the time dependent phase evolutions, i.e. Doppler shifts, of the plane waves incident at the receive array. The matrix **B** is the inverse spatial Fourier transform matrix, evaluated at receive antenna locations. This matrix depends explicitly on the receive antenna locations, and may be described as an inverse beam forming matrix. The antenna elements considered so far may be placed arbitrarily in space, and not necessarily in a regular linear array. In later numerical examples, the mobile array will be assumed to be a regular linear array. If the mobile array is not located in free space, but rather inside a car, then the matrix **B** used above changes form and becomes a coupling matrix, describing the response of each mobile antenna to a plane wave incident at the exterior of the vehicle. It will be assumed here that such near-field effects do not perturb the matrix **B** significantly, (viewing the car as mostly transparent), and **B** may therefore be described as an inverse Fourier transform matrix, as in (7).

The channel scattering matrix $\mathbf{H}_k$ is equivalent to a spatial Fourier transform of the Green's function of the channel with respect to the mobile receiver coordinates:

$$H_k(\mathbf{k},\mathbf{r}') = \int d\mathbf{r} G(\mathbf{r},\mathbf{r}') e^{-i\mathbf{k}\cdot\mathbf{r}} \tag{9}$$

The $\mathbf{H}_k$ matrix is thus characteristic of the channel and is independent of the receiver array design. It is assumed in this work that the scatterers are not moving, thus both the Green's function $G(\mathbf{r},\mathbf{r}')$ in (9) and the matrix $\mathbf{H}_k$ in (7) and (9) are not dependent on time. The Green's function in (9) is the same as the Green's function in (4), evaluated at $t = 0$. Furthermore, large scale variations in time due to, say, vehicle moving into a different environment in the next street are not included, leading to the angular spectra $A_m(\mathbf{k})$ in (7) remaining independent of time.

The representation of the **H** matrix used in (6) and (7) is simply a factorization of a general **H** matrix, with the noted restriction that the temporal variation is only a consequence of linear uniform motion, as described above. No other restrictions are imposed. In particular, the entries of the **H** matrix may be distributed as correlated complex Gaussian random variables, as is often assumed [11,12,13], or may be non-Gaussian as in the case of the keyhole channel [6, 20]. To generate a Gaussian distributed **H** matrix with correlations at the receiver stemming from a limited angular spread, the angular spectra A(**k**) may be generated as complex Gaussian *iid* variables, which are then weighted so as to produce an angular spectrum of desired shape. To introduce correlations at the transmitter, the **H** matrix may be multiplied from the right by an appropriate transformation [11,12,13, 22]. In this work, no assumptions of the distribution of the **H** matrix entries is made, and the results are independent of the spatial model.

The above model may be modified by carrying out a similar expansion of the field at the base transmitter. Such a representation, sometimes called a bi-directional channel model [15] is one of an infinite set of possible orthogonal expansions of the field. The resulting channel model is that of a coupling matrix, consisting of path coefficients relating outgoing plane waves at the transmitter to incoming plane waves at the receiver. In general, the coupling matrix is not diagonal, as a plane wave launched from the base results in multiple plane waves arriving at the terminal. Certain exceptional channels may be constructed, where the channel scattering matrix is diagonal, as for example, in the case of single scattering from specular scatterers. In general, multiple scattering takes place. As the base is normally stationary, plane wave expansions do not offer the same unique advantages as they do in the case of similar expansion at the mobile, where they form a set of temporal eigenfunctions. Angular spectrum representation at the base is therefore not considered in this work. In cases where particular clusters of scatterers are visible both from the base and from the mobile [19], beam forming at the base may prove beneficial. It should also be noted that in cases where there is a limited angle spread at the base, statistical description of the channel may be more naturally expressed in the angular spectrum basis.

III. Receiver processing for an infinite number of mobile antennas

As the number of mobile antennas N_R increases, the angular resolution of the mobile array increases, and the array begins to resolve individual plane waves in the limit of infinite number of antennas, and given a finite set of N_L distinct plane waves. In that limit the columns of the inverse Fourier transform matrix $\mathbf{B}$ become orthogonal:

$$\lim_{N_R\to\infty} \frac{1}{N_R}\mathbf{B}^{\dagger}\mathbf{B} = \lim_{N_R\to\infty} \frac{1}{N_R}\frac{N_R}{N_L}\begin{pmatrix} 1 & \frac{1}{N_R}\sum_{n=1}^{N_R} e^{i(\mathbf{k}_2-\mathbf{k}_1)\bullet\mathbf{r}_n} & \bullet \\ \frac{1}{N_R}\sum_{n=1}^{N_R} e^{i(\mathbf{k}_1-\mathbf{k}_2)\bullet\mathbf{r}_n} & 1 & \bullet \\ \bullet & \bullet & \bullet \end{pmatrix} = \frac{1}{N_L}\mathbf{I} \tag{10}$$

where $\mathbf{I}$ is the identity matrix of size N_L, and the dagger symbol denotes the Hermitian transpose. Here spacing between neighboring antennas remains fixed (and assumed to be 0.5 λ), thus the increase in the number of antennas leads to an increase in the mobile antenna effective aperture. The orthogonality of the rows of $\mathbf{B}$ holds in the limit of infinite number of mobile antennas, provided the number of distinct plane waves is large but finite while the number of mobile antennas approaches infinity. The restriction on the number of plane waves is used here but removed in the rest of the work. This causes the matrix $\mathbf{B}$ to have an increasing number N_R of rows, while keeping a constant number of columns N_L. The rank of $\mathbf{H}$ is therefore constrained not to exceed N_L by assumption. Alternatively, the set of wave vectors $\mathbf{k}$ used in defining $\mathbf{B}$ may be constructed so as to make $\mathbf{B}$ a DFT matrix, which is unitary by construction [14]. The Doppler matrix $\mathbf{D}$ in (7) is evidently unitary. Using (10) the time evolution of the $\mathbf{H}$ matrix in the limit of infinite number of mobile antennas may be represented as

$$\mathbf{H}(t) = \mathbf{B}\mathbf{D}(\mathbf{v}t)\mathbf{H}_k = \mathbf{B}\mathbf{D}(\mathbf{v}t)\frac{N_L}{N_R}\mathbf{B}^{\dagger}\mathbf{H}(0) = \mathbf{U}(\mathbf{v}t)\mathbf{H}(0) \tag{11}$$

where the temporal evolution operator

$$\mathbf{U}(\mathbf{v}t) = \frac{N_L}{N_R}\mathbf{B}\mathbf{D}(\mathbf{v}t)\mathbf{B}^{\dagger} \tag{12}$$

The connection between angular and temporal spectra is well known and has been used to derive fading statistics both theoretically [5] and in processing of measurements [8],[9]. In a practical receiver, the channel matrix **H**(t) must be estimated, usually from pilot or training signals. For a rapidly time varying channel, the channel must be determined at frequent intervals. While this is true in a general time-varying channel, the structure (7) of temporal evolution of the **H**(t) matrix may be exploited. In fact, if the channel is known at time t=0, it may be determined at a later time t through applying the evolution operators (12) in (11). Note that the evolution operators are not functions of the environment in a sense that the proper inverse beam forming matrix **B** is defined by the array geometry and the immediate vicinity of the mobile array (e.g. interior of a car). Similarly, the Doppler shift matrix **D** depends on vehicle velocity alone.

Alternatively, the receiver processing may include the evolution of the **H** matrix explicitly. The vector of received signals **r** in a MIMO system is related to the transmitted signal vector **s**:

$$\mathbf{r} = \mathbf{H}(t)\mathbf{s} + \mathbf{n} = \frac{N_L}{N_R}\mathbf{B}\mathbf{D}(\mathbf{v}t)\mathbf{B}^{\dagger}\mathbf{H}(0)\mathbf{s} + \mathbf{n} \tag{13}$$

where **n** is the vector of received noise signals. A vector of virtual receiver signals **r''** may be constructed by applying a sequence of operations to the received signal:

$$\mathbf{r}'' = \mathbf{D}^{\dagger}(\mathbf{v}t)\mathbf{B}^{\dagger}\mathbf{r} = \mathbf{B}^{\dagger}\mathbf{H}(0)\mathbf{s} + \mathbf{n}' \tag{14}$$

where $\mathbf{n}'$ is the equivalent receiver noise. The rank of both $\mathbf{B}^{\dagger}$ and **H**(0) may not exceed N_L. Note that the transformed received signal **r''** is related to the transmitted signal through a non-time varying channel. The capacity of such a channel is therefore constant with time as well. This is utilized below for the more practical case of a finite number of receivers at the mobile and no restriction on the number of plane waves present.

IV. Receiver processing for finite number of mobile antennas

In the case of a finite number of antennas, equation (7) still applies, but the transformation $\mathbf{B}^{\dagger}$ used in (13) does not generally have orthogonal rows. It is known, however, that any non-singular transformation of the received signal vector does not alter capacity, because the same transformation is applied both to the desired signal and noise. Applying the beam forming transformation $\mathbf{B}_1^{\dagger}$:

$$\mathbf{B}_1^{\dagger} = \frac{1}{\sqrt{N_R}}\begin{pmatrix} e^{-i\mathbf{K}_1\bullet\mathbf{r}_1} & e^{-i\mathbf{K}_1\bullet\mathbf{r}_2} & \bullet \\ e^{-i\mathbf{K}_2\bullet\mathbf{r}_1} & e^{-i\mathbf{K}_2\bullet\mathbf{r}_2} & \bullet \\ \bullet & \bullet & \bullet \end{pmatrix} \tag{15}$$

to the receive signal vector **r**, a vector of virtual received signals **r'** is constructed:

$$\mathbf{r}' = \frac{1}{\sqrt{N_L}} \begin{pmatrix} \sum_{\mathbf{k}} A_1(\mathbf{k}) G(\mathbf{k} - \mathbf{K}_1) e^{i\mathbf{k}\cdot\mathbf{v}t} & \sum_{k} A_2(\mathbf{k}) G(\mathbf{k} - \mathbf{K}_1) e^{i\mathbf{k}\cdot\mathbf{v}t} & \bullet \\ \sum_{\mathbf{k}} A_1(\mathbf{k}) G(\mathbf{k} - \mathbf{K}_2) e^{i\mathbf{k}\cdot\mathbf{v}t} & \sum_{\mathbf{k}} A_2(\mathbf{k}) G(\mathbf{k} - \mathbf{K}_2) e^{i\mathbf{k}\cdot\mathbf{v}t} & \bullet \\ \bullet & \bullet & \bullet \end{pmatrix} \mathbf{s} + \mathbf{B}_1^{\dagger}\mathbf{n} \quad (16)$$

where the complex amplitude response G of the receive array to the plane wave arriving from direction **k** is defined as

$$G(\mathbf{k} - \mathbf{K}_\mathbf{p}) = \frac{1}{\sqrt{N_R}} \sum_{n=1}^{N_R} e^{-i(\mathbf{k} - \mathbf{K}_p)\cdot\mathbf{r}_n} \quad (17)$$

when the receive antenna elements are phased so as to produce a maximum in the direction $\mathbf{K}_\mathbf{p}$. Note that $G(0) = \sqrt{N_R}$. The directions of $\mathbf{K}_\mathbf{p}$ may be chosen so as to make (15) unitary. $\mathbf{B}_1^{\dagger}$ is an $N_K \times N_R$ matrix, where N_K is the number of beam outputs. The number of information bearing receive signals is N_R and, therefore, the rank of the beam forming matrix $\mathbf{B}_1^{\dagger}$ must not be smaller than N_R, thus $N_K \geq N_R$. It should be noted, however, that as these beam signals are formed from a combination of N_R receive antenna signals, no additional information may be obtained if $N_K > N_R$. In dealing with a time-varying channel, however, there may be practical advantages to forming a number of receive beams greater than the number of receive antennas.

Changing the variables, (16)may be re-written as

$$\mathbf{r}' = \begin{pmatrix} e^{i\mathbf{K}_1\cdot\mathbf{v}t} & 0 & \bullet \\ 0 & e^{i\mathbf{K}_2\cdot\mathbf{v}t} & \bullet \\ \bullet & \bullet & \bullet \end{pmatrix} \frac{1}{\sqrt{N_L}}$$
$$\times \begin{pmatrix} \sum_{\mathbf{k}} A_1(\mathbf{k} + \mathbf{K}_1) G(\mathbf{k}) e^{i\mathbf{k}\cdot\mathbf{v}t} & \sum_{\mathbf{k}} A_2(\mathbf{k} + \mathbf{K}_1) G(\mathbf{k}) e^{i\mathbf{k}\cdot\mathbf{v}t} & \bullet \\ \sum_{\mathbf{k}} A_1(\mathbf{k} + \mathbf{K}_2) G(\mathbf{k}) e^{i\mathbf{k}\cdot\mathbf{v}t} & \sum_{\mathbf{k}} A_2(\mathbf{k} + \mathbf{K}_2) G(\mathbf{k}) e^{i\mathbf{k}\cdot\mathbf{v}t} & \bullet \\ \bullet & \bullet & \bullet \end{pmatrix} \mathbf{s} + \mathbf{B}_1^{\dagger}\mathbf{n} \quad (18)$$

One can approximate G(**k**) as being $G_0\sqrt{N_R}$ within the half-power beam width B_h, and zero outside this beam width, where G_0 is a scale factor chosen so that $\int d\Omega |G(\mathbf{k})|^2 = 1$ over all solid angles Ω. The equation (18) may then be re-written as

$$\mathbf{r}' = \overbrace{\begin{pmatrix} e^{i\mathbf{K}_1\bullet\mathbf{v}t} & 0 & \bullet \\ 0 & e^{i\mathbf{K}_2\bullet\mathbf{v}t} & \bullet \\ \bullet & \bullet & \bullet \end{pmatrix}}^{\mathbf{D}_1} \frac{G_0\sqrt{N_R}}{\sqrt{N_L}}$$

$$\times \begin{pmatrix} \sum_{k_x=-B_h/2}^{k_x=B_h/2} A_1(\mathbf{k}+\mathbf{K}_1)e^{i\mathbf{k}\bullet\mathbf{v}t} & \sum_{k_x=-B_h/2}^{k_x=B_h/2} A_2(\mathbf{k}+\mathbf{K}_1)e^{i\mathbf{k}\bullet\mathbf{v}t} & \bullet \\ \sum_{k_x=-B_h/2}^{k_x=B_h/2} A_1(\mathbf{k}+\mathbf{K}_2)e^{i\mathbf{k}\bullet\mathbf{v}t} & \sum_{k_x=-B_h/2}^{k_x=B_h/2} A_2(\mathbf{k}+\mathbf{K}_2)e^{i\mathbf{k}\bullet\mathbf{v}t} & \bullet \\ \bullet & \bullet & \bullet \end{pmatrix} \mathbf{s} + \mathbf{B}_1^{\dagger}\mathbf{n} \tag{19}$$

where k_x is the component of the wave vector $\mathbf{k}$ along an arbitrary x-axis, which may be chosen to be along the line of the array in the case of a uniform linear array.

The half-power beam width B_h near the broadside direction (in spatial frequency units m^{-1}) for a uniformly spaced linear array with antenna spacing d, is given by [7]:

$$B_h = 0.886k\frac{\lambda}{N_R d} = 0.886\frac{2\pi}{N_R d} \tag{20}$$

For example, a 4 element array with elements spaced $\lambda/2$ can generate a beam about 25^0 wide. Linear arrays with azimuthally omnidirectional elements will experience front-back ambiguity. It will be assumed here that the array elements are mounted on the back plane of the mobile terminal, thereby eliminating the front-back ambiguity and making equation (19) appropriate. For a large number of mobile antennas, the antennas may be placed in a planar grid array, where proper spacing will reduce the front-back ambiguity. The beam forming transformation (15) produces virtual directional antenna responses (19) , each pointing into a different direction, with generally some overlap between the beams. The channel induced time fluctuation of each beam is separated into a mean Doppler frequency shift, represented by the diagonal Doppler matrix $\mathbf{D}_1$ in (19) and a "slower" residual time fluctuations given by the $e^{i\mathbf{k}\bullet\mathbf{v}t}$ terms in the second matrix factor in (19), for $|k_x| < B_h/2$.

As is done in conventional receivers, the receiver on each beam would attempt to estimate the channel from the mth transmit antenna, perhaps by correlating the received signal with the corresponding pilot signal. Such a procedure produces a sequence in time of channel estimates of the actual channel coefficient

$$h_{nm}(t) = e^{i\mathbf{K}_n\bullet\mathbf{v}t} G_0\sqrt{N_R/N_L} \sum_{k_x=-B_h/2}^{k_x=B_h/2} A_m(\mathbf{k}+\mathbf{K}_1)e^{i\mathbf{k}\bullet\mathbf{v}t} = e^{i\omega_n t} h_{nm}^{lowpass}(t) \tag{21}$$

as follows from (19). The channel coefficient is seen to have the structure of a band pass signal, whose center frequency is the mean Doppler shift $\omega_n = \mathbf{K}_n \bullet \mathbf{v}$ suffered by the nth beam, which may be observed as a phase drift across the sequence of channel estimates on the nth receive beam. The estimation of the mean Doppler shift ω_n on each beam is then a problem of estimating the center frequency of a band pass process, which may be done, for example, by carrying out a linear least mean squares fit to the phase of the estimates of channel coefficient (21). Thus, while the Doppler shifts ω_n depend on the mobile velocity, the velocity need not be estimated explicitly. Given an estimate of the mean Doppler shift for each beam, a Doppler compensating matrix $\mathbf{D}_1^{\dagger}(\mathbf{v}t)$ may be formed and applied to the virtual received signal $\mathbf{r}'$:

$$\mathbf{r}'' = \mathbf{D}_1^{\dagger}(\mathrm{v}t)\mathbf{r}' = \frac{G_0\sqrt{N_R}}{\sqrt{N_L}} \begin{pmatrix} \sum\limits_{k_x=-B_h/2}^{k_x=B_h/2} A_1(\mathbf{k}+\mathbf{K}_1)e^{i\mathbf{k}\bullet\mathrm{v}t} & \sum\limits_{k_x=-B_h/2}^{k_x=B_h/2} A_2(\mathbf{k}+\mathbf{K}_1)e^{i\mathbf{k}\bullet\mathrm{v}t} & \bullet \\ \sum\limits_{k_x=-B_h/2}^{k_x=B_h/2} A_1(\mathbf{k}+\mathbf{K}_2)e^{i\mathbf{k}\bullet\mathrm{v}t} & \sum\limits_{k_x=-B_h/2}^{k_x=B_h/2} A_2(\mathbf{k}+\mathbf{K}_2)e^{i\mathbf{k}\bullet\mathrm{v}t} & \bullet \\ \bullet & \bullet & \bullet \end{pmatrix} \mathbf{s}$$

$$+\mathbf{D}^{\dagger}(\mathrm{v}t)\mathbf{B}_1^{\dagger}\mathbf{n}$$

$$= \mathbf{D}_1^{\dagger}(\mathrm{v}t)\mathbf{B}_1^{\dagger}\mathbf{r} \tag{22}$$

The above may be interpreted as an equivalent received signal, with "slower" time fluctuations, characterized by the Doppler bandwidth

$$\Delta f = \frac{B_h \mathrm{v}}{2\pi} = \frac{1.8\mathrm{v}}{N_R \lambda} \tag{23}$$

where (20) was used. Note that the Doppler bandwidth is inversely proportional to the number of mobile antennas. More frequently coherence time is used as a measure of temporal scale of channel fluctuations, as discussed in the next section.

As the "slowing" of the channel is achieved by a sequence of non-singular operators, the capacity of the effective channel experienced by the virtual received signal (22), has not been changed by the beam forming and Doppler compensation. This procedure has been demonstrated in computer simulations of a time-varying MIMO link in the presence of noise, as will be reported in a later publication.

The above "slowing down" procedure requires only an estimate of the mean Doppler shifts per beam to define the diagonal Doppler compensation matrix $\mathbf{D}_1^{\dagger}$ in (22). The beam forming matrix $\mathbf{B}_1^{\dagger}$ may be formulated for an arbitrary set of distinct beams, a procedure that requires only the knowledge of the relative disposition of the mobile array elements, independent both of the array orientation and the incident angular spectrum.

Due to the multiple scattering, beam forming at the base does not, in general, lead to a reduction in the width of the angular spectrum perceived at the terminal, and thus, no reduction in Doppler bandwidth.

The technique described above does not rely on any particular spatial distribution of scatterers. In particular, no assumption of clustering of scatterers is made. Furthermore, the beam forming weights are not adapted to the external angular spectrum but rather fixed. The actual signals received on the various beams are, of course, determined by the incident angular spectrum.

The channel "slowing" is demonstrated in Figures 1-4. In Figure 1 the beam patterns of four beams that may be formed by a linear array of 4 antennas are shown. The antenna elements are assumed separated by $\lambda/2$ and placed on a back plane (e.g. back of a laptop), restricting the beam patterns to a half space. The beams are mutually orthogonal, except for beam 4, pointing at 40^0, chosen so as not to point in the end-fire direction where the beam is naturally broader. The total Doppler spectrum observed at one antenna is plotted in Figure 2. The plane waves are assumed to be equiprobable over a half space, resulting in only positive frequencies. In this simulation the array is moving with the broadside direction along the direction of motion. The Doppler spectra on the four beam-formed signals are plotted in Figure 3, exhibiting the band limited property seen in (19). Beam 4 has the broadest frequency spread, as it is closest to the end-fire direction where the beams are naturally broader. Finally, the real part of the time-varying channel transfer coefficient as observed on one antenna is compared in Figure 4 to the real part of the "slowed down" channel transfer coefficient as seen on one of the beam signals after the Doppler compensation, as in (22). The rate of channel fluctuation is significantly reduced.

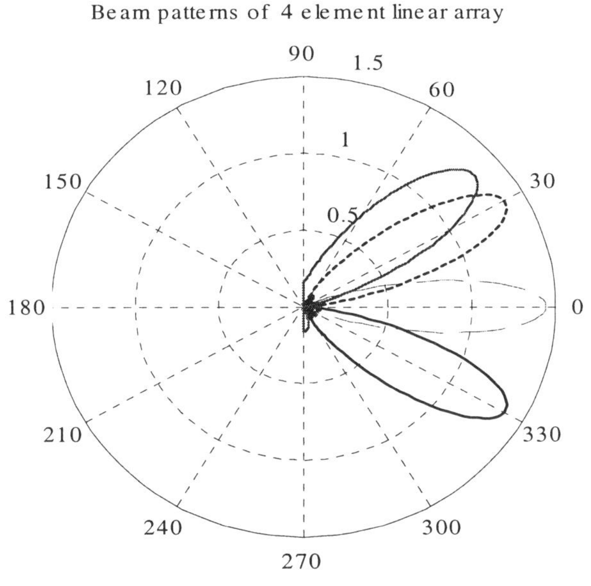

Figure 1. Beam patterns for a 4 antenna linear array, with antenna elements placed on a back plane and spaced by $\lambda/2$.

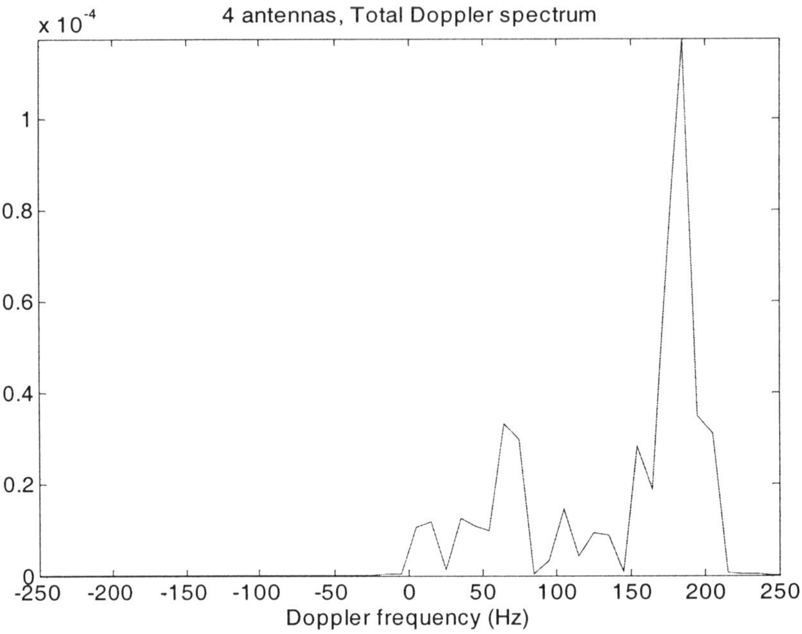

Figure 2. Total Doppler spectrum observed at 1 antenna at 60 mph.

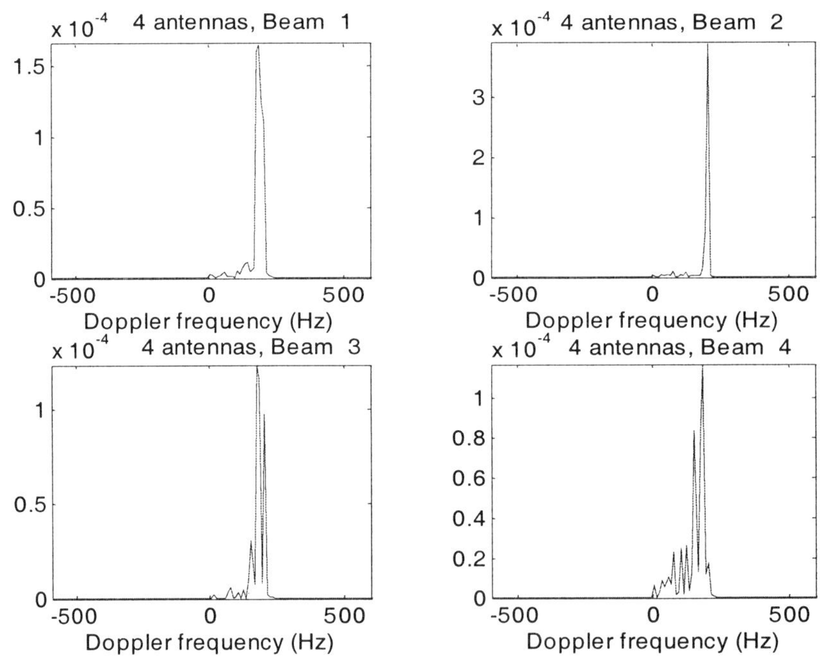

Figure 3. Doppler spectra on 4 beam signals at 60 mph.

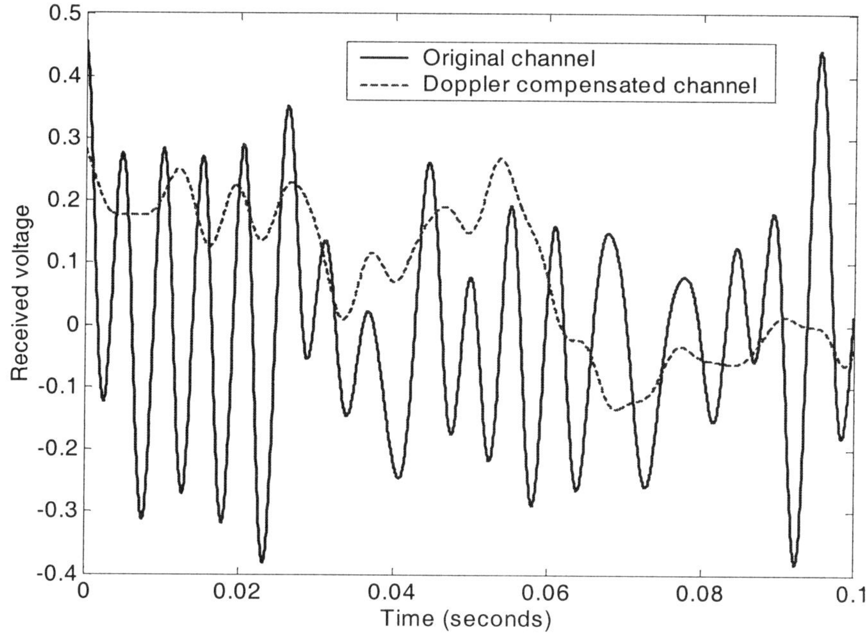

Figure 4. Channel time-fluctuation at 60 mph: original channel as seen on one antenna and beam-formed Doppler compensated channel on one of the beams.

V. Coherence time for a finite number of mobile antennas

It is of interest to determine the increase in coherence time experienced by a beam-formed signal as compared to the signal received on an omni-directional antenna. The correlation coefficient of a single **H** matrix entry $h(t)$ for a mobile moving at speed v in the direction indicated by angle ϕ is given by:

$$\langle h(t)h^*(t+\tau)\rangle = \rho(\tau) = \int_{-\pi}^{\pi} e^{ikv\tau\cos(\alpha-\phi)} p(\alpha)d\alpha \tag{24}$$

where the incident power angular spectrum p(α) is normalized : $\int_{-\pi}^{\pi} p(\alpha)d\alpha = 1$ and α is the angle of incidence.

Usually the coherence time is defined as the time τ_c at which the correlation coefficient has decreased to a prescribed value ρ_c from its maximum at τ=0. It may be shown that even a small error in the channel estimate leads to significant additional "channel mismatch noise". We are therefore interested in the case where ρ_c~1. For omnidirectional antennas, assuming the incident power spectrum is omnidirectional as well, $p(\alpha) = 1/2\pi$ and (24) leads to [5]

$$\rho(\tau_u) = \rho_c = J_0(kv\tau_u) \approx 1 - \left(kv\tau_u/2\right)^2 \tag{25}$$

Here the subscript u corresponds to the uniform angular distribution. The prescribed value of the correlation coefficient ρ_c is reached at time τ_u. Equation (25) also applies in the case of power spectrum distributed uniformly over an angular half space centered on the direction of motion. The angular receive pattern of a beam-formed phased array antenna of N_R elements is given by [7] $p(\alpha) = \dfrac{\sin^2\left(\dfrac{N_R kd\sin(\alpha)}{2}\right)}{N_R^2 \sin^2\left(\dfrac{kd\sin(\alpha)}{2}\right)}$. The main lobe of the antenna pattern may be shown to be closely approximated as a Gaussian beam of the same rms angular width for analytical convenience:

$$p(\alpha) = \frac{1}{\sqrt{2\pi}\sigma} \exp\left(-\frac{\alpha^2}{2\sigma^2}\right) \tag{26}$$

For narrow rms angular spread σ, the largest Doppler spread, i.e. the fastest fluctuation, occurs for motion perpendicular to the direction of beam maximum. Correspondingly, setting $\phi=90^0$ in (24), the correlation coefficient becomes

$$\rho(\tau) = \int_{-\pi}^{\pi} e^{ikv\tau\sin\alpha} p(\alpha)d\alpha \approx \int_{-\pi}^{\pi} e^{ikv\tau\alpha} p(\alpha)d\alpha = \exp\left(-(kv\tau\sigma)^2/2\right) \tag{27}$$

The prescribed value of the correlation coefficient ρ_c is reached at time τ_G, obtained by approximating (27) for small τ:

$$\rho_c = \exp\left(-(kv\sigma\tau_G)^2/2\right) \approx 1 - \left(kv\sigma\tau_G\right)^2/2 \tag{28}$$

The increase in coherence time due to beam forming may be obtained by equating (25) and (28), and using (20) to find the ratio of coherence times τ_G and τ_u:

$$\frac{\tau_G}{\tau_u} = \frac{1}{\sqrt{2}\sigma} = \frac{2.4}{\sqrt{2}\alpha_h} = \frac{2.4k}{\sqrt{2}B_h} \approx \frac{2.4N_R d}{\sqrt{2}\,0.886\lambda} \approx \frac{2N_R d}{\lambda} = N_R \tag{29}$$

where α_h is the half-power beam width in radians, B_h is approximated by (20), and the antenna separation d is set to $\lambda/2$. The above equation quantifies how the coherence time of a beam formed antenna increases as the beam narrows. Achievable beam width generally decreases with the number of mobile antennas and depends on the beam steering direction φ. For beams that are steered within 45^0 of broadside, the beam width is reasonably approximated by (20). It may be observed from (20) and (29) that the coherence time increases linearly with the number of mobile antennas. In Figure 5 the correlation coefficient for an omnidirectional antenna is compared to the correlation coefficient for a beam formed antenna with 25^0 half-power beam width pointing at 45^0 with respect to the direction of motion. A beam with this width may be formed by using a linear array of 4 antennas spaced $\lambda/2$ apart. Note from (24) that the coherence time τ appears in the correlation coefficient in a product with mobile speed. Consequently, for a particular value of ρ_c, the increase in coherence time due to beam forming and Doppler compensation at a particular speed may be used to increase the allowed vehicle speed.

This would allow the same level of channel estimation and training performance at higher vehicle speed. In the limit of an infinite number of mobile antennas, the effective channel becomes locally static.

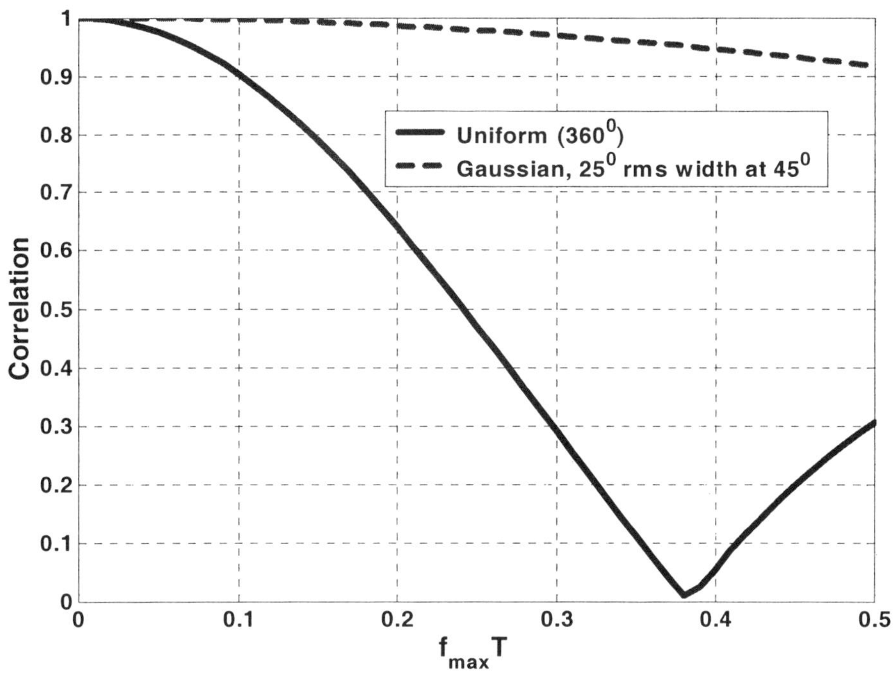

Figure 5. Plot of temporal correlation coefficient vs. normalized time $f_{max}T$, where $f_{max} = f_0 v / c$ is the maximum Doppler shift frequency of the mobile moving with speed v, and T is the actual time.

VI. Optimal number of transmit antennas

It has been pointed out by Marzetta [1], Marzetta and Hochwald [2], and Hassibi and Hochwald [3] that the optimal number of transmit antennas is limited by the coherence time of the channel. This result was obtained in the context of arbitrary form of fading, which was approximated as a block-fading process, of block duration T. It was further assumed that training symbols were time-multiplexed with message symbols. Training issues for continuously fading channels were explored by Sun, Lozano, and Huang [4], again for arbitrary forms of fading processes. This section explores how the fading process implied by (5), (6), and (7) modifies the conclusions reached in these works.

First the argument in [1] is reproduced. Marzetta has shown that the required training interval T_τ grows approximately linearly with the number of transmit antennas M, thus $T_\tau = \alpha M$, for some constant α, and this leaves $T_d = T - T_\tau = T - \alpha M$ time available for sending the data. Foschini [2] has shown that in complex Gaussian *iid* channels, the capacity C in units of bits per symbol is approximately proportional to $\min(M, N_R)$. Here we consider systems where $M \le N_R$, thus $C = \gamma M$ for some constant γ, which depends, among other things, on the SNR and N_R. Taking the view that the need for training imposes an overhead penalty by reducing the time available for sending the data, the total number of data bits that can be sent is equal to $C\frac{T_d}{T} = \gamma M \frac{(T - \alpha M)}{T}$. The optimum number of transmit antennas from the standpoint of maximizing the number of data bits is $M = T/2\alpha$, implying that the training interval is $T_\tau = T/2$.

The above analysis assumes that the coherence time of the channel is a fixed quantity, independent of the antenna array. It follows from (20) and (29) that the coherence time increases linearly with the number of mobile antennas, $T=\beta N_R$. This is approximately true for linear arrays, which are steered in directions within $\pm 45^0$ from broadside. The number of message bits that can be sent becomes

$$C\frac{T_d}{T} = \gamma M \frac{(T - \alpha M)}{T} = \gamma M \frac{(\beta N_R - \alpha M)}{\beta N_R} \tag{30}$$

Setting the number of receive antennas N_R as proportional to the number of transmit antennas M, $N_R = \kappa M$, (30) becomes

$$C\frac{T_d}{T} = \gamma M \frac{(\beta N_R - \alpha M)}{T} = \gamma M \frac{(\beta \kappa M - \alpha M)}{\beta \kappa M} = \gamma M \frac{(\beta \kappa - \alpha)}{\beta \kappa} \tag{31}$$

Equation (31) indicates that capacity increases linearly with number of transmit antennas *M*, provided the number of mobile antennas increases proportionally. In contrast to the result in [1], there is no finite optimum number of transmitting antennas. This argument is tempered by the fact that, in general, mobile motion is not the only source of temporal fading. Motion of scatterers, such as trees swaying in the wind, other vehicles, and pedestrians would induce channel fluctuations that are not represented by the model in (5)-(7). Another source of temporal channel fluctuation is the large-scale fading, experienced by the mobile as it traverses distances, which are larger than the spatial scale over which the angular spectrum may be assumed to be homogeneous (e.g. moving through a city block). These effects would still limit the optimum number of transmit antennas, as per [1], albeit on a much longer temporal scale. Nevertheless, it is generally recognized that mobile motion is the primary source of rapid temporal channel fluctuation, and MIMO channel estimation limits the maximum allowed vehicle speed. Reduction of perceived speed through beam forming and Doppler compensation would alleviate this constraint. As the number of mobile receive antennas increases, the number of channel coefficients that is estimated increases correspondingly. This presents at most a computational burden, but not a fundamental difficulty in estimation, as each beam space receiver needs to estimate channels from M transmitters and a single Doppler frequency. Note that for each mobile receive beam, all transmit antennas suffer the same mean Doppler shift, thus the need to estimate only a single Doppler frequency per mobile receive beam, regardless of the number of base transmit antennas. To illustrate the easing of the estimation difficulties, the number of parameters that needs to be estimated per unit time is examined. In the absence of beam forming, each receiver has to estimate M complex channel coefficients in the coherence time $T = \lambda/2\text{v}$ of the channel, so the "learning rate" is 2M/T real numbers/second for each receiver. In contrast, each of the beam space receivers needs to estimate the coefficient of the "slowed" channel at the rate of $2M/N_R T$ ($=2/T$, for $N_R = M$) real numbers/second and a single mean Doppler frequency. The total number of parameters to estimate per second is 2M/T, but additional receivers bring more received signals, thus increasing the computational burden but not the estimation error. The Doppler frequency changes at the rate of vehicle acceleration/deceleration, a slow rate neglected here.

VII. Conclusions

In this work, the MIMO channel is represented in terms of a sum of plane waves incident at the mobile. The channel transfer matrix is found to factor into time-dependent and time-independent parts. It is then proposed that the received signals are first pre-processed so as to produce signals that fluctuate on a much slower time scale. The pre-processing consists of beam forming and Doppler compensation for signals received on each beam. Both operations are non-singular and do not alter the capacity of the MIMO channel. This is shown to effectively "slow down" the channel fluctuation due to mobile motion. The coherence time of the pre-processed channel is found to increase by a factor on the order of the number of mobile receive antennas. In the limit of an infinite number of mobile receive antennas, the effective MIMO channel is shown to become locally static. These results remove constraints on the number of transmit antennas and allowed vehicle speeds imposed by training requirements. The "slowed down" channel may also be fed back to the transmitter, allowing an increase in information throughput.

VIII. Acknowledgments

I am grateful to Constantinos Papadias and Jerry Foschini for early discussions that proved invaluable to this work. Many thanks are due to Angel Lozano and Jack Saltz for their helpful comments on the draft of this work. Extensive discussions with Sivarama Venkatesan, Laurence Mailender and Reinaldo Valenzuela were crucial in assessing the applicability of this work. It is a pleasure to acknowledge discussions with Mike Gans, Ilya Korisch, Hao Xu, Jonathan Ling, and Dragan Samardzija, as well as useful suggestions by reviewers.

IX. References

[1] T. L. Marzetta, "BLAST Training: Estimating Channel Characteristics for High Capacity Space-Time Wireless", Proc. 37th Annual Allerton Conference on Communication, Control, and Computing, Monticello, IL, Sept.22-24, 1999.

[2] T. L. Marzetta, B. M. Hochwald, " Fundamental limitations on multiple-antenna wireless links in Rayleigh fading", *Proceedings. of the IEEE International Symposium on Information Theory*, 1998, p. 310

[3] B. Hassibi, B. M. Hochwald," How Much Training is Needed in Multiple-Antenna Wireless Links?", Submitted to IEEE Trans. Information Theory

[4] Q. Sun, D.C. Cox, A. Lozano and H.C. Huang, "Training-Based Channel Estimation for Continuous Flat Fading BLAST", ICC 2002, April 2002, New York City, NY.

[5] W.C. Jakes, Microwave Mobile Communications, IEEE Press, 1974

[6] D. Chizhik, G. J. Foschini and R. A. Valenzuela, "Capacities of Multi-element Transmit and Receive Antennas: Correlations and Keyholes", *IEE Electronic Lett.*, June 22, 2000, V. 36, No. 13, p. 1099-1100.

[7] W.L. Stutzman and G.A. Thiele, Antenna Theory and Design, J. Wiley and Sons, 1981.

[8] A. L. Moustakas, S. Arunachalam, Kam H. Wu, and H. Heller, "Wideband Spatio-Temporal Characterization of the Urban PCS Channel ", *International Symposium on 3G Infrastructure and Services*, Athens, Greece, July 2001.

[9] H. Xu, M. Gans, D. Chizhik, J. Ling, P. Wolniansky, and R. Valenzuela, "Measurements and Theory of Spatial and Temporal Variations of MIMO Channels", submitted to JSAC, 2001.

[10] D. Smardzija and N. Mandayam, "Pilot Assisted MIMO Channel Estimation and Achievable Rates", DIMACS Workshop on Signal Processing for Wireless Transmission, October 7-9, 2002, DIMACS Center, Rutgers University, Piscataway, NJ

[11] Da-Shan Shiu, G. J. Foschini, M. J. Gans, J. M. Kahn, "Fading correlation and its effect on the capacity of multielement antenna systems", *IEEE Transactions on Communications* ,vol. 48 Issue: 3 , March 2000, pp. 502 -513

[12] D. Chizhik, F. Rashid-Farrokhi, J. Ling, A. Lozano, "Effect of Antenna Separation on the Capacity of BLAST in Correlated Channels", *IEEE Communication Letters*, Vol. 4, No. 11, November 2000, pp. 337-339.

[13] K.I. Pedersen, J.B. Andersen, J.P. Kermoal, P. Mogensen, "A stochastic multiple-input-multiple-output radio channel model for evaluation of space-time coding algorithms",Vehicular Technology Conference, 2000. IEEE VTS Fall VTC 2000, 52nd , vol.2 , 2000 pp. 893 -897

[14] A. M. Sayeed, "Deconstructing multiantenna fading channels", *IEEE Transactions on Signal Processing*, vol.50 Issue: 10 , Oct. 2002, pp. 2563 -2579

[15] M. Steinbauer, A. F. Molisch, E. Bonek, "The double-directional radio channel", *IEEE Antennas and Propagation Magazine* , vol. 43, Issue: 4 , Aug. 2001, pp. 51 –63.

[16] O. Norklit, R. G. Vaughan, "Angular partitioning to yield equal Doppler contributions", *IEEE Transactions on Vehicular Technology*, vol. 48 Issue: 5 , Sept. 1999, pp. 1437 –1442.

[17] G. J. Foschini and M. J. Gans, "On Limits of Wireless Communications in a Fading Environment When Using Multiple Antennas", Wireless Personal Communications, Volume 6, No. 3, March 1998, p. 311.

[18] G. J. Foschini, "Layered Space-Time Architecture for Wireless Communication in a Fading Environment When Using Multiple Antennas", Bell Labs Technical Journal, Vol. 1, No. 2, Autumn 1996, pp 41-59.

[19] H. Asplund, A. F. Molisch, M. Steinbauer, N. B. Mehta, "Clustering of scatterers in mobile radio channels-evaluation and modeling in the COST259 directional channel model", ICC 2002. IEEE International Conference on Comm.,vol. 2, 2002, pp. 901–905.

[20] D. Gesbert, H. Bolcskei, D. Gore, A. Paulraj, "MIMO Wireless Channels: Capacity and Performance Prediction", CT10-5, *IEEE Globecom 2000*, Nov. 27-Dec. 1, 2000, San Fran., CA

[21] D. Chizhik, G. J. Foschini, M. J. Gans, R. A. Valenzuela, "Keyholes, correlations, and capacities of multielement transmit and receive antennas", *IEEE Transactions on Wireless Communications*, vol. 1(2), April 2002, pp. 361 –368.
[22] H. Xu, R. A. Valenzuela, D. Chizhik, and H. C. Huang, "Broadband MIMO Channel Model for Realistic Capacity and System Performance Evaluations", *DIMACS Series in Discrete Mathematics and Theoretical Computer Science*, 2002.

LUCENT TECHNOLOGIES, BELL LABORATORIES, HOLMDEL, NJ
E-mail address: chizhik@lucent.com

DIMACS Series in Discrete Mathematics
and Theoretical Computer Science
Volume **62**, 2003

A Spacetime Architectural Superstructure Enabling Efficient Multiple Antenna Communication

G. J. Foschini and M. Sellathurai

ABSTRACT: We consider an (M,N) wireless link, (M transmit antennas and N receive antennas), impaired by additive white Gaussian noise. The transmitter, which is subject to a power constraint, does not know the outcome of the random matrix channel which has a static and flat frequency characteristic. It does know the channel statistics. The link operates at a limit on the probability of outage. We describe the meaning of stratifying diagonal layers of spacetime to express a message for efficient communication with limited receiver complexity. The special message arrangement enables the receiver to substantially mute self-interference caused by multipath. Although the transmit signal is M-dimensional (M-D), an explosion of processing complexity in the spatial domain is avoided since the architecture uses only 1-D codecs.

We investigate examples in important downlink categories, showing that the message architecture can be extremely efficient. For all (M,1) systems it is maximally efficient. At 10% outage, and with matrix Rayleigh channels with 10dB average SNR, we see that an (8,3) and (4,2) system can, operate at over 97% and 88% of Shannon capacity respectively.

1. INTRODUCTION

1.1. OBJECTIVE. We explore narrowband (M,N) wireless communication links (M transmit antennas and N receive antennas) impaired by additive white Gaussian noise (AWGN). The transmitter, which is subject to a power limit, transmits with equal power out of each of its M transmit antennas. We also assume a "long burst" context. In this setting the view is that when the channel is used from time to time, its transfer characteristic holds constant for the communication burst, yet the channel can change significantly from one use to the next. It is assumed that, when using such a channel, the transmitter does not know the spectrally flat N by M matrix transfer characteristic. The transmitter only knows the channel matrix statistics. For each burst, a large number of symbols are sent, so that we use the standard infinite time horizon view common in information theory.

The channel is allowed to be in an outage state for a small percentage, say X%, of its realizations. Without knowledge of the individual channels, the (M,N) link, can, in principle [1,2] operate at the X% capacity level for the remaining channel outcomes. In this paper the (M,N) system will be constrained by the stipulation that despite the spatially M-dimensional (M-D) transmit signal, an explosion of processing complexity in the spatial domain must be avoided. We will investigate the extreme case of requiring communication with parallel spatially 1-D codecs. In Section 2.0, other basic constraints will be imposed as well.

We will suggest a means for expressing signals so that the encoded message is disposed in spacetime so that the receiver can substantially mute self-interference. After describing the communication architecture in Section 2.0, we analyze it 3.0 and 4.0 and then study it numerically for matrix Rayleigh channels in 5.0. We will see examples where such an architecture, when used in certain important downlink categories, like when M is substantially in excess of N, can operate at a specified outage level with near maximum efficiency.

The study of (M,N) systems is a very active area of research: [3-11] are sample references. Additional related papers are cited as we proceed. In contrast to many of the references, we are primarily concerned with a message's architectural superstructure so that the specific coding and modulation used in the fundamental 1-D space-time building blocks that we will introduce will not be our direct concern. A few of the references do deal with architectural superstructures: [12] is concerned with general (M,1), while references [13-14] and [15], investigate (2,1), and (4,1) respectively.

1.2. THE VECTOR CHANNEL. We take a baseband equivalent view of an *(M,N)* wireless communication link. The transmitted *M*-D vector signal has components denoted $s_i(t)$, $i = 1,\ldots M$ that are nominally complex, statistically independent white Gaussian signals of equal power, *P/M*. So the total radiated power is *P*. These *M* components of a vector process, $s(t)$, are transmitted over a noisy, spectrally flat, *N* by *M* complex matrix channel, *G*, that can cause the *M* transmitted signal components to interfere with each other. This self-interfering means of communication, is described by the following vector equation for the received *N*-D signal, $r(t)$ in terms of $s(t)$

$$r(t) = Gs(t) + \nu(t). \tag{1.1}$$

The *N*-D vector, $\nu(t)$, represents the complex AWGN impairment. We assume that $\nu(t)$ is both temporally and spatially white. For convenience we will often take time to be discrete, with each clock tick corresponding to the time it takes for exactly one coded vector symbol to be received (or alternately, sent).

When convenient, for simplifying our analysis we redefine the vectors in (1.1) to be normalized. Then, each component is divided by σ, the standard deviation of an additive noise component. With this normalization the noise variances are unity and the signal power radiated from each transmit antenna is $P/(M\cdot\sigma^2)$ instead of *P/M*. The signal $s(t)$ has spatially and temporally white Gaussian components representing the limit of a bandwidth efficient, forward error protected, encoded vector signal. We will architect the spacetime disposition of various basic building blocks that are put together to compose this white Gaussian vector process.

The channel matrix is perfectly known at the receiver: it takes vanishingly small rate to probe the channel, so that the receiver learns it with arbitrary accuracy without detracting from capacity. The Shannon capacity of the channel described by (1.1) is (1.2)

$$C = \log_2 \det\left[I_N + (P/(\sigma^2 M))GG^\dagger \right]. \tag{1.2}$$

In (1.2), I_N is the N by N identity matrix and det means determinant. The formula [1,2,8-10,16,17], termed the LogDet capacity formula can be derived from fundamental information theory considerations, as, e.g., appear in [18]. It holds when the transmitter only knows the value C, and not the actual entries in G. Then, with a suitably encoded signal, s, the capacity, C, can be "attained". More precisely, in the limit of a sequence of progressively more powerful codes, error free transmission at rate C can be approached to within an arbitrarily small deficit.

The LogDet capacity will be our target. Our primary interest is to explore how close we can come to the LogDet capacity of an (M,N) link in important downlink situations after we impose additional structure on the communication architecture.

2. STRATIFIED LAYERING TO REDUCE INTERFERENCE

2.1. DIAGONAL LAYERING OF SPACETIME. In our architectural superstructure for a received message, space is discrete with M components, $m = 1, 2, \ldots M$, corresponding to labeling of the transmit antennas. For the time it takes to receive (send) a coded message, according to equation (1.1) we can view spacetime as a rectangle. We will primarily be interested in the limit of messages of infinite duration.

Reference [1] offered a look at a means of communication processing that used the concept of imaginary "diagonals" passing downward and to the right through spacetime. Communication of this type, which effectively involves a spatially 1-D channel has been described as D-BLAST for diagonal Bell Labs layered space-time. The 1-D channel is an AWGN channel, but of the type whose noise power changes with time [19,20]. We will briefly highlight some of the ideas from [1]. Then we will introduce a more refined way to utilize spacetime.

Examples of diagonals are shown in Figure 1 for the case of five transmit antennas. These diagonals arise in describing the spacetime disposition of 1-D encoded message constituents that are cycled over the antennas that they radiate from. The transmitter, not knowing the channel, does not know which transmit antennas enjoy stronger propagation paths to the receive array and which suffer weaker paths, so, by cycling, the signal is hedged. These diagonals are referred to as diagonal layers (layers of spacetime). Importantly, these diagonals offer a means of muting mutual interference, while enabling 1-D processing at both transmit and receive ends. As explained in the reference, a diagonal layer can amount to one coded block with a downward and to the right time sense. Upon detection these diagonal layers are "peeled off" one by one, left to the right. At any time, signal constituents associated with previously peeled diagonals have been removed as a source of interference in detecting bits in subsequent diagonals. Interference from constituents in diagonals that will be peeled off later are muted using maximum signal to noise plus interference type processing. If the received signal to noise ratio is large enough, zero forcing is an option that is essentially as good.

The processing just outlined can be described as decision feedback in spacetime. As explained in reference [1], in many interesting cases C can be approached arbitrarily closely with 1-D processing in circumstances when the transmit array *does not know* the channel matrix. It was recently pointed out, [21], that this is so for N by M matrix channels at any SNR. The proof is an immediate consequence of combining D-BLAST, where we code over transmit antennas, with a method dealing with related geographically dispersed single antenna in a multiple access context. Unlike D-BLAST, in the multiple access case, bit-rate per transmit antenna information is fed back to each transmit antenna [22-24] as it makes no sense to code over the transmit antennas because they are dispersed.

Figure 1 shows two of many possibilities for diagonalization. There are two practical reasons for preferring the short over the long diagonal option. At the start of a message, [1,2] and once again at the end, there is wasted space-time and we see the waste is less for the case of shorter diagonals. Also, the channel coherence time can be a concern if the duration of a coded block begins to approach a significant fraction of the coherence time. This concern arises because the diagonal duration approaching the coherence time undermines the assumption of a constant channel during a message. We stress that these two concerns reflect practical, not theoretical, concerns. Strictly speaking, they are inconsequential under the idealizing assumptions that the channel is not time varying and the message is infinitely long.

For the two practical reasons just mentioned, it would appear that coding over a short diagonal is preferred. However, a countervailing practical issue arises. Namely, one cannot use too short a diagonal as that does not allow adequate time to support a powerful bandwidth efficient code with forward error correcting capabilities. Also, the signal to noise ratio changes as one moves down the diagonal and that makes for a nonstandard coding environment. We will confront these issues.

Figure 1: Two Examples of Diagonal Layering of Spacetime for the Same Message Duration. Space is composed of five transmit antennas. Every fifth diagonal layer is shown in both figures, including the first and last diagonal layer of a message. In the lower figure each diagonal has a longer time duration and therefore wastes more space-time both at the start and end of an encoded message.

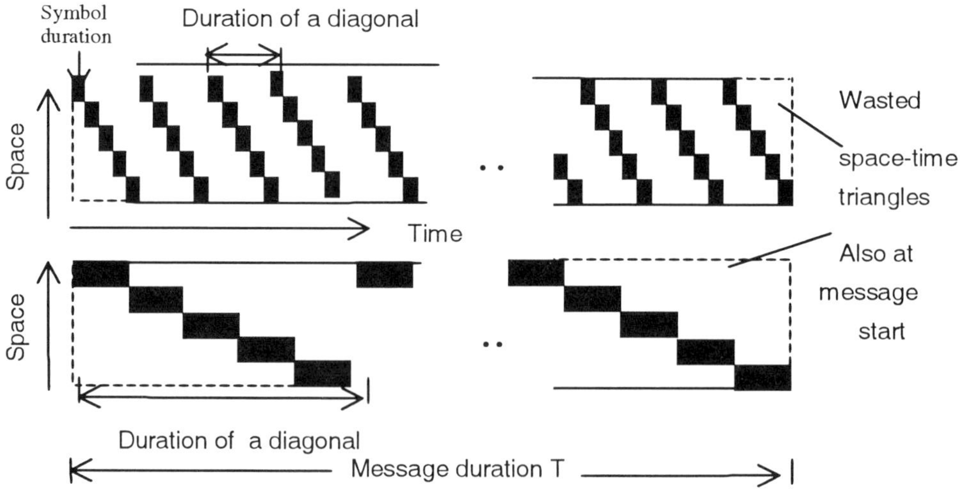

2.2. STRATIFIED DIAGONAL LAYERS TO MUTE INTERFERENCE. Toward overcoming these difficulties, we introduce a method of communication drawing on a more refined view of spacetime. We call the technique SD-BLAST for Stratified Diagonal BLAST.

As with D-BLAST, the method will involve cycling spatially 1-D coded/modulated signal constituents periodically over the M transmit antennas. In accord with this cycling we take the integer labels for transmit antennas to be the integers mod (M). It will also be convenient to sometimes view time, which is also finite, as modular. Since, we are interested in the limit of long messages, it will be harmless to assume that the number of vector symbols in a message, T, is a multiple of M. So the spacetime occupied by a message is viewed as the 2-D surface of a torus as shown in Figure 2. In place of successive diagonal layering of spacetime, as in [1], we think instead of layering toroidal spacetime with M helices perfectly winding around and covering the torus (two of the five helices are noted at the bottom of the figure. (In effect a long diagonal has been replaced by a helix that winds T/M times around the torus.) The M congruent helices are offset from each other in time, by the time to send one symbol. These M helices also have an internal n-fold stratified spatial structure as we next explain.

It will be useful to consider the toroidal space-time surface as plied so that the surface is actually n+1 nested, infinitely thin, tori. There are n+1 plies: n signal plies plus one AWGN ply. A message will be expressed as M·n 1-D coded blocks and each block occupies exactly one out of the M helices in one of the n signal plies. The M helices which, taken collectively, perfectly wind around and pave a toroidal ply are called strata. We can think of the plies as energized by received signal constituents, except that the last innermost ply is energized by AWGN. A received signal is composed of M diagonal layers and each diagonal layer is composed of n helices, or, what is the same, n strata.

Figure 3 shows, at a high level, the way that a 5-D transmitted message corresponding to the space-time structure of Figure 2 is composed. First, a primitive bit stream is demultiplexed into five separate streams of equal rate. Then each stream is further demultiplexed into four independently encoded substreams. Later we will see that there is flexibility in choosing the substream rates. We will assume each substream is modulated and encoded to compose four subsignals of equal power but different rates. The equal power assumption is to streamline the mathematical proof of our main result, reporting that in many cases, close to LogDet capacity can attained with 1-D codes.

For codes with a time sense, maximizing the number of helical windings around the torus, has all the strong – compensates – weak opportunities quickly (according to the processing clock) encountered and accomplished. This hastens the bit decision process. While diagonalization accomplishes hedging, as we shall see, the stratification feature enables muting of interference into any one of the inner plies.

The diagonals are themselves composed of diagonal layers having the same SE directed slope. These layers within the diagonal layers are the strata. Corresponding to Figure 3 there are four strata per diagonal layer. Within each of the five diagonal layers, each of the four strata has a different pattern. So there are a total of twenty distinct strata for the system in Figure 3. There are five additional strata within the AWGN ply.

At this point we suggest thinking of each basic strata constituent (i.e., smallest rectangular block) coordinatized by its shade of gray and pattern at time, t, as merely an accommodation in space-time for a basic signal constituent. As noted in Figure 2, will also keep time with a second separate clock called the processing clock.

FIGURE 2: VIEW OF SPACETIME AT RECEIVER

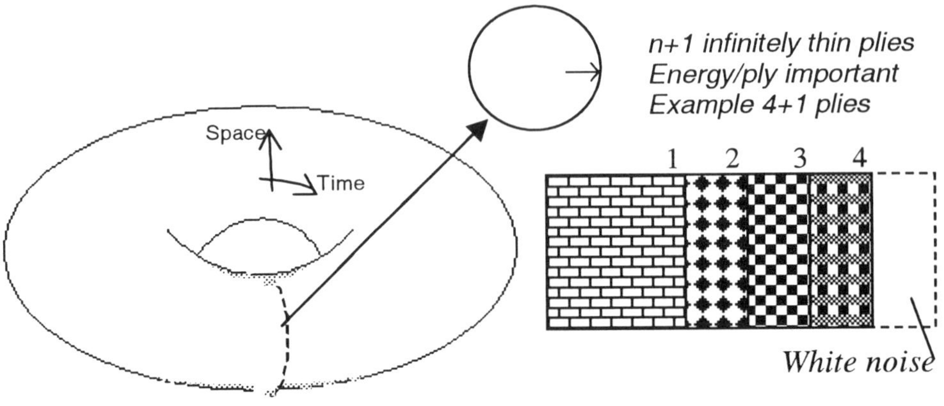

Each of the n plies is helically (diagonally) stratified. Loops taken by two of the five helices in the top ply are shown below for the case of five transmit antennas:

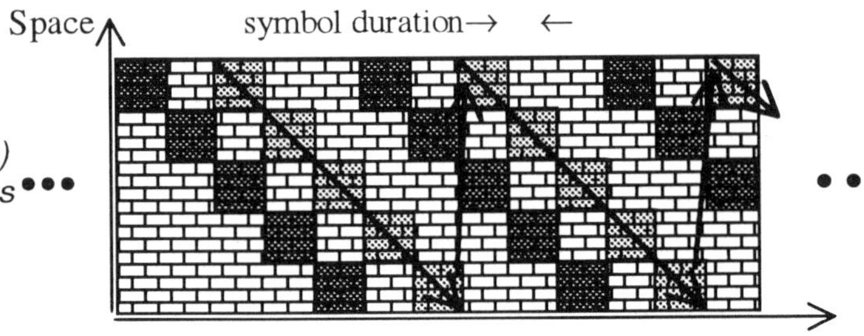

The helical (saw tooth) directed line expresses the time arrow for processing time.

3. DETECTION

In this section we will give a high level somewhat qualitative description of the form of processing that is involved in deciding bits. Then, in the next section, we give an information theory analysis showing that the stratified toroidal structure can, in certain important cases "attain" or come very close to "attaining" LogDet performance.

At the receive array, the encoded bits are detected as follows. Each ply is to be thought of as an annulus like the ring of an onion slice. See Figure 4. The outer brick patterned ply, is detected first. Detection is then followed by removal of the signals in this ply as a source of interference. Prior to removal, this outermost ply, which like all the four plies hold five strata, is said to be exposed. In the process of detecting the bits in the strata in this exposed ply, each brick patterned stratum can be, but need not be, processed simultaneously and separately. After all the bits in five strata that make up the outside ply are detected error free, the interference corresponding to this ply is then subtracted from the received signal vector. The diamond ply is then exposed. The bits in the five diamond strata are then detected and then the diamond strata signals are cancelled exposing the checkered ply, and so on, until finally the plaid patterned ply is exposed and the bits in its five strata detected.

FIGURE 3: STRATIFIED DIAGONAL BLAST (SD-BLAST) TRANSMITTER

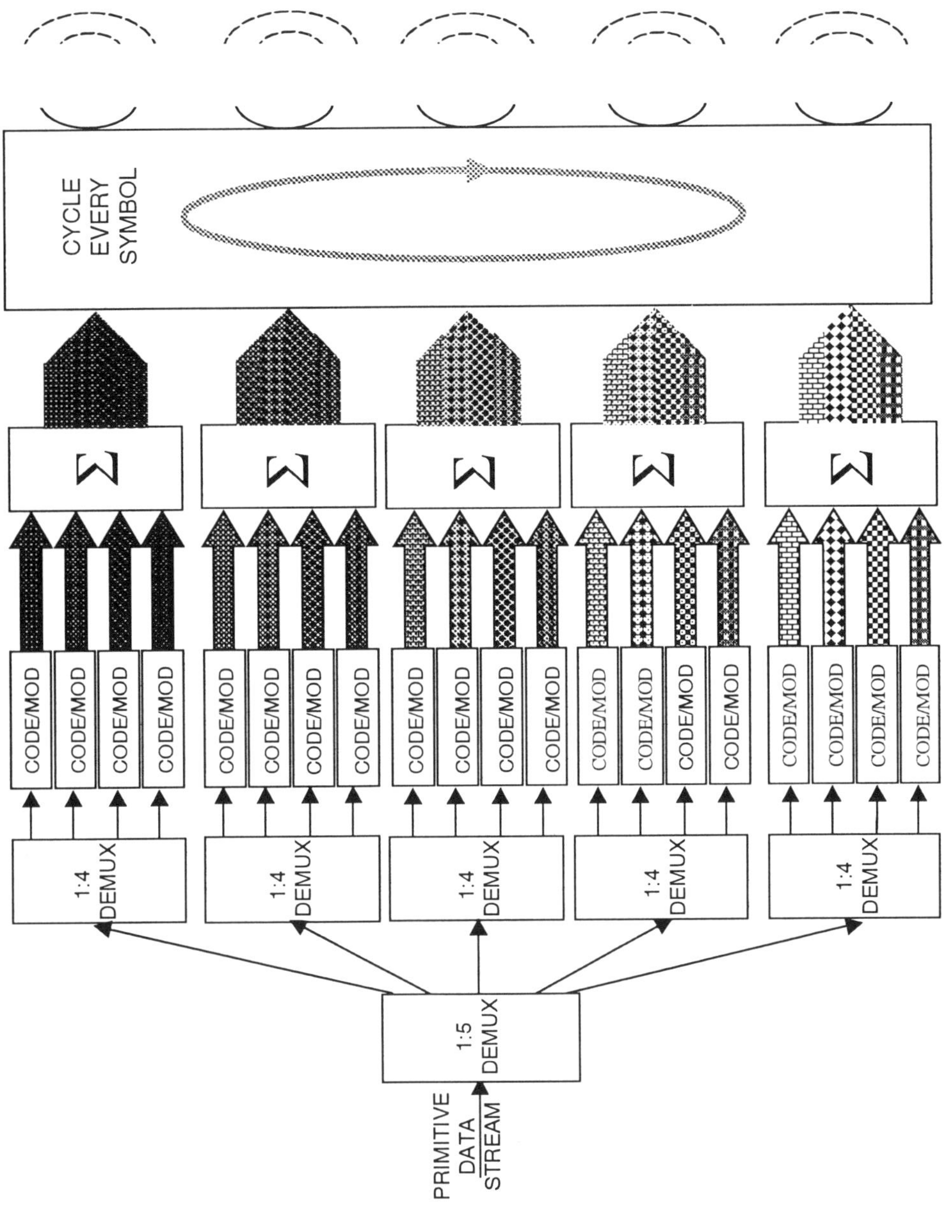

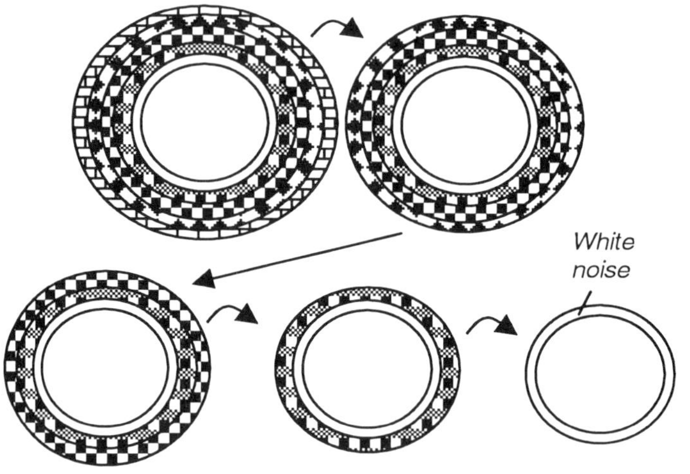

FIGURE 4: Peeling away of successive ply (five strata each) from outside-in. Here a layer is depicted as an "onion" with five "rings". In the sequence, peeling away another outer ring corresponds to removal of another ply of interference. Any stratum is interfered with by all the other strata on the same ply as well by the strata that make up relatively inner plies. The peeling process can be simultaneously conducted in all five of the gray shaded strata in a ply.

In preparing an exposed ply for detection, the N-D received vector, r(t), received at time t, has already been processed achieving perfect removal of the relatively outer ply. So each inside ply is free of interference from *all* outside plies. It may be helpful to think of a sequence of processed received vectors bootstrapped with $r(t)$. So, with each major step in the processing sequence, the original received vector process $r(t)$ sheds another ply's worth of interference

$$r(t) = r^1(t) \rightarrow r^2(t) \rightarrow r^3(t) \rightarrow r^4(t) \tag{3.1}$$

where each of the four vectors is an N-D function of time defined over the message duration. Adding another level of refinement, we point out that the detection process actually includes five different copies of (3.1), one is used for each layer in the example. Then a second superscript, m, is implicit for (3.1).

Each vector is then collapsed to a scalar using a weight vector, defined (up to an arbitrary complex scalar multiple), to maximize the signal to noise plus interference ratio (SINR) associated with this decision statistic. Maximization of SINR is a well known process, see, e.g., reference [2].

4. NEAR LogDet CAPACITY IN CERTAIN CASES

Next we derive a formula for the bit rate that each stratum can support if a genie informs the transmitter of the stratum capacities. Then we will obtain the asymptotic capacity for a large number of strata. These formulae will be useful in seeing what can be achieved when the genie is put back in the bottle: our main interest, is when the transmitter *is not* informed of the stratum capacities.

4.1. GENIE INFORMED STRATA CAPACITIES. We use n for the number of signal strata per diagonal layer. We derive the formula for the ultimate error free bit rate that each stratum is capable of supporting when it powered with power *P/(Mn)*. We investigate the received signal constituent, s_{mit}, on the ith strata of the mth component at time, *t*, and how it is impaired. For simplicity we proceed explicitly using only the first two coordinates. We do so for when the mth strata exposed on the ith ply is to be detected following the detection and removal of interference from the strata on plies numbered *1, 2, ... i-1*. Along with AWGN, this constituent, s_{mi}, is impaired by all of the simultaneously transmitted strata with equal or higher indices that are not yet detected. The constituents of these interfering strata are $\{s_{m'l}, i \leq l \leq n, 1 \leq m' \leq M, m' \neq m \text{ } when \text{ } l = i\}$.

It is convenient to express the *N* by *M* matrix, G, in terms of column vectors as follows:

$$G = (g_1, g_2, \cdots, g_M) \tag{4.1}$$

where g_m is an *N*-D vector corresponding to the mth transmit antenna. Drawing on (1.1) we see the vector expression involving the received scalar signal, s_{mi}, along with its additive impairment is therefore expressed as

$$(g_1, g_2, \cdots, g_M)\left[\begin{pmatrix} 0 \\ \vdots \\ 0 \\ s_{mi} \\ 0 \\ \vdots \\ 0 \end{pmatrix} + \sum_{l=i}^{n}{}^{(mi)} \begin{pmatrix} s_{1l} \\ s_{2l} \\ \vdots \\ \vdots \\ s_{Ml} \end{pmatrix}\right] + \nu. \tag{4.2}$$

The (mi) superscript means that the ith strata of the mth component is omitted in the sum, since, in (4.2), the scalar s_{mi} is signal, not interference. We are using the normalized form of equation (1.1), so that the *N*-D noise vector, ν, has complex i.i.d. Normal (0,1) components.

In equation (4.2), the impairment is the *N*-D vector, ξ_{mi}, defined by

$$\xi_{mi} = G \cdot \sum_{l=i}^{n} {}^{(mi)} \begin{pmatrix} s_{1l} \\ s_{2l} \\ \vdots \\ s_{Ml} \end{pmatrix} + \nu. \tag{4.3}$$

For each value of $m = 1, 2, \cdots, M$ the variance-covariance matrix of this impairment of s_{mi} is denoted by $\Sigma_{\xi_{mi}}$. In the next step we spatially whiten this vector impairment by multiplying the expression in (4.3) by the positive definite matrix $\Sigma_{\xi_{mi}}^{-1/2}$. We get the following expression for the spatially whitened i-th stratum of the m-th component

$$\sum\nolimits_{\xi_{mi}}^{-1/2} g_m s_{mi} + \mu^m \qquad (m = 1, 2, \cdots, M). \tag{4.4}$$

Each one of the vectors in the sequence of N-D vectors, $\{\mu^m\}_{m=1}^{M}$, has complex Normal(0,1) components that are statistically independent of each other. Indeed, for each value of m in a stratum, independent noise is encountered. For a *(1,N)* system, where the scalar signal s_{mi} is transmitted and the N-D noisy received signal vector is described by (4.4), we know from the LogDet formula for capacity that the capacity is

$$c_{mi} = \log_2\left[1 + \left\| \sum\nolimits_{\xi_{mi}}^{-1/2} g_m \right\|^2 \cdot \frac{P}{\sigma^2 Mn}\right] = \log_2\left[1 + g_m \bullet \sum\nolimits_{\xi_{mi}}^{-1} g_m \cdot \frac{P}{\sigma^2 Mn}\right] \tag{4.5}$$

where the heavy dot denotes scalar product. For this *(1,N)* system, maximum ratio combining of the received vector components is a key processing step for "achieving" Shannon capacity.

Next, employing E for expectation, we compute that

$$\begin{aligned} \sum\nolimits_{\xi_{mi}} &= E\left| \sum_{j=1}^{M} g_j \left(\sum_{l=i}^{n} {}^{(mi)} s_{jl} \right) + \nu \right|^2 \\ &= I + \left(\sum_{j=1}^{M} g_j g_j^\dagger \left(\frac{n-(i-1)}{n} \right) \left(\frac{P}{\sigma^2 \cdot M} \right) \right) - g_m g_m^\dagger \left(\frac{P}{\sigma^2 \cdot M} \right) \frac{1}{n}. \end{aligned} \tag{4.6}$$

We will also need the matrix identity

$$GG^\dagger = \sum_{j=1}^{M} g_j g_j^\dagger. \tag{4.7}$$

Later we will need the large n asymptotic form of (4.6). So, with z = i/n, we write

$$\sum\nolimits_{\xi_{mi}} = I + \left(P/(\sigma^2 M)\right) GG^\dagger (1-z) + o(1/n). \tag{4.8}$$

Therefore, the signal to interference plus noise (SINR) of the ith stratum of the mth component is explicitly

$$SINR_{mi} = g_m^\dagger \left(I_N + \left(P/(\sigma^2 M)\right) GG^\dagger (1-z)\right)^{-1} g_m \frac{P}{\sigma^2 nM} + o\left(\frac{1}{n}\right) \tag{4.9}$$

As time progresses, the ith stratum in any diagonal layer has an SINR of period M. The capacity added by the ith stratum in any diagonal layer, say, the dth, where $1 \le d \le M$, is then obtained by averaging over the capacity contributions from the M transmit antennas. This capacity, C_i^d, which by symmetry does not depend on d, is

$$(4.10) \qquad C_i^d = \frac{1}{M}\sum_{m=1}^{M}\log_2\left[1+g_m^{\dagger}\Sigma_{\xi_{mi}}^{-1}g_m\frac{P}{nM\sigma^2}\right].$$

This equation, draws on (4.6), a key equation to be used for computations in Section 5.

4.2. ASYMPTOTIC FORM OF GENIE AIDED STRATA CAPACITIES. Next we derive the large n asymptotic form of the strata capacities, still with the tentative assumption that the transmitter is made privy to these capacities. First we rewrite the previous equation as follows

$$(4.11) \quad C_i^d = \frac{1}{M}\sum_{m=1}^{M}\log_2\left[1+g_m^{\dagger}\left(I_N+\left(\frac{P}{M\sigma^2}\right)GG^{\dagger}(1-z)\right)^{-1}g_m\frac{P}{nM\sigma^2}\right]+o\left(\frac{1}{n}\right).$$

For the large n asymptote, we employ the small ε approximation for $\log_2(1+\varepsilon)$ so that differential capacity contribution of the ith stratum is expressed

$$(4.12) \qquad C_i^d = \frac{P}{\sigma^2\cdot n\cdot M^2\cdot\ln 2}\sum_{m=1}^{M}g_m^{\dagger}\left(I_N+\left(P/(\sigma^2M)\right)GG^{\dagger}(1-z)\right)^{-1}g_m.$$

Summing the capacities, C_i^d, from $i = 1, 2 \ldots n$, and letting $n \to \infty$, we get an integral over $[0,1]$. Substituting $\zeta \triangleq 1-z$, we get $\int_0^1 f(1-z)dz = \int_0^1 f(\zeta)d\zeta$.

So, in the large n limit, with $d\zeta$ associated with $1/n$ the capacity, $\sum_{i=1}^{n} C_i^d$, becomes the integral

$$(4.13) \quad \lim_{n\to\infty}\sum_{i=1}^{n}C_i^d = \frac{P}{\sigma^2\cdot M^2\cdot\ln 2}\sum_{m=1}^{M}\int_0^1 g_m^{\dagger}\left(I_N+\left(P/(\sigma^2M)\right)GG^{\dagger}\zeta\right)^{-1}g_m d\zeta.$$

Now, the detection of M such diagonal layers is executed in parallel so that the total capacity C is obtained by multiplying the right hand side of (4.13) by M. Employing tr for trace, after some minor linear algebra, the total capacity is seen to be

$$(4.14) \qquad C == \frac{P}{\sigma^2M\cdot\ln 2}\,tr\int_0^1 G^{\dagger}\left(I_N+\left(P/(\sigma^2M)\right)GG^{\dagger}\zeta\right)^{-1}Gd\zeta.$$

Using the singular value decomposition, [25], we write G as a triple product

$$(4.15) \qquad G = U\Lambda V^{\dagger}.$$

Let MIN = Min{M,N}. Λ is an N by M all zero matrix except that its NW corner is an MIN×MIN diagonal matrix with jj entry λ_j. V and U are M×M and N×N unitary matrices respectively. We obtain the following capacity formula, where the matrix with inverted diagonal entries depicted is a diagonal matrix and when j > MIN the λ_j vanish

$$C = \frac{P}{\sigma^2 \cdot M \cdot \ln 2} tr\left\{\int_0^1 \left(V\Lambda^\dagger U^\dagger\right)\left(UU^\dagger + \left(P/(\sigma^2 M)\right)U\Lambda\Lambda^\dagger U^\dagger \zeta\right)^{-1} U\Lambda V^\dagger d\zeta\right\}$$

(4.16)

$$= \frac{P}{\sigma^2 \cdot M \cdot \ln 2} tr\left\{\int_0^1 V\Lambda^\dagger \begin{bmatrix} \ddots & & \\ & \left(1+\frac{P\left|\lambda_j\right|^2 \zeta}{\sigma^2 \cdot M}\right)^{-1} & \\ & & \ddots \end{bmatrix} \Lambda V^\dagger d\zeta\right\}.$$

The trace of a square matrix is unitarily invariant, so carrying out the integration gives

$$C = \sum_{m=1}^{MIN} \log_2\left(1+\frac{P\left|\lambda_m\right|^2}{\sigma^2 \cdot M}\right). \tag{4.17}$$

Equation (4.17) can be expressed in the more a detailed form

$$C = \log_2 \det\left[UU^\dagger + \frac{P}{\sigma^2 \cdot M} U\Lambda V^\dagger V\Lambda^\dagger U^\dagger\right] = \log_2 \det\left[I_N + \frac{P}{\sigma^2 \cdot M} \cdot GG^\dagger\right]. \tag{4.18}$$

This is the same as the right hand side of equation (1.2). We have proved, that by knowing the strata rates, and using the communication means that we have described, in the limit of large n, LogDet capacity is "attained".

While our interest is when the strata rates are *not* known at the transmitter, we continue, for now, with the genie informing the transmitter of the ply rates. It will be useful to express the accumulation of capacity with the progressive peeling away of plies. We particularly want to do this for the asymptote of a large number of plies. We take i to be a fixed fraction, z, of n as n $\to \infty$. From equation (4.12) we see that the differential capacity added by the ith ply is

$$\frac{1}{\ln 2} \cdot \sum_{j=1}^{MIN} \frac{P\,|\,\lambda_j\,|^2\, z/(\sigma^2 M)}{1+\left(P\,|\,\lambda_j\,|^2/(\sigma^2 M)\right)(1-z)} dz. \tag{4.19}$$

Asymptotically for large n, the capacity accumulated through the normalized Z^{th} = $(i/n)^{th}$ ply is obtained by integrating from 0 to Z to give

$$\sum_{j=1}^{MIN} \log_2\left[1+\frac{P\,|\,\lambda_j\,|^2\, Z/(\sigma^2 M)}{1+\left(P\,|\,\lambda_j\,|^2/(\sigma^2 M)\right)(1-Z)}\right]. \tag{4.20}$$

As expected, setting Z = 1 gives equation (4.17).

4.3. PUTTING THE GENIE BACK IN THE BOTTLE WHERE IT BELONGS. As is well known, [16,26,27], a matrix channel possesses generalized eigenmodes. To access these noninterfering modes the transmitter needs to know the channel matrix. Using this knowledge, the transmit power can be spatially water poured over the modes to obtain a superior capacity than if it transmits equal power from each transmitter. Each of the summands in equation (4.17) is referred to as an eigenrate. We are interested in the eigenrates, the rates that the (generalized) eigenmodes can support when transmitting with power P/M on each of the antennas. When $M \geq N$ the channel has N eigenmodes, the interpretation of (4.17) is that they are being blindly accessed by the transmitter, each driven with power P/M. The wasted power, (M - N)P/M, with the inability to water pour, are the costs of channel blind operation.

We see that, if the entire genie does, is tell the transmitter the eigenrates, then LogDet capacity can be approached in the limit of an infinite number of strata per diagonal layer. The transmitter need not know how to access the eigenmodes. If there is only one eigenmode, as in (M, 1) systems, then knowing C is equivalent to knowing the eigenrate and the genie is unnecessary. (In (1,N) systems, the one eigenmode is also known without the help of a genie, but we are focusing on $M \geq N$.).

We now describe a Monte Carlo computation of a lower bound on what the stratified spacetime architecture can attain with the genie back in the bottle. The method uses only the channel's statistical characterization. We stress that the transmitter is assumed privy to that and nothing more. We discuss how to operate at a typically small, say X%, capacity outage for an arbitrary random channel ensemble. The transmitter, not knowing the set of n bit rates per ply, needs to make an educated guess to achieve a high value of the X%-tile outage capacity. The guessed ply rates must be low enough so that all n guesses are successful for at least (100-X)% of the channels.

Choose the guessed rates as those associated with any channel at an outage percentile Y% < X% of plies of a hypothetical channel population in which the ply rates are known. Enforce these somewhat less ambitious rates for each channel in a population where the rates are unknown and look at the percentage of the channels that have at least one of its n plies in violation of the demanded rates. Iterate this calculation by perturbing the starting outage percentile (all lower than X%) computed from the bit rates known case until an acceptably slightly less than X% violation free count occurs with rates unknown. This assures the desired X%-tile outage capacity is met for this guessing procedure. While the outage capacity is reduced compared to the genie-assisted case, in the next section we will see examples, where, it is fair to say, it can come close.

5. NUMERICAL EXAMPLES FOR MATRIX RAYLEIGH CHANNEL

We now report examples for the ideal matrix Rayleigh channel where the transmitter does not know what bit rate to use for each stratum, or equivalently, for each ply. When M >> N, the choice of rates/strata is aided by eigenrates becoming substantially hardened, [1,28-31], as they do when both M and N get large and M is a fixed fraction of N. In all cases, whether there is hardening or not, the iterative method of the last section can be applied.

We include results for a sampling of (M,N) systems using 10% outage, leaving comprehensive numerical studies for the future. The first examples, (8,1) and (8,3) are for the important downlink case where the end user has few antennas compared to the base(s). Also a (4,2) and a (2,2) case are reported. Heavy use will be made of the formulas from the last section. Since in all four examples $M \geq N$, we have that MIN = N.

The graphs that we present display 10% outage channel capacity versus average received signal to noise ratio (SNR). This average SNR, ρ, ($\rho_{dB} = 10\log_{10}\rho$) is calibrated as follows. If one makes a test measurement transmitting all the available power, P, out of any of the M transmit antennas, say the mth, and makes repeated statistically independent measurements at any receive antenna, say the jth, then ρ is given by $\rho = (P/\sigma^2)\cdot E|G_{jm}|^2$. The expectation averages over all the statistically independent instantiations of the entry G_{jm}. Note, ρ is independent of the index pair *jm*. In the SD-BLAST examples power P/M is transmitted from each of M antennas in the transmit array. All of the examples have SNRs in the range of $\rho_{dB} = -6$ dB to +24dB with 3dB steps.

A secondary feature of our figures is that we include some ultimately encoded Vertical-BLAST (V-BLAST) curves. We include these because the relative simplicity of the V-BLAST implementation has drawn interest. V-BLAST does not use the cycling (complication and advantage) of D-BLAST. See [32] for V-BLAST details. Instead, after an M:1 de-multiplexing, each of M independent data streams are individually encoded and mapped 1:1 to the M transmit antennas. Also V-BLAST offers the advantage of accommodating codes designed for a standard AWGN setting, in contrast to the nontrivial periodic SINR environment of D-BLAST. In our examples, V-BLAST reception is always based on maximum SINR, not zero forcing. It must be stressed that V-BLAST was designed for situations where N > M, so it cannot be expected to exhibit strong performance in our examples.

5.1. SD-BLAST EXAMPLE: (8,1). The SD-BLAST graphs for 1, 2, 4, 8, 16, 32 and 64 strata per layer are shown in Figure 5A for the (8,1) example (as they are in the next two examples). For an (8,1) system the SD-BLAST graphs evidence D-BLAST, or what is the same, LogDet, performance as an upper envelope as theory predicted. Since the transmitter knows the channel statistics, it knows the lower 10%-tile of $|\lambda_1|^2$. So, in this case, the transmitter knows the single eigenrate at 10% outage. (Note $|\lambda_1|^2$ is a chi-square variable with 16 degrees of freedom.) The bit rates per ply as $n \to \infty$ are known. Moreover, the guessing method was so effective that the counts of required rate/strata violations were negligible for all the n shown. So the ply bit rates are very close to being known. This will be less and less the case as we proceed with our examples.

The lower double dash curve in each figure gives the performance of the upper limit of encoded V-BLAST. As anticipated, V-BLAST exhibits inferior performance.

5.2. SD-BLAST EXAMPLE: (8,3). This (8,3) case, with M > N > 1, is more difficult then the previous one since the transmitter does not know the set of n bit rates per strata to use in the limit of large n. The transmitter's guessing of the ply rates is now crucial. We cannot expect to get the same 10%-tiles as if the ply rates were known. Again the "guessing" technique was used to produce the 10%-tile capacities. In Figure 5B, the degradation of these 10%-tiles from the 10%-tiles for when the n ply rates are known is shown and the degradation is now just about observable.

With V-BLAST, a smaller number of transmitter antennas than M can, depending on M, N and ρ_{dB} give superior performance than using all M antennas available. So, in the V-BLAST examples we optimized the number of transmit antennas actually used when M are available. The V-BLAST curves exhibit a peaking of the derivative since the optimum can change with the abscissa value. Such changes in the number of transmitters also occur in the other three V-BLAST curves, but in Figure 5A there was no noticeable peaking of the slope.

The curves that we have shown may make one wonder if the upper envelope of our genie unassisted examples is LogDet. The theory tells us that this cannot be for the (8,3), (4,2) and (2,2) examples. We give numerical confirmation of a deficit for this (8,3) example, at 18dB. We obtained results for a extremely large number of strata and indeed we noticed saturation below LogDet as the Figure 5B2 shows.

5.3. SD-BLAST EXAMPLE: (4,2). In the third example, a (4,2) case depicted in Figure 5C, the excess of M over N is not as great as in the previous examples. So we expect the relative performance to not be quite as good as the previous two cases. However, because of substantial eigenrate hardening, the deficit from known bit rates is still significantly less than one bps/Hz.

Figure 5A: (8,1) SD-BLAST, CAPACITY vs AVERAGE SNR

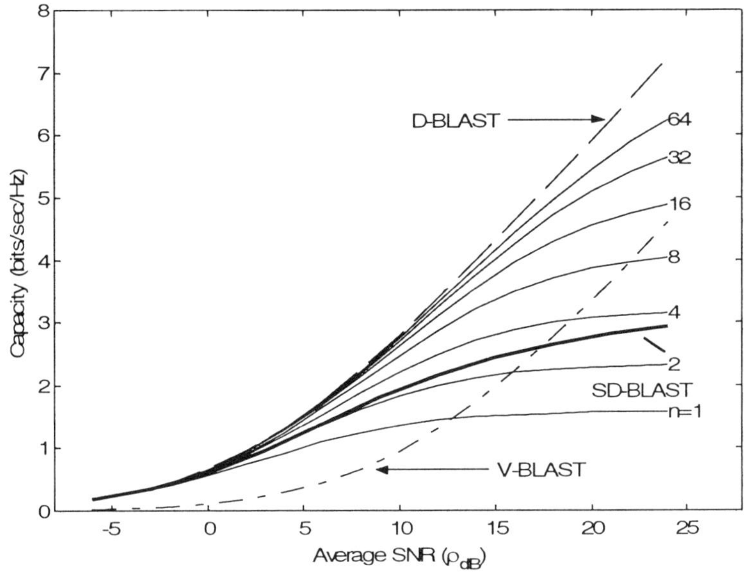

Figure 5B: (8,3) SD-BLAST, CAPACITY vs AVERAGE SNR

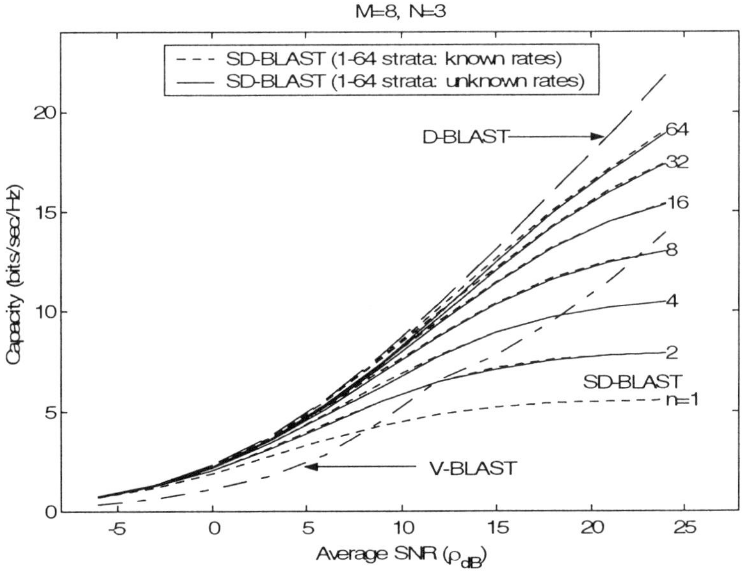

Figure 5B2: (8,3) SD-BLAST, CAPACITY AT 18dB vs No. OF STRATA. EXPECTED SATURATION BELOW D-BLAST ASYMPTOTE

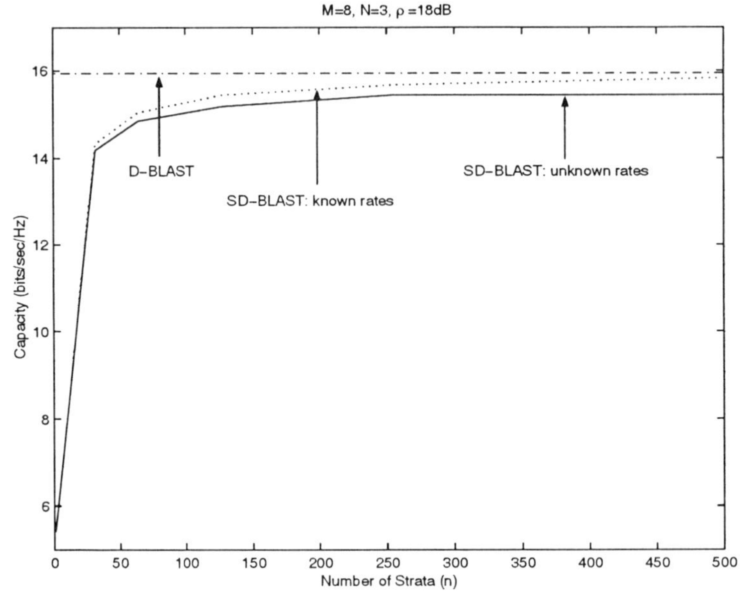

Figure 5C: (4,2) SD-BLAST, CAPACITY vs AVERAGE SNR

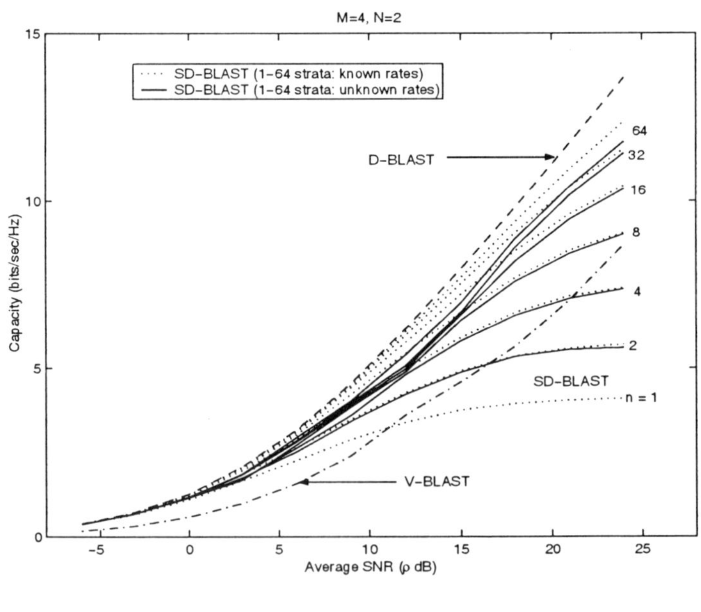

6. BRIEF COMMENTS ON VARIATIONS

There are straightforward variations on the covering of plies with helices. E.g. we can deviate from our assumption that the long message length, T, is a multiple of M, by perturbing, T slightly so that T and M are relatively prime. Then one can arrange to uniformly wrap a ply with just one (approximately M times longer) helix rather than M parallel helices. In the course of detection processing along the single longer helix, when repeating the time of symbol receipt coordinate, merges must have been forced on the simultaneously received, but previously processed helical section(s). In fact the entire message can be just one long helix by dropping down to the next inner ply to continue the ply wrapping after an outer ply is completely wrapped. In this case the received signal manifold can be taken to be like the snake that swallows its own tail along the lines depicted in [33]. As the number of plies increase, these design variations are easily shown to be of secondary importance in terms of capacity improvement relative to the M-fold simultaneous detection featured for simplicity in Section 4.

One can replace the spatially 1-D codec constraint with a spatially k-D codec constraint ($k \leq M$ and k dividing M). Our 1-D choice was the illustrative low dimensional extreme.

We focused a single one of many spectrally disjoint narrowband users. The numerical results presented in the last section go over immediately for any one of the orthogonal users in an idealized CDMA system based on orthogonal codes and operating over a flat channel. It is also possible to numerically assess performance of SD-BLAST CDMA systems using PN codes. Moreover, the SD-BLAST approach also extends to frequency selective systems.

7. SD-BLAST: SUMMARY AND FURTHER WORK

SD-BLAST is an architectural superstructure for an (M,N) communication system. Signal constituents are received symmetrically arranged in spacetime in nested tori called plies that are helically wound. They are so arranged to enable the receiver to substantially mute inevitable mutual interference of the M simultaneous transmissions in a multipath environment. At the same time SD-BLAST architecture uses 1-D codecs. It helps avoid problems associated with an earlier D-BLAST. Namely, it helps to avoid using long diagonals which force waste of the space-time resource and make it problematic to pack enough coded blocks into a message of limited duration.

The scope of this report does not include the lower level issue of 1-D codec design for the coded block occupying a stratum. With SD-BLAST, within a stratum, an SINR of period M is experienced. Reference [34,35] discusses coding for systems with periodically varying SNRs as arise in ADSL systems. The authors stress that the signaling alphabet must be sufficiently rich to take advantage of the better SNRs. We note that, with SD-BLAST, with more and more plies, even a code using a binary constellation meets this requirement because of the way that the bps/ply decrease with increasing n. Furthermore, the period M cycling at the transmitter has stronger transmissions compensating weaker ones at a maximum pace. This concentration of space diversity with the passage of processing time promotes quick delivery of bit decisions.

Information theory analysis showed that the SD-BLAST can often get close to LogDet performance. We highlighted downlink cases using matrix Rayleigh channels with M significantly greater than N, i.e., where the receiver has few antennas compared to the base(s). Namely, we looked at (8,1) and (8,3) cases. We also included (4,2) case. Under the assumption of equal energy per stratum, Figures 5A-C quantified how much stratification was needed to support a particular capacity level at 10% outage for SNRs spanning –6dB to +24dB. In these examples we quantified the proximity to LogDet performance.

In the figures in Section 5, the n signal plies were equally energized. It is worth noting that one can improve capacity at any outage %-tile by optimizing the energy distribution over the n plies. We illustrated for the (8,1) system that this complication can be a worthwhile refinement for future investigation. In Figure 5A, for the case n = 2 we used a one parameter search to optimize the energy distributed over the two plies. The capacity improvement is exhibited by the bold curve in Figure 5A. The upper LogDet envelope cannot be improved, but, one can certainly improve convergence to LogDet somewhat with increasing n by optimizing the energy distribution.

The simulation results in the literature, [36-39] (TURBO- BLAST architecture), [22] (Turbo SpaceTime Architecture) and [40], (Threaded SpaceTime Architecture) involving BLAST in conjunction with iterative receivers helped provide motivation for the work reported in our paper. These architectures do not include the element of stratification that we that enabled us to show that in some cases we can "attain" or come close to "attaining" LogDet performance in the limit of infinitely many strata.

Moreover, the SD-BLAST approach, analyzed here in the limit of zero error rate, was not iterative while the approaches in references of the previous paragraph were iterative. Of course, in practice, the SD-BLAST structure would operate under a specific bit or block error rate requirement. Using simulation, it would be interesting to quantify any advantages to including some iterative aspects of interstrata processing in SD-BLAST. Thus, one could see if it is worthwhile not to limit the processing to one pass through the various strata. Specific codes, like LDPC [41] and TURBO codes, can be tested in such SD-BLAST simulations.

From [22-24] we see that LogDet can also be "attained" with separately coded 1-D signals radiating from each transmit antenna using codes that are designed for a standard AWGN environment. Each transmitter transmits a channel dependent rate. However, unlike SD-BLAST, this requires that the M transmit rates be fed back to the transmitter.

ACKNOWLEDGEMENTS

Dr. S. Lek Ariyavisitakul's talks on his BLAST related research, later reported in [22], intensified interest in exploring improved performance, BLAST type systems. Thanks are also due to Dr. C. Papadias and Dr. J. Salz.

REFERENCES

[1] G. J. Foschini, "Layered Space-Time Architecture for Wireless Communication in a Fading Environment when Using Multi-Element Antennas", *Bell Labs Technical Journal,* Vol. 1, No. 2, pp 41-59, Autumn 1996

[2] G. J. Foschini and M. J. Gans, "On Limits of Wireless Communication a Fading Environment when Using Multiple Antennas", *Wireless Personal Comm,* Vol. 6, pp 311-335, No. 3, 1998

[3] V. Tarokh, N. Seshardri and A. R. Calderbank, "Space-time Codes for High Data Rate Wireless Communications: Performance analysis and Code Construction", *IEEE Trans. on Inform. Theory*, Vol. 44, No. 2, pp 744-765, March 1998

[4] V. Tarokh, A. F. Naguib, N. Seshardri and A. R. Calderbank, "Combined Array Processing and Space-Time Coding", *IEEE Trans. on Inform. Theory*, vol 45, No. 4, pp 1121-1128, May 1999

[5] A. F. Naguib, N. Seshardri and A. R. Calderbank, "Applications of Space-Time Codes and Interference Suppression for High Capacity and High Data Rate Wireless Systems", *Proceedings 32nd Asilomar Conference on Computers, Signals and Systems*, Pacific Grove, CA, 1998, pp 1803-1810

[6] B. M. Hochwald and T. L. Marzetta, "Unitary Space-Time Modulation for Multiple-Antenna Communications in Rayleigh Flat Fading", *IEEE Transa. on Inform. Theory*, vol 44, no. 2, pp 744-765, March 1998

[7] Babak Hassibi and Bertrand Hochwald, High-Rate Codes that are Linear in Space and Time, Submitted to *IEEE Trans. on Inform. Theory*

[8] P. F. Driessen and G. J. Foschini, On the Capacity Formula for Multiple Input-Multiple Output Wireless Channels: A Geometric Interpretation, *IEEE Trans. On Comm.* Vol. 47, No. 2 pp 173-176, February 1999

[9] C-N Chuah, J. M. Kahn, G. J. Foschini, R. A. Valenzuela, D. Chizhik, J. Ling, "Capacity Growth of Multi-Element Arrays in Indoor and Outdoor Wireless Channels, *Proceedings of WCNC 2000*, Chicago, Il, Sept 23-28

[10] D.-S. Shiu, J, M, Kahn, G. J. Foschini and M. J. Gans, "Fading Correlation and Its Effect on the Capacity Multielement Antenna Systems", *IEEE Transactions On Communications,* Vol. 48 No. 3, pp 502-213, March 2000

[11] E. Biglieri, G. Caire and G. Taricco, "Limiting Performance of Block Fading Channels with Multiple Antennas", Accepted for *IEEE Trans. on Inform. Theory*

[12] Narula, M. D. Trott and G. W. Wornell, "Performance Limits of Coded Diversity Methods for Transmitter Antenna Arrays", *IEEE Transactions on Information Theory*, Vol 45, pp 2418-2433, Nov. 1999

[13] S. Alamouti, "A Simple Transmitter Diversity Scheme for Wireless Communication", *IEEE Journal On Selected Areas in Comm.,* Vol. 17, No. 10, pp 1451 1458, October 1998

[14] C. Papadias, "On the Spectral Efficiency of Space-Time Spreading Schemes for Multiple Antenna Systems", *33rd Asilomar Conference on Signals, Systems and Computers,* Pacific Grove, CA, Oct 24-27, 1999, pp 639-643

[15] C. Papadias and G. J. Foschini, "An Open-Loop Transmit Diversity Technique for Systems Employing Four Transmitter Antennas", BL0113K.11.09.00-TM#11

[16] G. J. Foschini and R. K. Mueller, "The Capacity of Linear Channels with Additive Gaussian Noise", *Bell System Technical Journal*, pp 81-94, January 1970

[17] S. Verdu, "Capacity Region of Gaussian CDMA Channels: The Symbol-Synchronous Case, *Proceedings 37th Annual Allerton Conference, on Communication, Control and Computing,* Allerton, IL, October 1986, pp1025-1034

[18] M. S. Pinsker, *Information and Information Stability of Random Processes*, Holden Bay, San Fransisco, 1964, Chapter 10

[19] E. Csiszar and J. Korner, Information Theory: Coding Theorems for Discrete Memoryless Systems, Academic Press, New York, 1981

[20] S. Z. Stambler, "Shannon's Theorems for a Complete Class of Discrete Channels Whose State is Known At the Output," *Problems of Information Transmission,* No. 11, Plenum Press, New York, pp 263-270, November 1976

[21] G. J. Foschini, D. Chizhik and M. J. Gans, "Universal Optimality of Signalling Along Diagonals in Space-Time in BLAST Systems", *Talk at Lucent BLAST Workshop,* Feb. 2000

[22] S. L. Ariyavisitakul, "Turbo Space-Time Processing to Improve Wireless Channel Capacity", *IEEE Transactions on Communications*, Vol. 48, No. 8, pp 1347-1359, August 1999

[23] S. Verdu and S. Shamai, "Spectral Efficiency of CDMA with Random Spreading" *IEEE Transactions on Information Theory* , Vol. 45, No. 2, pp 622-640, March 1999

[24] M. K. Varanasi and T. Guess, "Optimum Decision Feedback MultiuserEqualization with Successive Decoding Achieves the Total Capacity of the Gaussian Multiple-Access Channel,"Proc. Asilomar Conf. on Signals, Systems and Computers, pp. 1405-1409, Montery, CA, Nov. 1997.

[25] P. Lancaster and M. Tismenetsky, *The Theory of Matrices*, Academic Press, New York, 2nd Edition, pg 192, 1985

[26] G. G. Rayleigh and J. M. Cioffi, "Spatio-Temporal Coding for Wireless Channels", *IEEE Transactions on Communications*, Vol. 44, No.3 pp 357-366, March 1998

[27] D. Chizhik, G. J. Foschini, R. A. Valenzuela, "Capacities of Multielement Transmit and Receive. Antennas: Correllations and Keyholes", *IEE Electronic Letters,* 22/6/00, Vol. 36(13), pp. 1099-1100

[28] D. Josson, "Some Limit Theorems for the Eigenvalues of a Sample Covariance Matrix", *Journal of Multivariate Analysis*, Vol. 12, 1982, pp 1- 18

[29] V. Grenander and J. W. Silverstein, "Spectral Analysis of Networks with Random Topologies", *SIAM Journal of Applied Mathematics*, Vol. 32, No. 2, pp 499-519, March 1997

[30] A. L. Moustakas, H. U. Baranger, L. Balents, A. M. Sengupta and S. H. Simon, ``Communication Through a Diffusive Medium: Coherence and Capacity", Science, Vol 287, page 287 (2000).

[31] A. M. Sengupta and P. P. Mitra, ``Capacity of Multivariate Channels with Multiplicative Noise: I. Random Matrix Techniques and Large-N Expansions for Full Transfer Matrices", Available at http://xxx.lanl.gov/abs/physics/0010081

[32] G. J. Foschini, G. D. Golden, R. A. Valenzuela and P. W. Wolniansky, "Simplified Processing for High Spectral Efficiency Wireless Communications Employing Multi-Element Arrays", *IEEE Journal On Selected Areas in Communications*, Vol. 17, No. 11, pp 1841-1852, November 1999

[33] J. Gribbin and M. Rees, Cosmic Coincidences, Bantam Books, New York, 1989, Fig. 3.1

[34] R. D. Wesel, X. Liu, and W. Shi, "Trellis Codes for Periodic Erasures," *IEEE Transactions on Communications*, Vol. 48, No. 6, pp 938-947, June 2000.

[35] R. D. Wesel and J. M. Cioffi, "Trellis Code Design for Periodic Interleavers," *IEEE Communications Letters*, Vol. 3, No. 4, pp 103-105, April 1999.

[36] M. Sellathurai and S. Haykin, "A Nonlinear Iterative Beamforming Technique for Wireless Communications", *33rd ASILOMAR Conference on Signals Systems and Computers,* Vol. 2, pp 957-961, November, 1999

[37] M. Sellathurai and S. Haykin, ``Turbo-BLAST for wireless communications: theory and experiments", *IEEE Trans. on Signal Processing*, vol. 50, no. 10, pp. 2538-2546, October 2002.

[38] M. Sellathurai and S. Haykin, "Turbo-BLAST for High-speed Wireless Communications", in *Wireless Communications and Network Conference Proceedings,* September 2000, Chicago.

[39] M. Sellathurai and S. Haykin, ``A simplified diagonal BLAST architecture with iterative parallel-interference cancellation", *Proceedings of IEEE Int. conf. on communications*, vol. 10, pp. 3067-3071, June 2001.

[40] H. El Gamal and A. R. Hammons Jr., "New Approach to Space-Time Transmitter/Receiver Design," *Proceedings 37th Allerton Conference, on Communication, Control and Computing,* September, 1999

[41] Gallager, R. G., "Low Density Parity Check Codes", Monograph, M.I.T. Press, 1963

G. J. FOSCHINI IS WITH LUCENT TECHNOLOGIES, BELL LABS, CRAWFORD HILL LAB, HOLMDEL, NJ *E-dress: gjf@.lucent.com*

M. SELLATHURAI IS WITH THE COMMUNICATIONS RESEARCH CENTRE OF CANADA, OTTAWA, ON K2H 8S2 *E-dress: msellath@crc.ca*

DIMACS Series in Discrete Mathematics
and Theoretical Computer Science
Volume **62**, 2003

Space–Time Coding with Iterative Receiver Interfaces

Ezio Biglieri Alessandro Nordio Giorgio Taricco

ABSTRACT. We compare suboptimum linear and nonlinear interfaces to be used for decoding space-time codes transmitted over a multiple-antenna Rayleigh fading channel with perfect channel-state information available at the receiver. The codes we consider are obtained by apportioning evenly among the transmit antennas the symbols of off-the-shelf convolutional codes. We observe how the introduction of an interleaver can be beneficial here. We introduce a new simple iterative linear interface, based on hard Viterbi decoding and offering a performance considerably improved with respect to noniterative receivers.

1. Introduction

We study the behavior of error probabilities of space-time codes over the Rayleigh fading channel when a limited-complexity receiver interface is used in lieu of a maximum-likelihood (ML) receiver. We first compare the ML receiver performance with those of linear and nonlinear (BLAST) minimum mean-square error (MMSE) receivers. Next, we introduce a new iterative receiver, based on hard decoding and interference cancellation, and we show that it outperforms considerably the linear MMSE receiver. The simulations presented show how a performance close to ML can be achieved with simpler receiver interfaces.

This paper is organized as follows. Section 2 describes the channel model and the general problem of designing a suboptimum (linear or nonlinear) interface. Coding schemes are described in Section 3, along with the effect of a row interleaver on the receiver performance. The receivers complexities are discussed in Section 4. Numerical results are contained in Chapter 5, which compares by simulation the error probabilities of several receivers.

2. Decoding with multiple antennas

We consider a radio system with t transmit and r receive antennas, $r \geq t$. Assume that a space-time code [**25**] $\mathcal{X}$ is used with block length N. Specifically, the transmitted signal is represented by the $t \times N$ matrix $\mathbf{X} \triangleq (\mathbf{x}_0, \ldots, \mathbf{x}_{N-1})$. The row index of $\mathbf{X}$ indicates space, while the column index indicates time: that is,

1991 *Mathematics Subject Classification.* Primary 94A05, 94A14; Secondary 94B60.
Key words and phrases. Space–time codes; multiple-antenna systems; radio receivers.
This work was supported by Lucent Technologies, EU, and CERCOM.

$\mathbf{x}_i$ is the t-tuple of channel symbols transmitted simultaneously at discrete time i, $0 \le i \le N-1$. The received signal is modeled by the $r \times N$ matrix

$$\mathbf{Y} = \mathbf{H}\mathbf{X} + \mathbf{Z} \tag{2.1}$$

where $\mathbf{Z}$ is a matrix of complex Gaussian random variables (RV) with zero mean and independent real and imaginary parts having the same variance $N_0/2$ (i.e., circularly-distributed). Thus, the noise affecting the received signal is spatially and temporally independent, with $\mathbb{E}[\mathbf{Z}\mathbf{Z}^\dagger] = NN_0\mathbf{I}_r$, where $\mathbf{I}_r$ denotes the $r \times r$ identity matrix and $(\cdot)^\dagger$ denotes Hermitian transposition. The channel is described by the $r \times t$ complex Gaussian random matrix $\mathbf{H}$ whose entries h_{ij} are circularly distributed with variance of their real and imaginary parts equal to $1/2$. Equivalently, each entry of $\mathbf{H}$ has uniformly-distributed phase and Rayleigh-distributed magnitude, with expected magnitude square equal to 1. $\mathbf{H}$ is independent of both $\mathbf{X}$ and $\mathbf{Z}$, it remains constant during the transmission of an entire code word (the "quasi-static," or "block-fading," assumption [**4**], and its realization (the "channel state information," or CSI) is known at the receiver. Under the assumptions above, the maximum-likelihood receiver chooses the matrix $\mathbf{X} \in \mathcal{X}$ which minimizes the squared Euclidean (or "Frobenius") norm:

$$\|\mathbf{Y} - \mathbf{H}\mathbf{X}\|^2 \tag{2.2}$$

2.1. Suboptimum receiver interfaces. As ML detection might be too intensive computationally, one may resort to a suboptimum receiver interface. This may be either linear or nonlinear.[1]

2.1.1. *Linear interfaces.* The linear interface we consider here generates first a "soft estimate" of the transmitted matrix $\mathbf{X}$ in the form

$$\widetilde{\mathbf{Y}} = \mathbf{A}\mathbf{Y} \approx \mathbf{X} \tag{2.3}$$

with $\mathbf{A}$ a $t \times r$ matrix, chosen to make $\widetilde{\mathbf{Y}}$ close to $\mathbf{X}$, so that the metric

$$\|\widetilde{\mathbf{Y}} - \mathbf{X}\|^2 \tag{2.4}$$

can be used. Here we restrict ourselves to the choice of $\mathbf{A}$ that minimizes the mean-square error $\mathbb{E}[\|\mathbf{A}\mathbf{Y} - \mathbf{X}\|^2]$. By assuming independent coded symbols with average energy E_s we have

$$\mathbf{A} = (\mathbf{H}^\dagger\mathbf{H} + \delta_s\mathbf{I}_t)^{-1}\mathbf{H}^\dagger \tag{2.5}$$

where $\delta_s \triangleq N_0/E_s$.

2.1.2. *Nonlinear interfaces.* The general form of nonlinear interface we consider here is shown in Fig. 1. It is based again on the generation of a soft estimate $\widetilde{\mathbf{Y}}$ of $\mathbf{X}$ and on the use of the metric (2.4). Specifically, the soft estimate is obtained by subtracting, out of a linearly-filtered version of the observed signal $\mathbf{Y}$, a linearly-filtered version of the coded symbols that have been previously decoded:

$$\widetilde{\mathbf{Y}} = \mathbf{G}\mathbf{Y} - \mathbf{L}\widehat{\mathbf{X}} \tag{2.6}$$

[1]The reader should be able to perceive the similarity of the problem of choosing a suboptimum interface here with the problems of equalizing a dispersive channel or of designing a suboptimum multiuser detector. Their common feature is the need to control interference, which is spatial interference for multiple-antenna systems, intersymbol interference for the dispersive channel, and multiple-access interference for multiuser detectors. Spatial interference is characterized, as we shall see below, by the off-diagonal elements of $\mathbf{H}^\dagger\mathbf{H}$.

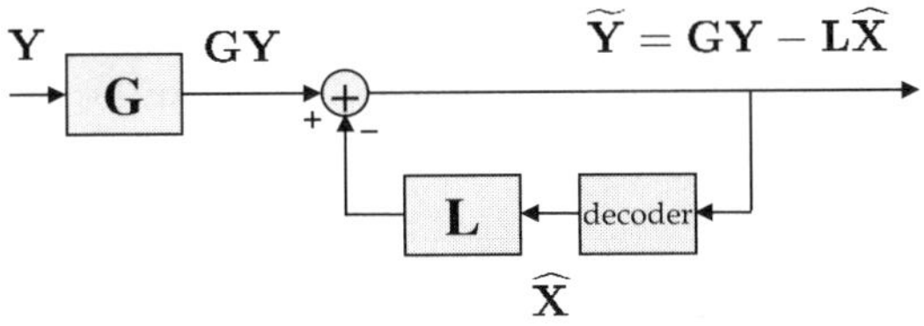

FIGURE 1. *General form of a nonlinear interface.*

where $\widehat{\mathbf{X}}$ is a hard estimate of $\mathbf{X}$, and $\mathbf{L}$ is assumed to have zero diagonal entries because $\mathbf{L}\widehat{\mathbf{X}}$ is intended to be an approximation of the spatial interference affecting $\mathbf{GY}$. The choice of the $t \times r$ matrix $\mathbf{G}$ and the $t \times t$ matrix $\mathbf{L}$ categorizes the interface structure.

Vertical BLAST interface. If $\mathbf{L}$ is assumed to be strictly upper triangular, that is, $[\mathbf{L}]_{i,j} = 0$ unless $j > i$, then this "causality" constraint (which makes $(\widetilde{\mathbf{Y}})_k$ independent of $(\widehat{\mathbf{X}})_j$, $j \leq k$) allows (2.6) to be implemented sequentially in t steps.[2] The proposed decoding algorithm is described by the following two steps.

(1) For $i = t, \ldots, 1$, we decode the ith row of (2.6)

$$(\widetilde{\mathbf{Y}})_i = (\mathbf{GY})_i - \sum_{k=t-i+1}^{t} [\mathbf{L}]_{i,k} (\widehat{\mathbf{X}})_k \tag{2.7}$$

over the subcode $\mathcal{X}_i$ obtained as a projection the space-time code $\mathcal{X}$ (whose words are the matrices $\mathbf{X}$) over the positions corresponding to the ith row of the matrix code words. Notice that the ith subcode decoding step is feasible since the ith row of the soft estimate $\widetilde{\mathbf{Y}}$ depends only on already available rows of $\widehat{\mathbf{X}}$ because of the causality constraint. The decoded word is assigned to the $(t-i)$th row of $\widehat{\mathbf{X}}$.

(2) Next we use $\widehat{\mathbf{X}}$ to obtain the soft estimate $\widetilde{\mathbf{Y}}$ which is used to implement an ML algorithm based on the metric $\|\widetilde{\mathbf{Y}} - \mathbf{X}\|^2$. The final output is

$$\arg\min_{\mathbf{X}} \|\mathbf{GY} - \mathbf{L}\widehat{\mathbf{X}} - \mathbf{X}\|^2 \tag{2.8}$$

Clearly, the validity of this approach depends on how well the soft estimate $\widetilde{\mathbf{Y}}$ approximates the transmitted code word $\mathbf{X}$. We can regard the difference $\widetilde{\mathbf{Y}} - \mathbf{X}$ as a disturbance, and select the matrices $\mathbf{G}$ and $\mathbf{L}$ so as to minimize its effect. We choose to minimize the MSE of the spatial interference plus noise: this approach leads to minimum mean-square error (MMSE) BLAST [**6, 15**].

Minimization of the mean square error $\mathbb{E}[\|\mathbf{GY} - \mathbf{LX} - \mathbf{X}\|^2]$ leads to the choice

$$\begin{cases} \mathbf{G} = \operatorname{diag}^{-1}(\mathbf{S})\mathbf{S}^{-\dagger}\mathbf{H}^{\dagger} \\ \mathbf{L} = \operatorname{diag}^{-1}(\mathbf{S})\mathbf{S} - \mathbf{I}_t \end{cases} \tag{2.9}$$

[2] We denote the ith row of a matrix $\mathbf{A}$ as $(\mathbf{A})_i$.

where $\mathbf{S}$ is an upper triangular matrix obtained from the Cholesky factorization $\mathbf{H}^\dagger\mathbf{H} + \delta_s\mathbf{I}_t = \mathbf{S}^\dagger\mathbf{S}$. As a result, the soft estimate $\widetilde{\mathbf{Y}}$ can be written as

$$\begin{aligned}\widetilde{\mathbf{Y}} &= \operatorname{diag}^{-1}(\mathbf{S})\mathbf{S}^{-\dagger}\mathbf{H}^\dagger\mathbf{Y} - (\operatorname{diag}^{-1}(\mathbf{S})\mathbf{S} - \mathbf{I}_t)\widehat{\mathbf{X}} \\ &= (\mathbf{I}_t - \operatorname{diag}^{-1}(\mathbf{S})\mathbf{S}^{-\dagger})\mathbf{X} + (\operatorname{diag}^{-1}(\mathbf{S})\mathbf{S} - \mathbf{I}_t)(\mathbf{X} - \widehat{\mathbf{X}}) \\ &\quad + \operatorname{diag}^{-1}(\mathbf{S})\mathbf{S}^{-\dagger}\mathbf{H}^\dagger\mathbf{Z}\end{aligned} \tag{2.10}$$

where the three terms in the last expression play the role of *i)* the (biased) useful term; *ii)* the interference due to past wrong decisions; and *iii)* colored noise.

Iterative interface. An alternative to BLAST consists of performing an iterative spatial interference cancellation along the lines of [**21**]. At iteration k, $k = 0, 1, \ldots$, we generate an estimate of the spatial interference in the form

$$\mathbf{W}^{(k)} = (\mathbf{A}\mathbf{H} - \operatorname{diag}(\mathbf{A}\mathbf{H}))\widehat{\mathbf{X}}^{(k)} \tag{2.11}$$

Here $\widehat{\mathbf{X}}^{(k)}$ is the decoded word at iteration k, computed by minimizing the metric $\|\widetilde{\mathbf{Y}}^{(k)} - \mathbf{X}\|^2$, where

$$\begin{aligned}\widetilde{\mathbf{Y}}^{(k+1)} &= \widetilde{\mathbf{Y}} - \mathbf{W}^{(k)} \\ &= \widetilde{\mathbf{Y}} - (\mathbf{A}\mathbf{H} - \operatorname{diag}(\mathbf{A}\mathbf{H}))\widehat{\mathbf{X}}^{(k)}\end{aligned} \tag{2.12}$$

and for $k = 0$ we define $\widehat{\mathbf{X}}^{(0)} \triangleq \mathbf{0}$. It can be easily seen that, if decoding is perfect (that is, if $\widehat{\mathbf{X}}^{(k)} = \mathbf{X}$ for some k), then

$$\widetilde{\mathbf{Y}}^{(k)} = \operatorname{diag}(\mathbf{A}\mathbf{H})\,\mathbf{X} + \mathbf{A}\mathbf{Z} \tag{2.13}$$

which shows that the spatial interference is completely removed.

The main difference between the multiuser iterative receiver of [**21**] and our receiver is that in the former the rows of $\mathbf{X}$ (each corresponding to one user's message) are independent, and decoding is performed independently for each user at each iteration. In our receiver the rows of $\mathbf{X}$ are jointly decoded at each iteration, because there is correlation among the signals transmitted by each antenna. Both systems employ Viterbi hard decoding in lieu of computationally intensive soft-input, soft-output decoding: it was shown in [**21**] that this choice offers a good tradeoff between complexity and performance.

3. Coding schemes for nonlinear interfaces

A categorization of coding schemes for nonlinear interfaces, which are based on preliminary decoding procedures made to the purpose of canceling the spatial interference from the received signal $\mathbf{Y}$, was described in [**1, 18**](for coding schemes matched to BLAST architectures, see [**2, 11, 22, 23**]): one is "vertical coding," which consists of using a single encoder and of spreading the coded symbols across the transmit antennas. The other one, "horizontal coding," separately encodes the signals transmitted by each antenna (see Fig. 2). A difficulty with vertical coding is in interference cancellation, because no preliminary decisions are available to increase the reliability of the symbols detected and used for canceling. On the other hand, the performance of horizontal coding, which does interference cancellation using previously decoded symbols, is limited by the weakest code. Two possible improvements of these coding schemes consist of using iterative decoding in conjunction with vertical coding [**1, 18**], or a concatenated code in a combination of vertical and horizontal coding [**18**].

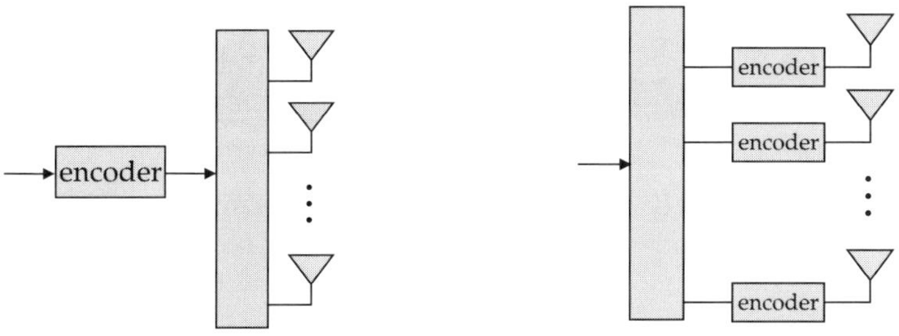

(a) Vertical coding (b) Horizontal coding

FIGURE 2. *Space–time coding.*

The approach we advocate here, and which is especially attractive when simplicity is at a premium, consists of using, for vertical coding, subcodes of a single larger code. The role of the individual subcodes is to improve the cancellation process by providing more reliable decisions, while the larger code, to be decoded as a final step, compensates for uncanceled interference and noise by introducing a coding gain. Our decoding scheme matches to "vertical" BLAST architectures any code whose words can be evenly split among the transmit antennas (for a discussion of the rationale behind the choice of an off-the-shelf code for multiple-antenna systems, see [**5**]).

3.1. Effect of the interleaver. Consider again ML detection, and the pairwise error probability [**5**]

$$P(\mathbf{X} \to \widehat{\mathbf{X}}) = Q\left(\frac{\|\mathbf{H}\boldsymbol{\Delta}\|}{\sqrt{2N_0}}\right) = Q\left(\sqrt{\frac{\mathrm{Tr}\ (\mathbf{H}^\dagger\mathbf{H}\boldsymbol{\Delta}\boldsymbol{\Delta}^\dagger)}{2N_0}}\right) \tag{3.1}$$

where $\boldsymbol{\Delta} \triangleq \mathbf{X} - \widehat{\mathbf{X}}$. The previous equation shows that if $\mathbf{H}^\dagger\mathbf{H}$ is a scalar matrix, i.e., a scalar multiple of the identity matrix $\mathbf{I}_t$, then the pairwise error probability depends only on $\mathrm{Tr}\ (\boldsymbol{\Delta}\boldsymbol{\Delta}^\dagger)$, the squared Euclidean distance between $\mathbf{X}$ and $\widehat{\mathbf{X}}$. More generally, the off-diagonal elements of $\mathbf{H}^\dagger\mathbf{H}$ introduce "spatial interference," which cause a loss of performance because some degrees of freedom in the receiver must be used to remove or reduce it. Specifically, let us rewrite $\mathbf{H}$ and $\boldsymbol{\Delta}$ in the partitioned form

$$\mathbf{H} = [\mathbf{h}_1 \cdots \mathbf{h}_t] \qquad\qquad \mathbf{X}^\dagger = [\mathbf{X}_1^\dagger \cdots \mathbf{X}_t^\dagger]$$

and observe that

$$\mathbf{H}\boldsymbol{\Delta} = \sum_{i=1}^{t} \mathbf{h}_i \boldsymbol{\Delta}_i \tag{3.2}$$

so that

$$\text{(3.3)} \quad \operatorname{Tr}(\mathbf{H}^\dagger\mathbf{H}\boldsymbol{\Delta}\boldsymbol{\Delta}^\dagger) = \sum_{i=1}^{t}\sum_{j=1}^{t} \boldsymbol{\Delta}_i\boldsymbol{\Delta}_j^\dagger \mathbf{h}_j^\dagger \mathbf{h}_i^\dagger = \underbrace{\sum_i \|\mathbf{h}_i\|^2\|\boldsymbol{\Delta}_i\|^2}_{\xi_0} + \underbrace{\sum_{i\neq j} \boldsymbol{\Delta}_i\boldsymbol{\Delta}_j^\dagger \mathbf{h}_j^\dagger \mathbf{h}_i}_{\xi_1}$$

Eq. (3.3) exhibits explicitly the spatial interference term ξ_1 (note that this is not independent of ξ_0). Its deleterious effect can be understood by observing the application of Jensen's inequality to the convex $\cup$ function $Q(\sqrt{\cdot})$:

$$\begin{aligned} \mathbb{E}\,[Q(\sqrt{\xi_0+\xi_1})] &= \mathbb{E}_{\xi_0}\left[\mathbb{E}_{\xi_1|\xi_0}[Q(\sqrt{\xi_0+\xi_1})]\right] \\ &\geq \mathbb{E}_{\xi_0}\left[Q\left(\sqrt{\mathbb{E}_{\xi_1|\xi_0}[\xi_0+\xi_1]}\right)\right] \\ &= \mathbb{E}[Q(\sqrt{\xi_0})] \end{aligned}$$

A way of removing ξ_1 consists of reducing $\mathbf{H}^\dagger\mathbf{H}$ to a scalar matrix by letting r tend to infinity while t remains constant [**5**]. If r and t are finite, the spatial interference term can be removed by choosing a proper coding scheme which makes $\mathbf{H}^\dagger\mathbf{H}$ a scalar matrix (see [**17**]; Alamouti scheme is one of these). Another way of removing ξ_1, which we advocate here, consists of increasing the code length N while introducing a row interleaver as described below.

Consider a row interleaver whose action is described by the multiplication of each row $\mathbf{X}_i$ of $\mathbf{X}$ by a permutation matrix $\mathbf{P}_i$, $i = 1, \ldots, t$. We write $\mathbf{X}_i^\pi \triangleq \mathbf{X}_i\mathbf{P}_i$, and consequently

$$\mathbf{X}^{\dagger\pi} = [\mathbf{X}_1^{\dagger\pi} \cdots \mathbf{X}_t^{\dagger\pi}]$$

With interleaving, the rows of $\boldsymbol{\Delta}$ are permuted as well, and we have, with obvious notations,

$$\text{(3.4)} \qquad \xi_1^\pi = \sum_{i\neq j} \boldsymbol{\Delta}_i^\pi \boldsymbol{\Delta}_j^{\dagger\pi} \mathbf{h}_j^\dagger \mathbf{h}_i$$

while ξ_0^π is left unchanged by the interleaver. We have $\mathbb{E}\,\xi_0 = r\|\boldsymbol{\Delta}\|^2$ and $\mathbb{E}\,\xi_1^\pi = 0$. After some algebra, it can be shown that the variance of ξ_1, with an average taken over the fading as well as over the set of random interleavers, is given by $\mathbb{E}\,[|\xi_1^\pi|^2] = O(1/N)$. This shows that as $N \to \infty$ the spatial interference tends to disappear.

With the introduction of a row interleaver, the received signal takes the form

$$\text{(3.5)} \qquad \mathbf{Y} = \mathbf{H}\mathbf{X}^\pi + \mathbf{Z}$$

where $\mathbf{X}^\pi$ denotes the code word after interleaving. In this case the receiver computes

$$\text{(3.6)} \qquad \widetilde{\mathbf{Y}}^\pi = \mathbf{A}\mathbf{Y}$$

for the linear interface, or

$$\text{(3.7)} \qquad \widetilde{\mathbf{Y}}^\pi = \mathbf{G}\mathbf{Y} + \mathbf{L}\widehat{\mathbf{X}}^\pi$$

for the BLAST interface. The matrices $\widetilde{\mathbf{Y}}^\pi$ and $\widehat{\mathbf{X}}^\pi$ denote the interleaved versions of $\widetilde{\mathbf{Y}}$ and $\widehat{\mathbf{X}}$, respectively. In both cases the metrics are computed according to (2.4). Simulation results will show that interleaving yields nonnegligible benefits to the receiver performance.

4. Receivers complexity

To evaluate and compare the complexities of the receiver interfaces described above, namely ML, linear MMSE, iterative MMSE, and BLAST receivers, we consider a rate-κ/n convolutional code with 2^M states. Coded bits are mapped onto symbols belonging to a modulation alphabet $\mathcal{A}$ with cardinality $|\mathcal{A}| = 2^m$. The overall complexity of an interface is measured as the number of real operations required to decode a frame, and depends on the complexity of each one of the following items:

- Computation of the soft-estimate (linear and nonlinear filtering)
- Computation of the metrics
- Viterbi algorithm

We define a *complex operation* as the combination of a complex product and a complex addition; a *real operation* is the combination of a real product and a real addition. We assume that a complex operation has a complexity equivalent to 4 real operations.

4.1. Computation of the soft estimates.

4.1.1. *MMSE.* The receiver computes $\widetilde{\mathbf{Y}} = \mathbf{AY}$, where $\mathbf{A}$ is a $t \times r$ matrix and $\mathbf{Y}$ is an $r \times N$ matrix. This product requires Ntr complex operations.

4.1.2. *Iterative MMSE.* At each iteration k, $k = 1, \dots, K$, the receiver computes $\widetilde{\mathbf{Y}}^{(k)} = \widetilde{\mathbf{Y}} - (\mathbf{FH} - \mathrm{diag}\{\mathbf{FH}\})\widehat{\mathbf{X}}^{(k-1)}$ where $\mathbf{FH} - \mathrm{diag}\{\mathbf{FH}\}$ is a $t \times t$ matrix with zeros on its main diagonal. The computation of $\widetilde{\mathbf{Y}}^{(k)}$ requires $Nt(t-1)$ complex operations per iteration, which yields an overall cost of $Nt(t-1)K$ complex operations.

4.1.3. *BLAST.* The soft estimate of the transmitted codeword is computed as $\widetilde{\mathbf{Y}} = \mathbf{GY} - \mathbf{L}\widehat{\mathbf{X}}$. The product $\mathbf{GY}$ involves Ntr operations, since $\mathbf{G}$ is a $t \times r$ matrix. The product $\mathbf{L}\widehat{\mathbf{X}}$ requires $Nt(t-1)/2$ operations, because $\mathbf{L}$ is upper triangular with zeros on its diagonal. The overall cost is $Nt(r + t/2 - 1/2)$ complex operations.

4.1.4. *ML.* No soft estimates of the transmitted code word $\mathbf{X}$ are computed.

4.2. Computation of the metrics.

4.2.1. *MMSE, iterative MMSE, and BLAST.* At each iteration these receivers compute the metrics

$$\|\widetilde{\mathbf{Y}} - \mathbf{X}\|^2 = \sum_{i=0}^{N-1} \sum_{j=0}^{t-1} |(\widetilde{\mathbf{Y}})_{ij} - (\mathbf{X})_{ij}|^2 \tag{4.1}$$

For each i and j the norm $|(\widetilde{\mathbf{Y}})_{ij} - (\mathbf{X})_{ij}|^2$ is computed for all possible symbols $(\mathbf{X})_{ij}$, thus requiring $Nt2^m$ complex operations. Here only $t2^m$ values need storing at each step.

4.2.2. *ML.* The ML receiver computes the metrics

$$\|\mathbf{Y} - \mathbf{HX}\|^2 = \sum_{i=0}^{N-1} |\mathbf{y}_i - \mathbf{H}\mathbf{x}_i|^2 \tag{4.2}$$

where $\mathbf{y}_i$ and $\mathbf{x}_i$ denote the ith column of $\mathbf{Y}$ and $\mathbf{X}$, respectively. Since for each i there are 2^{mt} possible symbol vectors $\mathbf{x}_i$, the computation of (4.2) requires $Nr2^{mt}$ complex operations. Here only $r2^m$ values need storing at each step.

System	*Soft estimates*	*Metrics*	*Viterbi*
ML	0	$4r2^{mt}$	$\epsilon\beta$
MMSE	$4tr$	$4t2^m$	$\epsilon\beta$
itMMSE	$4t(r+K(t-1))$	$4t(K+1)2^m$	$\epsilon(K+1)\beta$
BLAST	$2t(2r+t-1)$	$4t2^m$	$\epsilon t\beta$

TABLE 1. Normalized complexity $\mathcal{C}/N$ of some receivers in terms of real operations per frame. Here $\beta \triangleq 2^{\kappa+M}$.

4.3. Viterbi algorithm. The Viterbi algorithm operates on a trellis with 2^M nodes, with 2^κ transitions entering each node. The Viterbi algorithm finds, for each trellis state, the transition that minimizes the accumulated metric. This involves 2^κ real additions and comparisons per state. The complexity of a real addition and a comparison generally depends on the hardware used: in some cases the cost of an addition is negligible with respect to that of a multiplication, while in some other cases (and especially when the hardware allows pipelining) the two costs are equivalent. If we introduce the (hardware-dependent) coefficient $\epsilon \in [0,1]$ denoting the ratio between the cost of an addition and that of a multiplication, the Viterbi algorithm requires $\epsilon N 2^{\kappa+M}$ real operations per frame. The Viterbi algorithm is called once per frame by the ML and MMSE receivers, $K+1$ times per frame (K the number of iterations) by the iterative MMSE receiver, and t times per frame by the BLAST receiver.

4.4. Comparisons. Table 1 summarizes the complexities of the four interfaces considered here. $\mathcal{C}$ is the number of real operations. We first compare the complexity of the iterative MMSE receiver with that of the BLAST receiver. The ratio ρ_1 between the two complexities is

$$\begin{aligned}\rho_1 &\triangleq \frac{\mathcal{C}_{\text{itMMSE}}}{\mathcal{C}_{\text{BLAST}}} \\ &= \frac{4t(r+K(t-1)+(K+1)2^m)+\epsilon(K+1)2^{\kappa+M}}{4t(r+t/2-1/2+2^m)+\epsilon t 2^{\kappa+M}}\end{aligned}$$

The ratio between the complexity of the iterative MMSE receiver and the ML receiver is instead

$$\begin{aligned}\rho_2 &\triangleq \frac{\mathcal{C}_{\text{itMMSE}}}{\mathcal{C}_{\text{ML}}} \\ &= \frac{4t(r+K(t-1)+(K+1)2^m)+\epsilon(K+1)2^{\kappa+M}}{4r2^{mt}+\epsilon 2^{\kappa+M}}\end{aligned}$$

The ratio ρ_2/ρ_1 compares the complexities of BLAST and itMMSE.

As a special case of the above, consider 2PSK modulation (i.e., $m=1$) and three system configurations that will be examined below in terms of their error performance:

(1) $t=2$, $r=2$, rate-1/2 convolutional code with generators $(5,7)$ ($\kappa=1$, $n=2$, $M=2$)
(2) $t=2$, $r=4$, rate-1/2 convolutional code with generators $(5,7)$ ($\kappa=1$, $n=2$, $M=2$)

System	1	2	3
ρ_1, $K = 0$	0.89	0.92	0.80
ρ_1, $K = 1$	1.55	1.38	1.46
ρ_2, $K = 0$	1.00	0.75	0.37
ρ_2, $K = 1$	1.75	1.12	0.68

TABLE 2. Complexity comparisons for $\epsilon = 0$.

System	1	2	3
ρ_1, $K = 0$	0.77	0.82	0.60
ρ_1, $K = 1$	1.38	1.29	1.13
ρ_2, $K = 0$	1.00	0.78	0.41
ρ_2, $K = 1$	1.80	1.22	0.76

TABLE 3. Complexity comparisons for $\epsilon = 1$

(3) $t = 4$, $r = 4$, rate-1/4 convolutional code with generators $(13, 13, 15, 17)$ ($\kappa = 1$, $n = 4$, $M = 3$)

The relative complexities ρ_1 and ρ_2 are shown in Tables 2 and 3 for $\epsilon = 0$ and $\epsilon = 1$, respectively.

From these Tables we see that MMSE ($K = 0$) has a complexity lower than BLAST ($\rho_1 < 1$), and that iterative MMSE with one iteration ($K = 1$) has a complexity that exceeds that of (noniterative) BLAST by about $15\% - 50\%$, depending on system parameters. Moreover, we notice that with systems 1 and 2 ML has approximately the same complexity of the noniterative MMSE receiver, but a better performance. With system 3 iterative MMSE seems to provide the best trade-off between complexity and performance.

5. Numerical results

One of the results of [**5**] is that, for moderate signal-to-noise ratios, good space-time codes can be obtained by choosing codes that exhibit a good performance on the (unfaded) additive white Gaussian noise channel, and splitting evenly each code word among the t transmit antennas. Here we examine the performance of the receiver interfaces discussed above with space-time codes obtained from simple off-the-shelf convolutional codes.

Fig. 3 compares the simulated performances, in terms of frame-error rate, of the space-time code obtained by choosing a good binary, 4-state, rate-1/2 convolutional code, mapping its symbols onto 2PSK, and splitting them evenly between two transmit antennas. Here $t = r = 2$. The frame length is 130 symbols, including 1 symbol for trellis termination.

The lines with no markers show the performance of the ML receiver (solid) and that of an uncoded system with $t = 1$ and $r = 2$ (dashed). The dashed line with a $\circ$ marker refers to the noniterative linear MMSE receiver without interleaving. The two dashed lines with a $*$ and $\square$ marker refer to the MMSE-BLAST receiver with and without interleaving. The three lines with a $\diamond$ marker refer to the MMSE iterative receiver with 0, 1, and 2 iterations. Finally, the two lowest lines show the outage probability lower bound [**4**]: the upper assumes 2PSK inputs (i.e., the signal set considered here) and the lower assumes Gaussian inputs.

The effect of interleaving can be seen by comparing the uppermost solid line with a $\diamond$ marker (noniterative MMSE with interleaving) and the dashed line with a $\circ$ marker (noniterative MMSE without interleaving). The gain of the former reaches 5 dB at a FER of 10^{-3}. The interleaver is also effective on BLAST, which performs slightly better than the linear MMSE interface. The performance of the latter interface improves considerably when 1 or 2 iterations are performed. The gain of the iterative MMSE receiver and 2 iterations over the MMSE-BLAST receiver (both with interleaving) is of 6 dB at a FER of 10^{-3}. Increasing the number of iterations beyond 2 does not improve performance significantly, and leaves a gap with respect to ML that increases with SNR. The ML performance is about 2.5 dB away from the outage lower bound at a FER of 10^{-3}.

Fig. 4 refers to a system with the same parameters as in Fig. 3, except that now $r = 4$. Here we can see that the performance of linear MMSE interfaces is much closer to ML (without interleaving), with a single iteration yielding about the same error rate as ML. Here the noniterative linear MMSE interface outperforms BLAST, albeit slightly. The ML performance is about 2.5 dB away from the outage lower bound at a FER of 10^{-3}.

Fig. 5 refers to a system with the same parameters as in Fig. 4, except that now $t = 4$, with a rate-1/4 convolutional code. Without interleaving, the noniterative linear MMSE interface performs much better than BLAST. We interpret this result by observing that the iterative procedure intrinsic in BLAST need strong codes on each row of $\mathbf{X}$, which is not the case here. We also observe that the presence of an interleaver makes the BLAST performance similar to that of MMSE. Moreover, interleaved MMSE performs even better than noninterleaved ML. The beneficial effect of interleaving on ML can be more directly observed from Fig. 6, which refers to the same code as in Fig. 5, but with a shorter block length $N = 10$ (the shortening of block length is made necessary because interleaved ML is decoded by exhaustive search through all code words, which entails a complexity $O(2^N)$). From this figure we also observe that with a shorter block length the performance improvement caused by one iteration in MMSE is smaller and the iterative MMSE receiver performance does not exceed the ML receiver (both with and without interleaving). Finally, the ML performance is about 4 dB away from the outage lower bound at a FER of 10^{-3}.

6. Conclusions

We have examined the performance of linear MMSE and nonlinear (BLAST) interfaces used to decode space-time codes with a limited number of transmit and receive antennas. In particular, we have introduced a simple iterative linear interface, based on hard Viterbi decoding and offering a performance considerably improved with respect to noniterative receivers. A complexity analysis has been performed, showing the relative merits of linear and nonlinear interfaces.

Acknowledgment

This work was supported by EU through a FET contract and by CERCOM.

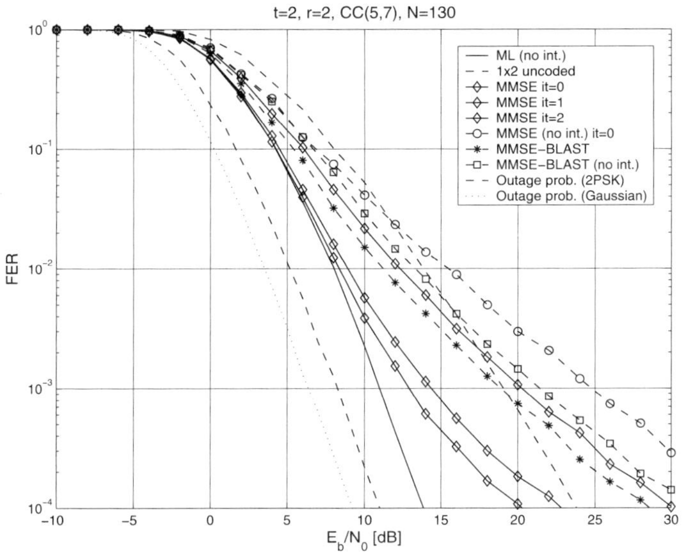

FIGURE 3. *Frame error probability of a convolutional code with different interfaces vs. the signal-to-noise ratio* E_b/N_0. *Here* $t = r = 2$.

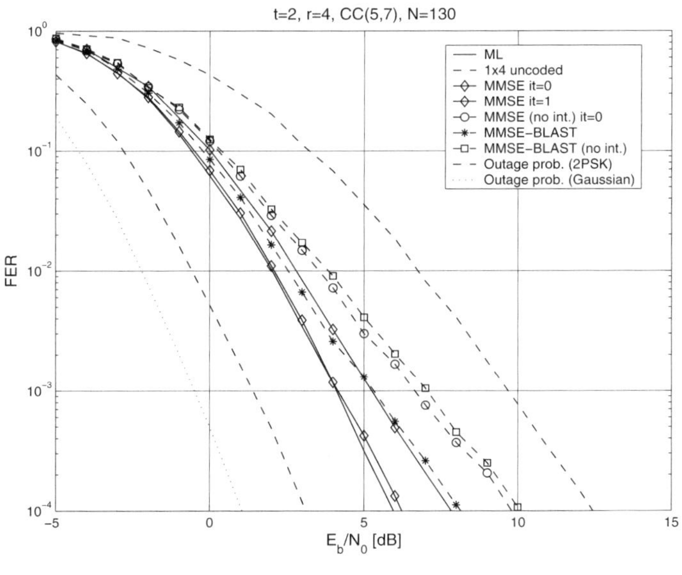

FIGURE 4. *Frame error probability of a convolutional code with different interfaces vs. the signal-to-noise ratio* E_b/N_0. *Here* $t = r/2 = 2$.

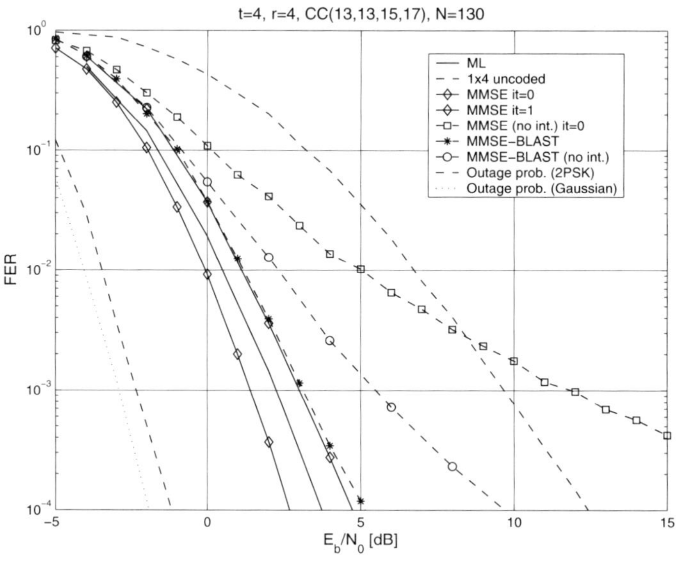

FIGURE 5. *Frame error probability of a convolutional code with different interfaces vs. the signal-to-noise ratio* E_b/N_0. *Here* $t = r = 4$.

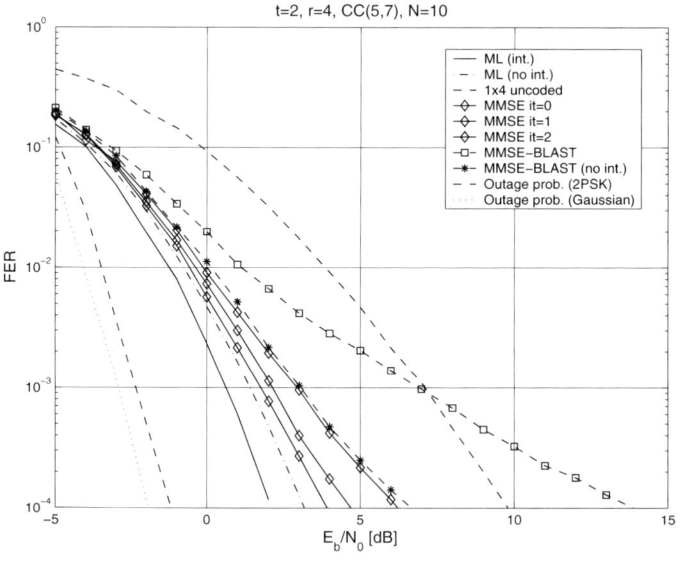

FIGURE 6. *Frame error probability of a convolutional code with different interfaces vs. the signal-to-noise ratio* E_b/N_0. *Here* $t = r = 4$.

References

[1] S. L. Ariyavisitakul, *Turbo space-time processing to improve wireless channel capacity.* IEEE Trans. Commun., **48** (2000), pp. 1347–1359.

[2] S. Bäro, G. Bauch, A. Pavlic, and A. Semmler, *Improving BLAST performance using space-time block codes and turbo decoding.* Proc. IEEE Globecom, San Francisco, CA, 2002, pp. 1067–1071.

[3] S. Benedetto and E. Biglieri, *Digital Communication Principles with Wireless Applications.* Plenum, New York, 1999.

[4] E. Biglieri, J. Proakis, and S. Shamai (Shitz), *Fading channels: Information-theoretic and communications aspects.* IEEE Trans. Inform. Theory **44** (1998), pp. 2619–2692.

[5] E. Biglieri, G. Taricco, and A. Tulino, *Performance of space-time codes for a large number of antennas.* IEEE Trans. Inform. Theory **48** (2002), pp. 1794–1803.

[6] E. Biglieri, G. Taricco, and A. Tulino, *Decoding space-time codes with BLAST architectures.* IEEE Trans. Signal Processing **50** (2002), pp. 2547–2552.

[7] I. M. Boyarinov and G. L. Katsman, *Linear unequal error protection codes.* IEEE Trans. Inform. Theory, **27** (1981), pp. 168–175.

[8] G. Buch and F. Burkert, *Concatenated Reed-Muller codes for unequal error protection.* IEEE Commun. Letters, **3** (1999), pp. 202–204.

[9] G. Caire and E. Biglieri, *Parallel concatenated codes with unequal error protection.* IEEE Trans. Commun., **46** (1998), pp. 565–567.

[10] W.-J. Choi, R. Negi, and J. M. Cioffi, *Combined ML and DFE decoding for the V-BLAST system.* Proc. IEEE Int. Conf. Commun. , 2000, pp. 1243–1248.

[11] H. El Gamal and R. Hammons, *A new approach to layered space-time coding and signal processing.* IEEE Trans. Inform. Theory, **47** (2001), pp. 2321–2334.

[12] G. J. Foschini, *Layered space-time architecture for wireless communication in a fading environment when using multi-element antennas.* Bell Labs Technical Journal, (Autumn 1996), pp. 41–59.

[13] G. J. Foschini and M. J. Gans, *On limits of wireless communications in a fading environment when using multiple antennas.* Wireless Personal Communications, **6** (1998), pp. 311–335.

[14] G. J. Foschini, G. D. Golden, R. A. Valenzuela, and P. W. Wolniansky, *Simplified processing for high spectral efficiency wireless communication employing multi-element arrays. IEEE Journal on Selected Areas in Communications*, **17** (1999), pp. 1841–1852.

[15] G. Ginis and J. M. Cioffi, *On the relation between V-BLAST and the GDFE.* IEEE Commun. Letters, **5** (2001), pp. 364–366.

[16] B. Hassibi, *An efficient square-root algorithm for BLAST.* Proc. IEEE Int. Conf. on Acoustics, Speech, and Signal Processing (ICASSP 2000), Istanbul, Turkey (2000), pp. II737–II740.

[17] B. Hassibi and B. Hochwald, *High-rate codes that are linear in space and time.* IEEE Trans. Inform. Theory, **48** (2002), pp. 1804–1824.

[18] X. Li, H. Huang, G. J. Foschini, and R. A. Valenzuela, *Effects of iterative detection and decoding on the performance of BLAST.* Proc. IEEE Globecom, San Francisco, CA, (2000), pp. 1061–1066.

[19] B. Masnick and J. K. Wolf, *On linear unequal error protection codes.* IEEE Trans. Inform. Theory, **3** (1967), pp. 600–607.

[20] R. H. Morelos-Zaragoza and H. Imai, *Binary multilevel convolutional codes with unequal error protection capabilities.* IEEE Trans. Commun., **46** (1998), pp. 850–853.

[21] A. Nordio, M. Hernandez, and G. Caire, *Design and performance of a low-complexity iterative multuser joint decoder based on Viterbi decoding and parallel interference cancellation.* Proc. IEEE ICC 2002, New York, 2002, pp. 298–302.

[22] N. Prasad and M. K. Varanasi, *Optimum efficiently decodable layered space-time block codes.* Proc. 35th Asilomar Conf. Signals, Systems, and Computers, (2001).

[23] M. Sellathurai and S. Haykin, *Turbo-BLAST for high-speed wireless communications.* Proc. IEEE WCNC, (2000), pp. 315–320.

[24] P. Stevens, *On decoding unequal error protection product codes.* IEEE Trans. Inform. Theory, **36** (1990), pp. 890–895.

[25] V. Tarokh, N. Seshadri, and A. R. Calderbank, *Space-time codes for high data rate wireless communication: Performance criterion and code construction.* IEEE Trans. Inform. Theory, **44** (1998), pp. 744–765.

[26] W. J. Van Gils, *Linear unequal error protection codes from shorter codes.* IEEE Trans. Inform. Theory, **30** (1984), pp. 544-546.
[27] A. Vardy, *Trellis structure of codes.* In: V. S. Pless *et al.*, *Handbook of Coding Theory.* Elsevier Sc. Publ., Amsterdam, 1998.
[28] S. Verdú, *Multiuser Detection.* Cambridge University Press, 1998.
[29] P. W. Wolniansky, G. J. Foschini, G. D. Golden, and R. A. Valenzuela, *V-BLAST: An architecture for realizing very high data rates over the rich-scattering wireless channel.* Proc. ISSSE-98, Pisa, Italy, (1988).
[30] D. Wuebben, R. Boehnke, J. Rinas, V. Kuehn, and K. D. Kammeyer, *Efficient algorithm for decoding layered space-time codes.* Electronics Letters, **37** (2001), pp. 1348–1349.

Current address: Dipartimento di Elettronica, Politecnico di Torino, Italy
E-mail address: `<name>@polito.it`

DIMACS Series in Discrete Mathematics
and Theoretical Computer Science
Volume **62**, 2003

Diversity Order of Space-Time Block Codes in Inter-Symbol Interference Multiple-Access Channels

S. N. Diggavi, N. Al-Dhahir, and A. R. Calderbank

ABSTRACT. In this paper we study the multiple-access channel where users employ space-time block codes (STBC). The problem is formulated in the context of an inter-symbol interference (ISI) channel which occurs for transmission over frequency-selective channels. Each of the K users transmits using $M_t = 2$ transmit antennas and a time-reversed space-time block code suitable for frequency-selective channels. We show that a diversity order of $2M_r(\nu+1)$ is achievable at full transmission rate for each user, when we have M_r receive antennas, channel memory of ν and an optimal multiuser maximum-likelihood (ML) decoder is used.

1. Introduction

In wireless communication networks, frequency spectrum is a scarce resource that should be efficiently utilized. Since their invention, space-time block codes (STBC) [**4**] have been shown to have the potential to significantly increase the rates and spectral efficiency of wireless transmissions. Given the limited spectral resources, in this paper we consider multiple co-channel users each equipped with two transmit antennas sharing a frequency-selective channel. The goal is to design space–time transmitter and receiver techniques that allow for efficient detection of the co-channel users while realizing rate and diversity gains without bandwidth expansion.

In this paper, we consider time-reversed space–time transmitter and receiver techniques in multiuser environments. Multiuser detection has been a rich area of research with many results related to Code Division Multiple Access (CDMA) systems (see for example [**10**] for more information on this topic). Our focus in this paper is on multiple-antenna transmitters and receivers which employ STBC at the transmitter. The interference cancellation technique presented in [**8**] for flat–fading channels can be directly extended to frequency–selective channels by combining it with either orthogonal frequency division multiplexing (OFDM) or with a single–carrier frequency–domain equalizer (SC–FDE) [**2**]. There are three main reasons for considering the time-reversal (TR) STBC single–carrier technique introduced in [**7**]. The first and perhaps the most important reason is that there is a simple technique to ensure both spatial and multipath diversity gains for two transmit antennas without rate loss.[1] Second, OFDM suffers from the problems of high peak-to-average power ratio and increased sensitivity to frequency synchronization errors. Single–carrier techniques do not suffer from this problem. The last reason is that the performance of the time-reversal technique is better than OFDM and SC–FDE for uncoded systems [**3**] and the three schemes have comparable performance for coded systems.

Previous related work includes space–time interference cancellation techniques for flat–fading channels in [**8**], joint frequency–domain zero–forcing interference cancellation and equalization for frequency–selective channels in [**2**], and extensions to the case of more than two users and more than two transmit antennas in [**1**]. All these techniques realized spatial diversity alone and did not harvest the multipath diversity available.

The main result of this paper is to prove that a diversity order of $2M_r(\nu+1)$ is achievable for each of the K users transmitting using TR-STBC. This code is transmitted at at full rate using $M_t = 2$ transmit antennas over ISI channels of memory ν when we use M_r receive antennas. This paper is organized as follows. In Section 2 we introduce the transmission technique used and set up the notation. In Section 3, we develop the joint ML decoder and present the diversity order result for the optimal decoder. The paper is concluded in Section 4 with the detailed proof given in Appendix.

2. Data Model

We transmit information by encoding over two transmission blocks each of length N (see Figure 1) over which the channel is assumed to be quasi-static. In addition, ν zero symbols are inserted as guard between data blocks to eliminate inter–block interference.

Our starting point is the single–user scenario and the representation of the input–output relationship in the following matrix form

$$\mathbf{y}_1 = \mathbf{H}_{1,1}\mathbf{I}_{zs}\mathbf{c}_{1,1} + \mathbf{H}_{1,2}\mathbf{I}_{zs}\mathbf{c}_{2,1} + \eta_1 \ , \tag{2.1}$$

where $\mathbf{H}_{1,1}$ and $\mathbf{H}_{1,2}$ are square $(N+\nu)$–dimensional lower–triangular Toeplitz matrices whose first columns are equal to the $(\nu+1)$ impulse response coefficients of the channels from the first and second transmit antennas to the first receive antenna appended by $(N-1)$ zeros, respectively. The output and noise vectors $\mathbf{y}_{1,1}$ and $\eta_{1,1}$ are $(N+\nu)$–dimensional while the data vectors $\mathbf{c}_{1,1}$ and $\mathbf{c}_{2,1}$ are N–dimensional. This matrix model assumes the insertion of ν zeros at the end of each data vector to eliminate inter–block interference. This **zero-stuffing** operation is represented in (2.1) by the matrix $\mathbf{I}_{zs} \stackrel{def}{=} \begin{bmatrix} \mathbf{I}_N \\ \mathbf{0}_{\nu\times N} \end{bmatrix}$. According to the TR–STBC scheme [**7**], the output of the second sub–block is given by

$$\mathbf{y}_2 = \mathbf{H}_{1,1}\mathbf{I}_{zs}\mathbf{J}_N\mathbf{c}_{1,2}^* - \mathbf{H}_{1,2}\mathbf{I}_{zs}\mathbf{J}_N\mathbf{c}_{1,1}^* + \eta_2 \ , \tag{2.2}$$

where[2] $\mathbf{J}_N$ is the N–dimensional reversal matrix that consists of ones on the anti–diagonal and zeros everywhere else and $(.)^*$ denotes complex conjugation.

[1] Except for the rate loss associated with the guard sequence which is common to all block transmission schemes over ISI channels.

[2] In this paper, for a complex matrix (or vector) $\mathbf{A}$ we denote by $\mathbf{A}^T$ its transpose, by $\bar{\mathbf{A}}$ its Hermitian transpose, and by $\mathbf{A}^*$ its complex conjugate.

By applying a linear transformation which performs a partial reversal of the second sub-block, it is shown in [**11**] that the following finite-block length form is obtained.

$$\begin{bmatrix} \mathbf{y}_1 \\ \mathbf{P}\mathbf{y}_2^* \end{bmatrix} = \begin{bmatrix} \mathbf{H}_{1,1}^{(c)} & \mathbf{H}_{1,2}^{(c)} \\ -\bar{\mathbf{H}}_{1,2}^{(c)} & \bar{\mathbf{H}}_{1,1}^{(c)} \end{bmatrix} \begin{bmatrix} \mathbf{I}_{zs}\mathbf{c}_1 \\ \mathbf{I}_{zs}\mathbf{c}_2 \end{bmatrix} + \begin{bmatrix} \eta_1 \\ \mathbf{P}\eta_2^* \end{bmatrix}$$

$$\Rightarrow \mathbf{r}_1^{(c)} \stackrel{def}{=} \mathbf{H}_1\mathbf{c} + \eta_1^{(c)}, \tag{2.3}$$

where the matrices $\mathbf{H}_{1,k}^{(c)}$, $k = 1, 2$ represent the $(N + \nu)$-dimensional square circulant matrices derived from $\mathbf{H}_{1,k}$, $k = 1, 2$, $\mathbf{P}$ is a partial permutation matrix[3] [**11**], and $\bar{(.)}$ denotes the complex conjugate transpose operation.

For the case of M_r receive antennas and K users each using $M_t = 2$ transmit antennas, (2.3) can be generalized as follows

$$\mathbf{r} = \begin{bmatrix} \mathbf{r}_1^{(c)} \\ \mathbf{r}_2^{(c)} \\ \vdots \\ \mathbf{r}_{M_r}^{(c)} \end{bmatrix} = \underbrace{\begin{bmatrix} \mathbf{H}_1^{(1)} & \dots & \mathbf{H}_1^{(K)} \\ \mathbf{H}_2^{(1)} & \dots & \mathbf{H}_2^{(K)} \\ \vdots & \vdots & \vdots \\ \mathbf{H}_{M_r}^{(1)} & \dots & \mathbf{H}_{M_r}^{(K)} \end{bmatrix}}_{\mathbf{H}} \underbrace{\begin{bmatrix} \mathbf{c}^{(1)} \\ \mathbf{c}^{(2)} \\ \vdots \\ \mathbf{c}^{(K)} \end{bmatrix}}_{\mathbf{c}} + \underbrace{\begin{bmatrix} \eta_1 \\ \eta_2 \\ \vdots \\ \eta_{M_r} \end{bmatrix}}_{\eta} \tag{2.4}$$

where $\mathbf{H}_m^{(k)}$, $k = 1, \dots, K$, $m = 1, \dots, M_r$ is the channel matrix from the kth user to the mth receive antenna and $\mathbf{c}^{(k)}$ is the data vector of the kth user.

3. Diversity Order of ML Decoder

Given the multiuser input-output model in (2.4), we can develop the optimal joint multiuser detector based on maximum-likelihood (ML) decoding [**10**]. We will illustrate this using the matrix model of (2.4). The ML decoding metric is

$$\{\mathbf{c}^{(1)}, \mathbf{c}^{(2)}, \dots, \mathbf{c}^{(K)}\} = \text{argmin}_{\mathbf{c}} ||\mathbf{r} - \mathbf{Hc}||^2 \tag{3.1}$$

which is computed using a joint trellis implementing the Viterbi algorithm [**10**].

The notion of diversity order for space-time codes has been defined in [**9**] as follows.

DEFINITION 3.1. A coding scheme which has an average error probability $P_e(SNR)$ as a function of SNR that behaves as

$$\lim_{SNR \to \infty} \frac{\log(P_e(SNR))}{\log(SNR)} = -d \tag{3.2}$$

is said to have a diversity order of d.

In words, a scheme with diversity order d has an error probability at high SNR behaving as $P_e(SNR) \approx SNR^{-d}$. The notion of full transmission rate (as defined in [**9**]) implies that if we use a constellation size of 2^b for transmission, the space time code sends b bits/sec/Hz information symbols. For example the STBC defined by Alamouti [**4**] has full transmission rate since 2 information symbols are sent over two time units.

[3]For $\mathbf{a}^T = [\mathbf{a}(0) \cdots \mathbf{a}(N + \nu - 1)]$, $\mathbf{Pa} = [\mathbf{a}(N-1) \cdots \mathbf{a}(0)\mathbf{a}(N + \nu - 1) \cdots \mathbf{a}(N)]^T$, where $(.)^T$ denotes the transpose.

For the multiple-access channel defined in (2.4), using the ML decoding metric given in (3.1), we can prove the following result on the diversity order of TR-STBC transmissions.

THEOREM 3.2. *A multiple-access system with M_r receive antennas and K synchronous users each transmitting TR-STBC signals using $M_t = 2$ antennas over ISI channels with memory ν achieves a diversity order of $2M_r(\nu + 1)$ at full transmission rate for each user.*

The proof of Theorem 3.2 is given in Appendix A. This result implies that if optimum decoding is used then the performance observed by any individual user is equivalent to the system where only that user is transmitting. This is quite satisfying since we know (see [**7**],[**11**]) that for a single-user system with $M_r = 2$, the TR-STBC achieves the maximal order of diversity $2M_r(\nu + 1)$.

In order to achieve the diversity order predicted in Theorem 3.2, we would need to do joint multiuser ML decoding of the K-user multiple-access ISI channel. This is computationally expensive with the decoding complexity being exponential in the channel length ν, the number of users K, and the spectral efficiency of the signal constellation. This motivates the suboptimal reduced-complexity multiuser linear detector structures studied in [**5**].

4. Conclusion

In this paper we presented a space-time combined interference suppression, equalization, and decoding scheme for multiple synchronous co-channel users each equipped with two transmit antennas over frequency–selective channels. We demonstrated that a diversity order of $2M_r(\nu + 1)$ is achievable when optimum ML decoding is applied to detect users transmitting using TR-STBC. The extensions of this idea to other orthogonal designs and reduced-complexity detectors are natural questions to ask next.

Appendix A. Proof of Theorem 3.2

We give this proof for the case of two users ($K = 2$) and $M_r = 2$ receive antennas. The steps can be very easily generalized to the case with $K > 2$ users and $M_r > 2$ receive antennas. The idea of the proof is that the pairwise error probability (PEP) of the ML decoder can be derived in terms of the error vectors of the different users of the multiple-access channel. By using the derived expression of the PEP, we can show that the diversity order is achievable. The proof relies quite heavily on the quaternionic structure that the TR-STBC imposes on the equivalent channels in (2.4).

We begin with the following well-known observation [**6**] on circulant matrices.

FACT A.1. A circulant matrix $\mathbf{C}$ of size M has an eigendecomposition $\mathbf{C} = \mathbf{Q}\Lambda\bar{\mathbf{Q}}$ with its eigenvectors as the Fourier matrix $\mathbf{Q}$ whose elements are given by $[\mathbf{Q}]_{p,q} = exp(-j\frac{2\pi}{M}(p-1)(q-1))$. Moreover, $\Lambda = \text{diag}(\mathbf{Q}\mathbf{c}_1)$, where $\mathbf{c}_1$ is the first column of $\mathbf{C}$, and $\text{diag}(\cdot)$ creates a diagonal matrix from the elements of a vector.

Using this fact, we can represent (2.4) (for the two-user case) in the frequency domain as

$$\mathbf{R} = \begin{bmatrix} \Lambda_{\mathbf{H_1}} & \Lambda_{\mathbf{G_1}} \\ \Lambda_{\mathbf{H_2}} & \Lambda_{\mathbf{G_2}} \end{bmatrix} \begin{bmatrix} \tilde{\mathbf{C}} \\ \tilde{\mathbf{S}} \end{bmatrix} + \begin{bmatrix} \tilde{\eta}_1 \\ \tilde{\eta}_2 \end{bmatrix}, \tag{A.1}$$

where $\mathbf{R} = [\mathbf{R}_1^T, \mathbf{R}_2^T]^T$, with $\mathbf{R}_k = \breve{\mathbf{Q}}\mathbf{r}_k$ and

$$\breve{\mathbf{Q}} = \begin{bmatrix} \mathbf{Q} & \mathbf{0} \\ \mathbf{0} & \mathbf{Q} \end{bmatrix}. \tag{A.2}$$

Furthermore,

$$\Lambda_{H_k} = \begin{bmatrix} \Lambda_{H_{k,1}} & \Lambda_{H_{k,2}} \\ -\bar{\Lambda}_{H_{k,2}} & \bar{\Lambda}_{H_{k,1}} \end{bmatrix}, \tag{A.3}$$

with $\Lambda_{H_{k,l}}$, $l = 1, 2$ contains the eigenvalues of $\mathbf{H}_{k,l}^{(1)}$. Similarly,

$$\Lambda_{G_k} = \begin{bmatrix} \Lambda_{G_{k,1}} & \Lambda_{G_{k,2}} \\ -\bar{\Lambda}_{G_{k,2}} & \bar{\Lambda}_{G_{k,1}} \end{bmatrix}, \tag{A.4}$$

with $\Lambda_{G_{k,l}}$, $l = 1, 2$ contains the eigenvalues of $\mathbf{H}_{k,l}^{(2)}$. The data vectors are written as $\tilde{\mathbf{C}} = \breve{\mathbf{Q}}\mathbf{c}^{(1)}$ and $\tilde{\mathbf{S}} = \breve{\mathbf{Q}}\mathbf{c}^{(2)}$. Given that the Fourier transformation is orthonormal, the noise is still white Gaussian noise with variance 1.

The PEP for (A.1) can be bounded as [**9**],

$$\mathbb{P}(\mathbf{c} \to \mathbf{c}' | \mathbf{H}^{(1)}, \mathbf{H}^{(2)}) \leq exp(-d^2(\mathbf{R}, \mathbf{R}')/4), \tag{A.5}$$

where

$$d^2(\mathbf{R}, \mathbf{R}') = \begin{bmatrix} \bar{\mathbf{e}}_c & \bar{\mathbf{e}}_s \end{bmatrix} \begin{bmatrix} \bar{\Lambda}_{H_1} & \bar{\Lambda}_{H_2} \\ \bar{\Lambda}_{G_2} & \bar{\Lambda}_{G_2} \end{bmatrix} \begin{bmatrix} \Lambda_{\mathbf{H_1}} & \Lambda_{\mathbf{G_1}} \\ \Lambda_{\mathbf{H_2}} & \Lambda_{\mathbf{G_2}} \end{bmatrix} \begin{bmatrix} \mathbf{e}_c \\ \mathbf{e}_s \end{bmatrix}, \tag{A.6}$$

and $\mathbf{e}_c = \mathbf{c}^{(1)} - \mathbf{c}^{(1)'} = [\mathbf{e}_{c_1}^T, \mathbf{e}_{c_2}^T]^T$. After re-arranging terms, we can rewrite the quadratic form in (A.6) as

(A.7)

$$d^2(\mathbf{R}, \mathbf{R}') = \begin{bmatrix} \bar{\lambda}_{H_1} & \bar{\lambda}_{H_2} & \bar{\lambda}_{G_1} & \bar{\lambda}_{G_2} \end{bmatrix} \begin{bmatrix} \mathbf{I}_2 \otimes (\mathbf{E}_c\bar{\mathbf{E}}_c) & \mathbf{I}_2 \otimes (\mathbf{E}_c\bar{\mathbf{E}}_s) \\ \mathbf{I}_2 \otimes (\mathbf{E}_s\bar{\mathbf{E}}_c) & \mathbf{I}_2 \otimes (\mathbf{E}_s\bar{\mathbf{E}}_s) \end{bmatrix} \begin{bmatrix} \lambda_{H_1} \\ \lambda_{H_2} \\ \lambda_{G_1} \\ \lambda_{G_2} \end{bmatrix},$$

where $\bar{\lambda}_{H_k} = [\bar{\lambda}_{H_{k,1}}, \bar{\lambda}_{H_{k,2}}]$, $\lambda_{H_{k,1}} = vec(\Lambda_{H_{k,1}})$, and $\bar{\lambda}_{G_k}$ is defined similarly. Applying the $vec(.)$ operation to a diagonal matrix constructs a vector from its diagonal elements. Also

$$\mathbf{E}_c = \begin{bmatrix} \mathbf{E}_{c_1} & -\bar{\mathbf{E}}_{c_2} \\ \mathbf{E}_{c_2} & \bar{\mathbf{E}}_{c_1} \end{bmatrix}, \quad \mathbf{E}_s = \begin{bmatrix} \mathbf{E}_{s_1} & -\bar{\mathbf{E}}_{s_2} \\ \mathbf{E}_{s_2} & \bar{\mathbf{E}}_{s_1}^* \end{bmatrix}, \tag{A.8}$$

with $\mathbf{E}_{c_k} = diag(\mathbf{e}_{c_k})$, $\mathbf{E}_{s_k} = diag(\mathbf{e}_{s_k})$, $k = 1, 2$. Using Fact A.1, we can write $\lambda_{H_k} = \breve{\mathbf{Q}}\mathbf{h}_k$, with $\mathbf{h}_k = [\mathbf{h}_{k,1}^T, \mathbf{h}_{k,1}^T]^T$ being the first columns of the circulant matrices $\mathbf{H}_{k,l}^{(1)}$. Since we have FIR channels, only the first $\nu + 1$ elements of $\mathbf{h}_{k,1}$ are non-zero, and we denote this by the length-$(\nu + 1)$ vector $\tilde{\mathbf{h}}_{k,1}$ and form $\tilde{\mathbf{h}}_k = [\tilde{\mathbf{h}}_{k,1}^T, \tilde{\mathbf{h}}_{k,2}^T]^T$. The definition for λ_{G_k} is identical and we do not repeat it here. Hence the quadratic form in (A.7) can be rewritten as

(A.9)

$$d^2(\mathbf{R}, \mathbf{R}') = \begin{bmatrix} \bar{\tilde{\mathbf{h}}}_1 & \bar{\tilde{\mathbf{h}}}_2 & \bar{\tilde{\mathbf{g}}}_1 & \bar{\tilde{\mathbf{g}}}_2 \end{bmatrix} \underbrace{\begin{bmatrix} \mathbf{I}_2 \otimes (\bar{\tilde{\mathbf{Q}}}\mathbf{E}_c\bar{\mathbf{E}}_c\tilde{\mathbf{Q}}) & \mathbf{I}_2 \otimes (\bar{\tilde{\mathbf{Q}}}\mathbf{E}_c\bar{\mathbf{E}}_s\tilde{\mathbf{Q}}) \\ \mathbf{I}_2 \otimes (\bar{\tilde{\mathbf{Q}}}\mathbf{E}_s\bar{\mathbf{E}}_c\tilde{\mathbf{Q}}) & \mathbf{I}_2 \otimes (\bar{\tilde{\mathbf{Q}}}\mathbf{E}_s\bar{\mathbf{E}}_s\tilde{\mathbf{Q}}) \end{bmatrix}}_{\mathcal{E}} \begin{bmatrix} \tilde{\mathbf{h}}_1 \\ \tilde{\mathbf{h}}_2 \\ \tilde{\mathbf{g}}_1 \\ \tilde{\mathbf{g}}_2 \end{bmatrix},$$

where $\tilde{\mathbf{Q}}$ is a $2(N+\nu) \times 2(\nu+1)$ block diagonal matrix with the blocks being the first $(\nu+1)$ columns of the Fourier matrix $\mathbf{Q}$. Therefore, inserting this in (A.5) and averaging the Gaussian quadratic form over the channel parameters in a standard manner (see for example [**9**]), we obtain

$$\text{(A.10)} \qquad \mathbb{P}(\mathbf{c} \to \mathbf{c}') \leq \prod_{l=1}^{8(\nu+1)} \frac{1}{1 + \frac{\gamma(l) SNR}{4}},$$

where $\gamma(l)$ represents the eigenvalues of $\mathcal{E}$ in the quadratic form of (A.9). Hence the diversity order depends on the rank of $\mathcal{E}$, and we next show that this matrix has rank $4(\nu+1)$ for any error sequence on each user yielding the diversity order result claimed in Theorem 3.2.

First note that in (A.9), since we are examining the diversity order for any user, the maximal rank of $\mathcal{E}$ is $4(\nu+1)$. This is true because we can set, for example, $\mathbf{E}_c \neq 0, \mathbf{E}_s = 0$ when there is an error only in the first user's transmission. The main step is to show that this maximal rank is achieved and therefore giving us the result claimed in Theorem 3.2. Observe that by row and column interchanges of $\mathcal{E}$, it can be written as

$$\text{(A.11)} \quad \mathbf{I}_2 \otimes \begin{bmatrix} \bar{\tilde{\mathbf{Q}}}\mathbf{E}_c\bar{\mathbf{E}}_c\tilde{\mathbf{Q}} & \bar{\tilde{\mathbf{Q}}}\mathbf{E}_c\bar{\mathbf{E}}_s\tilde{\mathbf{Q}} \\ \bar{\tilde{\mathbf{Q}}}\mathbf{E}_s\bar{\mathbf{E}}_c\tilde{\mathbf{Q}} & \bar{\tilde{\mathbf{Q}}}\mathbf{E}_s\bar{\mathbf{E}}_s\tilde{\mathbf{Q}} \end{bmatrix} = \mathbf{I}_2 \otimes \left(\begin{bmatrix} \bar{\tilde{\mathbf{Q}}}\mathbf{E}_c \\ \bar{\tilde{\mathbf{Q}}}\mathbf{E}_s \end{bmatrix} \begin{bmatrix} \bar{\mathbf{E}}_c\tilde{\mathbf{Q}} & \bar{\mathbf{E}}_s\tilde{\mathbf{Q}} \end{bmatrix} \right).$$

Next we will show that $\bar{\tilde{\mathbf{Q}}}\mathbf{E}_c$ has rank $2(\nu+1)$ if $\mathbf{E}_c \neq 0$. This will prove that $\mathcal{E}$ achieves a rank of $4(\nu+1)$. Let the $(N+\nu) \times (\nu+1)$ matrix containing the first $(\nu+1)$ columns of $\mathbf{Q}$ be denoted by $\mathbf{T}_Q$ and $\bar{\mathbf{T}}_Q = [\psi_1, \ldots, \psi_{N+\nu}]$ with $[\psi_k]_p = e^{j\frac{2\pi}{N+\nu}(k-1)(p-1)}$, $k = 1, \ldots, N+\nu$, $p = 1, \ldots, \nu+1$. Since $\mathbf{T}_Q$ is the truncated Fourier matrix it has full rank of $(\nu+1)$ and hence any $(\nu+1)$ rows of $\mathbf{T}_Q$ are linearly independent. Therefore, this allows to write

$$\text{(A.12)} \qquad \bar{\tilde{\mathbf{Q}}}\mathbf{E}_c = \begin{bmatrix} \bar{\mathbf{T}}_Q\mathbf{E}_{c_1} & -\bar{\mathbf{T}}_Q\bar{\mathbf{E}}_{c_2} \\ \bar{\mathbf{T}}_Q\mathbf{E}_{c_2} & \bar{\mathbf{T}}_Q\bar{\mathbf{E}}_{c_1} \end{bmatrix}.$$

It is easy to show that if $\mathbf{c}_k - \mathbf{c}_k' \neq 0,\ k = 1, 2$ then the diagonal matrix $\mathbf{E}_{c_k}$ has at least $(\nu+1)$ non-zero entries using the property that $\mathbf{QI}_{zs}$ has full rank[4]. Without loss of generality we can assume $\mathbf{c}_1 - \mathbf{c}_1' \neq 0$, then $\mathbf{E}_{c_1}$ has at least $(\nu+1)$ non-zero entries and let us choose the $2(\nu+1)$ columns corresponding to those entries in (A.12), and denote this set of columns by $f(1), \ldots, f(\nu+1)$. If these columns are linearly dependent, then there exist scalars $\{\alpha_l, \beta_l\}_{l=1}^{\nu+1}$ not all zero such that

$$\text{(A.13)} \quad \sum_{l=1}^{\nu+1} \left\{ \alpha_l \begin{bmatrix} \mathbf{e}_{c_1}(f(l))\psi_{f(l)} \\ \mathbf{e}_{c_2}(f(l))\psi_{f(l)} \end{bmatrix} + \beta_l \begin{bmatrix} -\bar{\mathbf{e}}_{c_2}(f(l))\psi_{f(l)} \\ \bar{\mathbf{e}}_{c_1}(f(l))\psi_{f(l)} \end{bmatrix} \right\} = \mathbf{0}_{2(N+\nu)\times 1}.$$

Since $\mathbf{T}_Q$ is full rank of $\nu+1$ this reduces to the simpler set of equations

$$\text{(A.14)} \quad \begin{bmatrix} \tilde{\mathbf{e}}_{c_1}(f(l)) & -\bar{\tilde{\mathbf{e}}}_{c_2}(f(l)) \\ \tilde{\mathbf{e}}_{c_2}(f(l)) & \bar{\tilde{\mathbf{e}}}_{c_1}(f(l)) \end{bmatrix} \begin{bmatrix} \alpha_l \\ \beta_l \end{bmatrix} = \mathbf{0}_{2\times 1},\ l = 1, \ldots, \nu+1$$

which cannot be true since we have chosen $f(l)$ such that $\tilde{\mathbf{e}}_{c_1}(f(l)) \neq 0,\ \forall l$. Hence, there do not exist scalars $\{\alpha_l, \beta_l\}_{l=1}^{\nu+1}$ not all zero such that (A.13) holds. This proves that $\bar{\tilde{\mathbf{Q}}}\mathbf{E}_c$ has rank $2(\nu+1)$. To conclude the proof of Theorem 3.2 for $K = 2, M_r = 2$, we can use the standard union bound argument to bound the

[4]Remember that $\mathbf{e}_{c_k} = \mathbf{QI}_{zs}(\mathbf{c}_k - \mathbf{c}_k')$, $k = 1, 2$ and $\mathbf{E}_{c_k} = diag(\mathbf{e}_{c_k})$.

average probability in terms of the pairwise error probability. Since we are using a constant rate code and the rank of $\mathcal{E}$ is $4(\nu+1)$, the PEP decays at a rate $\frac{1}{SNR^{4(\nu+1)}}$, therefore the diversity order according to Definition 3.1 is $4(\nu+1)$.

For the general case, the proof structure is identical. It can be easily verified that the equivalent $M_rK(N+\nu)$ square matrix $\mathcal{E}$ in (A.9) has the (p,q)th square block components as $\mathbf{I}_{M_r} \otimes (\mathbf{E}_{\mathbf{C}^{(p)}}\bar{\mathbf{E}}_{\mathbf{C}^{(q)}})$ of size $M_r2(N+\nu)$. The rank of $\mathcal{E}$ would again be determined by the rank of $\tilde{\bar{\mathbf{Q}}}\mathbf{E}_c$ which is found by the above argument to be $2(\nu+1)$, yielding the rank of $\mathcal{E}$ as $2M_r(\nu+1)$. This will allow us to show that the diversity order is $2M_r(\nu+1)$.

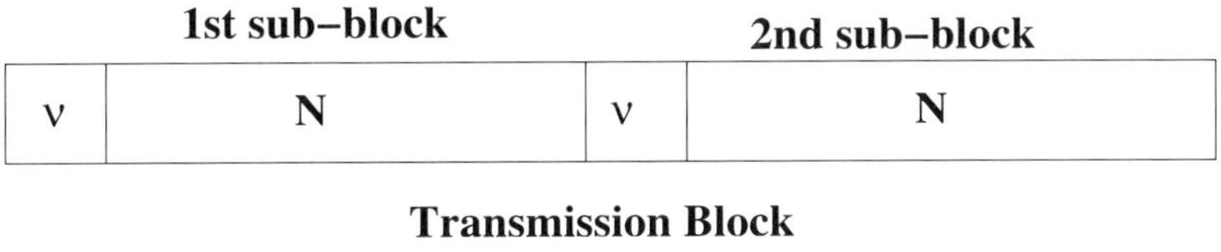

FIGURE 1. Block Transmission Format

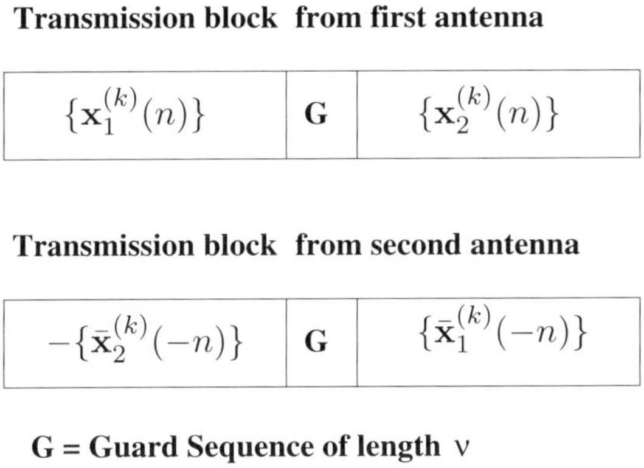

FIGURE 2. Transmission Format for Time-Reversal Space-Time Block Coding

References

[1] A. Stamoulis and N. Al-Dhahir and A.R. Calderbank. Further Results on Interference Cancellation and Space-Time Block Codes. In *Asilomar*, pages 257–262, October 2001.

[2] N. Al-Dhahir. Single-Carrier Frequency-Domain Equalization for Space-Time Block-Coded Transmissions over Broadband Wireless Channels. In *PIMRC*, pages 143–146, September 2001.

[3] N. Al-Dhahir, M. Uysal, and C. Georghiades. Three Space-Time Block-Coding for Frequency-Selective Fading Channels with Application to EDGE. In *VTC*, pages 1834–1838, October 2001.

[4] S.M. Alamouti. A simple transmit diversity technique for wireless communications. *IEEE Journal on Selected Areas in Communications*, 16(8):1451–1458, October 1998.

[5] S. N. Diggavi, N. Al-Dhahir, and A.R. Calderbank. Algebraic properties of multiuser joint equalization and decoding of space-time block codes. AT&T Technical Report, Tech Rep # TD-5FSQX5, 2002.

[6] R. Gray. On the Asymptotic Eigenvalue Distribution of Toeplitz Matrices. *IEEE Transactions on Information Theory*, 18(6), November 1972.

[7] E. Lindskog and A. Paulraj. A transmit diversity scheme for channels with intersymbol interference. In *ICC 2000*, volume 1, pages 307–311, 2000.

[8] A.F. Naguib, N. Seshadri, and A.R. Calderbank. Applications of Space-Time Block Codes and Interference Suppression for High Capacity and High Data Rate Wireless Systems. In *Thirty-Second Asilomar Conference on Signals, Systems and Computers*, pages 1803 –1810, 1998.

[9] V. Tarokh, N. Seshadri, and A.R. Calderbank. Space-time codes for high data rate wireless communications: Performance criterion and code construction. *IEEE Transactions on Information Theory*, 44(2):744–765, March 1998.

[10] S. Verdu. *Multiuser Detection.* Cambridge University Press, United Kingdom, 1998.

[11] A. Zhou and G. B. Giannakis. Space-time coding with maximum diversity gains over frequency-selective fading channels. *IEEE Signal Processing Letters*, 8(10):269–272, October 2001.

AT&T Shannon Laboratory, 180 Park Avenue, Bldg. 103, Florham Park, New Jersey, NJ 07932, USA.

E-mail address: suhas@research.att.com

AT&T Shannon Laboratory, 180 Park Avenue, Bldg. 103, Florham Park, New Jersey, NJ 07932, USA.

E-mail address: naofal@research.att.com

AT&T Shannon Laboratory, 180 Park Avenue, Bldg. 103, Florham Park, New Jersey, NJ 07932, USA.

E-mail address: rc@research.att.com

DIMACS Series in Discrete Mathematics
and Theoretical Computer Science
Volume **62**, 2003

Pilot Assisted Estimation of MIMO Fading Channel Response and Achievable Data Rates

Dragan Samardzija and Narayan Mandayam

ABSTRACT. In this paper we analyze the effects of pilot assisted channel estimation error of a frequency-flat time-varying channel on achievable data rates. We consider multiple-input multiple-output (MIMO) wireless systems and emphasize differences with respect to single-input single-output (SISO) systems. The analysis connects results of information theory with practical wireless communication systems. Using a block-fading channel model, where the block length corresponds to the channel coherence time, we consider two pilot based approaches for the estimation. Per transmit antenna, the first approach uses a single pilot symbol per block with different power than data symbols, while the second approach uses more than one pilot symbol per block with same power as data symbols. In the MIMO case, the orthogonality between the pilots assigned to different transmit antennas is assumed. The effects of the estimation error are evaluated in the case of the estimates being available at the receiver only (open loop). The presented analysis is a study of mismatched receiver algorithms in MIMO systems.

1. Introduction

Fading channels are an important element of any wireless propagation environment [9]. Different aspects of fading channels have been studied and publicized. It has been recognized that the inherent temporal and spatial variations of wireless channels impose stringent demand on design of a communication system to allow it approach the data rates that are achievable in, for example, wire-line systems. A number of different solutions exploit variations in wireless channels. For example, a transmitter optimization scheme (using power control), known as the water pouring algorithm, maximizes the capacity for the constrained average transmit power [12] (see also [3]). In addition to power control, recent applications of variable coding rate and modulation formats illustrate a wide range of resource allocation techniques used to exploit and combat effects of fading channels in multiuser wireless systems [24]. An extensive review of the information theoretical aspects of communications in fading channels is given in [1]. Furthermore, modulation and channel coding for fading channels is also being studied (see [1] and references therein).

Multiple-transmit multiple-receive antenna systems represent an implementation of the MIMO concept in wireless communications [6]. This particular multiple

---

Key words and phrases. Estimation, Ergodic Capacity, MIMO.

antenna architecture provides high capacity wireless communications in rich scattering environments. It has been shown that the theoretical capacity (approximately) increases linearly as the number of antennas is increased [6, 7]. This and related results point to the importance of understanding all aspects of MIMO wireless systems. For example, the studies regarding propagation [4, 5, 23], detection [8, 21], space-time coding and implementation aspects [11, 20, 27] are well publicized.

Next generation wireless systems and standards are supposed to operate over wireless channels whose variations are faster and/or further pronounced. For example, using higher carrier frequencies (e.g., 5 GHz for 802.11a) results in smaller scale of spatial variations of the electro-magnetic field. Also, compared to SISO channels, MIMO channels have greater number of parameters that a receiver and/or transmitter has to operate with, consequently pronouncing the channel variations. In addition, there has been a perpetual need for supporting higher mobility within wireless networks. These are just a few motivations for studying the implications of channel variations on achievable data rates in wireless systems.

In this paper we analyze how the estimation error of the channel response affects the performance of a MIMO wireless system. Considering the practical importance of single-input single-output (SISO) systems, we analyze them as a subset of MIMO systems. Considering terminology in literature (see [1] and references therein), the channel response estimate corresponds to *channel state information* (CSI). We assume a frequency-flat time-varying wireless channel with additive white Gaussian noise (AWGN). The above system may also correspond to one subchannel (i.e., carrier) of an OFDM wireless system [25]. We consider two pilot arrangement schemes in this paper. The first scheme uses a single pilot symbol per block with the different power than the data symbol power. The second scheme uses more than one pilot symbol per block, whose power is the same as the data symbol power. For the given pilot schemes, in both cases, *maximum-likelihood* (ML) estimation of the channel response is considered [18]. In the MIMO case, the orthogonality between the pilots assigned to different transmit antennas is assumed. The effects of the estimation error are evaluated in the case of the estimates being available at the receiver only. The presented analysis may be viewed as a study of mismatched receiver algorithms in MIMO systems. The analysis connects results of information theory (see [16, 22] and references therein) with practical wireless communication systems (employing pilot assisted channel estimation) and generalizing it to MIMO systems. We believe that the presented results are directly applicable to current and next generation wireless systems. Furthermore, the results may be used as baseline benchmarks for performance evaluation of more advanced estimation, such as anticipated in future systems.

2. System Model

In the following we present a MIMO communication system that consists on M transmit and N receive antennas (denoted as a $M \times N$ system). At the receiver we assume sampling with the period $T_{smp} = 1/B$, where B is the signal bandwidth, thus preserving the sufficient statistics. The received signal is a spatial vector $\mathbf{y}$

$$\mathbf{y}(k) = \mathbf{H}(k)\mathbf{x}(k) + \mathbf{n}(k), \quad \mathbf{y}(k) \in \mathcal{C}^N, \mathbf{x}(k) \in \mathcal{C}^M, \mathbf{n}(k) \in \mathcal{C}^N, \mathbf{H}(k) \in \mathcal{C}^{N\times M} \tag{2.1}$$

where $\mathbf{x}$ is the transmitted vector, $\mathbf{n}$ is AWGN ($\mathrm{E}[\mathbf{n}\mathbf{n}^{\mathrm{H}}] = N_0\,\mathbf{I}_{N\times N}$), and $\mathbf{H}$ is the MIMO channel response matrix, all corresponding to the time instance k. We assign index $m = 1, \cdots, M$ to denote the transmit antennas, and index $n = 1, \cdots, N$ to denote the receive antennas. Thus, $h_{nm}(k)$ is the n-th row and m-th column element of the matrix $\mathbf{H}(k)$. Note that it corresponds to a SISO channel response between the transmit antenna m and the receive antenna n. The n-th component of the received spatial vector $\mathbf{y}(k) = [y_1(k) \cdots y_N(k)]^{\mathrm{T}}$ (i.e., signal at the receive antenna n) is

$$y_n(k) = \sum_{m=1}^{M} h_{nm}(k) g_m(k) + n_n(k) \tag{2.2}$$

where $g_m(k)$ is the transmitted signal from the m-th transmit antenna (i.e., $\mathbf{x}(k) = [g_1(k) \cdots g_M(k)]^{\mathrm{T}}$), and $\mathbf{n}(k) = [n_1(k) \cdots n_N(k)]^{\mathrm{T}}$. To perform estimation of the channel response $\mathbf{H}(k)$, the receiver uses a pilot (training) signal that is a part of the transmitted data. The pilot is sent periodically, every K sample periods. We consider the transmitted signal to be comprised of two parts: one is the data bearing signal and the other is the pilot signal. Within the pilot period consisting of K symbols, L symbols (i.e., signal dimensions) are allocated to the pilot, per transmit antenna. As a common practical solution (see [11,20,27]), we assume that the pilot signals assigned to the different transmit antennas, are mutually orthogonal. For more details on signal design for multiple transmit antenna systems see also [13,14]. Consequently we define a K-dimensional temporal vector $\mathbf{g}_m = [g_m(1) \cdots g_m(K)]^{\mathrm{T}}$, whose k-th component is $g_m(k)$ (in (2.2)), is defined as

$$\mathbf{g}_m = \underbrace{\sum_{i=1}^{K-LM} a_{im}^{d} d_{im}^{d} \mathbf{s}_i^{d}}_{Data} + \underbrace{\sum_{j=1}^{L} a_{jm}^{p} d_{jm}^{p} \mathbf{s}_{jm}^{p}}_{Pilot} . \tag{2.3}$$

In the above the first sum is the information, i.e., data bearing signal and the second corresponds to the pilot signal, corresponding to the transmit antenna m. Superscripts "d" and "p" denote values assigned to the data and pilot, respectively. d_{im}^{d} is the unit-variance circularly symmetric complex data symbol. The pilot symbols (d_{jm}^{p}, $j = 1, \cdots, L$) are predefined and known at the receiver. Without loss of generality, we assume that $|d_{jm}^{p}|^2 = 1$. We also assume that the amplitudes are $a_{im}^{d} = A$, and $a_{jm}^{p} = A_P$, and they are known at the receiver. Further, the amplitudes are related as $A_P = \alpha A$. Note that the amplitudes are identical across the transmit antennas (because we assumed that the transmit power is equally distributed across them).

Furthermore, $\mathbf{s}_i^{d} = [s_i^{d}(1) \cdots s_i^{d}(K)]^{\mathrm{T}}$, $(i = 1, \cdots, (K - LM))$ and $\mathbf{s}_{jm}^{p} = [s_{jm}^{p}(1) \cdots s_{jm}^{p}(K)]^{\mathrm{T}}$, $(j = 1, \cdots, L$, and $m = 1, \ldots M)$ are waveforms, denoted as *temporal signatures*. The temporal signatures are mutually orthogonal. For example, $\mathbf{s}_i^{d}$ (or $\mathbf{s}_{jm}^{p}$) could be a canonical waveform (i.e., could be a TDMA-like waveform, where $\mathbf{s}_i^{d}$ (or $\mathbf{s}_{jm}^{p}$) is the unit-pulse at the time instance i). Alternately, $\mathbf{s}_i^{d}$ (or $\mathbf{s}_{jm}^{p}$) could also be a K-dimensional Walsh sequence spanning all K sample intervals (i.e., similar to a CDMA-like waveform [15]). As said earlier, we assume that the pilot signals are orthogonal between the transmit antennas. The indexing and summation limits in (2.3) conform to this assumption, i.e, temporal signatures $\mathbf{s}_{jm}^{p}(j = 1, \cdots, L)$ are uniquely assigned to the transmit antenna m. In other words, transmit antenna m must not use the temporal signatures that are assigned as pilots to other antennas and assigned to data, which is consequently lowering the achievable data rates (will be revisited in the following sections). Unlike the pilot temporal signatures, the data bearing temporal signatures $\mathbf{s}_i^d$ $(i = 1, \ldots, (K - LM))$ are reused across the transmit antennas, which is an inherent property of MIMO systems, potentially resulting in high achievable data rates.

We rewrite the received spatial vector in (2.1) as

$$\mathbf{y}(k) = \mathbf{H}(k)(\mathbf{d}(k) + \mathbf{p}(k)) + \mathbf{n}(k), \quad \mathbf{d}(k) \in \mathcal{C}^M, \mathbf{p}(k) \in \mathcal{C}^M \tag{2.4}$$

where $\mathbf{d}(k)$ is the information, i.e., data bearing signal and $\mathbf{p}(k)$ is the pilot portion of the transmitted spatial signal, at the time instance k. The m-th component of the data vector vector $\mathbf{d}(k) = [d_1(k) \cdots d_M(k)]^{\mathrm{T}}$ (i.e., data signal at the transmit antenna m) is

$$d_m(k) = \sum_{i=1}^{K-LM} a_{im}^d d_{im}^d s_i^d(k). \tag{2.5}$$

The m-th component of the pilot vector $\mathbf{p}(k) = [p_1(k) \cdots p_M(k)]^{\mathrm{T}}$ (i.e., pilot signal at the transmit antenna m) is

$$p_m(k) = \sum_{j=1}^{L} a_{jm}^p d_{jm}^p s_{jm}^p(k). \tag{2.6}$$

Let us now describe the assumed properties of the MIMO channel $\mathbf{H}(k)$. The channel coherence time is assumed to be greater or equal to KT_{smp}. This assumption approximates the channel to be constant over at least K samples ($h_{nm}(k) \approx h_{nm}$, for $k = 1, \cdots, K$, for all m and n), i.e., approximately constant during the pilot period. In the literature, channels with the above property are known as block-fading channels [25]. Furthermore, we assume that the elements of $\mathbf{H}$ are independent identically distributed (*iid*) random variables, corresponding to highly scattering channels. In general, the MIMO propagation measurements and modeling have shown that the elements of $\mathbf{H}$ are correlated (i.e., not independent) [4,5,23]. In addition, assuming independence is a common practice because the information about correlation is usually not available at the receiver and/or the correlation is time varying (not stationary) and hard to estimate. Based on the above, the received temporal vector at the receiver n, whose k-th component is $y_n(k)$ (in (2.2)), is

$$\mathbf{r}_n = [y_n(1) \cdots y_n(K)]^{\mathrm{T}} = \sum_{m=1}^{M} h_{nm}\mathbf{g}_m + \mathbf{n}_n, \ \ \mathbf{r}_n \in \mathcal{C}^K \tag{2.7}$$

where $\mathbf{n}_n = [n_n(1) \cdots n_n(K)]^{\mathrm{T}}$ and $\mathrm{E}[\mathbf{n}_n \mathbf{n}_n^{\mathrm{H}}] = N_0 \, \mathbf{I}_{K \times K}$.

Note that when applying different number of transmit antennas, the total average transmitted power must remain the same, i.e., conserved. This is a common assumption in MIMO systems [6, 7]. Also, the power is equally distributed across the transmit antennas. The average transmit power (from all transmit antennas) is

$$P_{av} = M \frac{\left(\sum_{i=1}^{K-LM} (a_{im}^d)^2 + \sum_{j=1}^{L} (a_{jm}^p)^2\right)}{K} = M \frac{((K - LM) + L\alpha^2)A^2}{K}. \tag{2.8}$$

Thus

$$A = \sqrt{\frac{K}{((K - LM) + \alpha^2 L)} \frac{P_{av}}{M}}. \tag{2.9}$$

As seen from the above, we assume that the total average transmitted energy (within the pilot period) is the same, but differently distributed between the data bearing portion of the signal and the pilot. Consequently, we observe the performance of the system with respect to the amount of transmitted energy that is allocated to the pilot (percentage wise). This percentage is denoted as μ and is given as

$$\mu = \frac{L\alpha^2}{(K - LM) + L\alpha^2} 100 \ [\%]. \tag{2.10}$$

As said earlier, in this study we consider two different pilot arrangements:

(1) $L = 1$ and $A_P \neq A$. The amplitude is

$$A_1 = \sqrt{\frac{K}{((K - M) + \alpha^2)} \frac{P_{av}}{M}}. \tag{2.11}$$

In the remainder of the paper, the above pilot arrangement is referred to as case 1. For example, in SISO systems the above pilot arrangement is applied in CDMA wireless systems (e.g., IS-95 and WCDMA [15]). In MIMO systems, it is applied in narrowband MIMO implementations described in [11, 20, 27]. Also, it is applied in a wideband MIMO implementation based on 3G WCDMA [2].

(2) $L \geq 1$ and $A_P = A \, (\alpha = 1)$. The amplitude is

$$A_2 = \sqrt{\frac{K}{(K - L(M - 1))} \frac{P_{av}}{M}}. \tag{2.12}$$

In the remainder of the paper, the above pilot arrangement is referred to as case 2. Note that the above pilot arrangement is frequently used in SISO systems, e.g., wire-line modems [19] and some wireless standards (e.g., IS-136 and GSM [25]). To the best to our knowledge, this arrangement is not used in MIMO systems.

3. Estimation of Channel Response

Due to the orthogonality of the pilots and assumption that the elements of $\mathbf{H}$ are *iid*, it can be shown that to obtain the maximum likelihood estimate of $\mathbf{H}$ it is sufficient to estimate h_{nm} (for $m = 1, \cdots, M$, $n = 1, \cdots, N$), independently. This is identical to estimating a SISO channel response between the transmit antenna m and receive antenna n. The estimation is based on averaging the projections of the received signal on $d_{jm}^p \mathbf{s}_{jm}^p$ (for $j = 1, \cdots, L$ and $m = 1, \cdots, M$) as

$$\begin{aligned} \hat{h}_{nm} &= \frac{1}{LA_P} \sum_{j=1}^{L} (d_{jm}^p \mathbf{s}_{jm}^p)^{\mathrm{H}} \mathbf{r}_n \\ &= h_{nm} + \frac{1}{LA_P} \sum_{j=1}^{L} (d_{jm}^p \mathbf{s}_{jm}^p)^{\mathrm{H}} \mathbf{n}_n \end{aligned} \tag{3.1}$$

where $\hat{h}_{nm}$ denotes the estimate of the channel response h_{nm}. It can be shown that for a frequency-flat AWGN channel, given pilot signal [18] and assumed properties of $\mathbf{H}$, (3.1) is the maximum-likelihood estimate of the channel response h_{nm}. The estimation error is

$$n^e_{nm} = \frac{1}{LA_P}\sum_{j=1}^{L}(d^p_{jm}\mathbf{s}^p_{jm})^{\mathrm{H}}\mathbf{n}_n. \tag{3.2}$$

Note that n^e_{nm} corresponds to a white Gaussian random process with distribution $\mathcal{N}(0, N_0/(L\,(\alpha A)^2))$. Thus, the channel matrix estimate $\widehat{\mathbf{H}}$ is

$$\widehat{\mathbf{H}} = \mathbf{H} + \mathbf{H_e} \tag{3.3}$$

where $\mathbf{H}_e$ is the estimation error. It can be shown that each component of the error matrix $\mathbf{H}_e$ is an independent and identically distributed random variable n^e_{nm} given in (3.2) (where n^e_{nm} is the n-th row and m-th column element of $\mathbf{H}_e$).

Having the channel response estimated, the estimate of the transmitted data that is associated with the temporal signature $\mathbf{s}^d_i$ is obtained starting from the following statistics

$$z_{ni} = \frac{1}{A}(\mathbf{s}^d_i)^{\mathrm{H}}\mathbf{r}_n \tag{3.4}$$

where the amplitude A is assumed to be known at the receiver. The above procedure is the matched filtering operation and z_{ni} corresponds to the n-th component of the vector

$$\mathbf{z}_i = [z_{1i}\cdots z_{Ni}]^{\mathrm{T}} = \mathbf{H}\,\mathbf{d}_i + \frac{1}{A}\mathbf{n}_i,\ \ i = 1,\ldots,K - LM \tag{3.5}$$

where the m-th component of $\mathbf{d}_i = [d^d_{i1}\cdots d^d_{iM}]^{\mathrm{T}}$ is d^d_{im} (data transmitted from the antenna m and assigned to the temporal signature $\mathbf{s}^d_i$). Also, $\mathrm{E}[\mathbf{n}_i\mathbf{n}_i^{\mathrm{H}}] = N_0\,\mathbf{I}_{N\times N}$. It can be shown that $\mathbf{z}_i$ is a sufficient statistic for detecting the transmitted data.

As a common practice, the detection procedure assumes that the channel response is perfectly estimated, and that $\widehat{\mathbf{H}}$ corresponds to the true channel response. Let us rewrite the expression in (3.5) as

$$\mathbf{z}_i = (\mathbf{H} + \mathbf{H_e})\,\mathbf{d}_i + \frac{1}{A}\mathbf{n}_i - \mathbf{H_e}\,\mathbf{d}_i = \widehat{\mathbf{H}}\,\mathbf{d}_i + \left(\frac{1}{A}\mathbf{n}_i - \mathbf{H_e}\,\mathbf{d}_i\right). \tag{3.6}$$

The effective noise in the detection procedure (as a spatial vector) is

$$\bar{\mathbf{n}}_i = \left(\frac{1}{A}\mathbf{n}_i - \mathbf{H_e}\,\mathbf{d}_i\right). \tag{3.7}$$

For the given $\widehat{\mathbf{H}}$, the covariance matrix of the effective noise vector is

$$\boldsymbol{\Upsilon} = \boldsymbol{\Upsilon}(A) = \mathrm{E}_{\bar{\mathbf{n}}_i|\widehat{\mathbf{H}}}[\bar{\mathbf{n}}_i\bar{\mathbf{n}}_i^{\mathrm{H}}] = \frac{N_0}{A^2}\mathbf{I} + \mathrm{E}_{\mathbf{H_e}|\widehat{\mathbf{H}}}[\mathbf{H_e}\mathbf{H_e}^{\mathrm{H}}] \tag{3.8}$$

and it is a function of the amplitude A. As said earlier $\mathbf{H_e}$ is a matrix of *iid* Gaussian random variables with distribution $\mathcal{N}(0, N_0/(L\,(\alpha A)^2))$.

4. Estimates Available to Receiver: Open Loop Capacity

Assuming that the channel response estimate is available to the receiver, only, we determine the lower bound for the open loop ergodic capacity as follows.

$$C \geq R = \frac{K - LM}{K} \mathrm{E}_{\widehat{\mathbf{H}}} \left[\log_2 \det \left(\mathbf{I}_{M \times M} + \widehat{\mathbf{H}} \widehat{\mathbf{H}}^{\mathrm{H}} \boldsymbol{\Upsilon}^{-1} \right) \right]. \tag{4.1}$$

The term $(K - LM)/K$ is introduced because L temporal signature per each transmit antenna are allocated to the pilot. Also, the random process $\widehat{\mathbf{H}}$ has to be stationary and ergodic (this is a common requirement for fading channel and ergodic capacity [1, 17]). We assume that the channel coding will span across great number of channel blocks (i.e., we use the well known infinite channel coding time horizon, required to achieve error-free data transmission with rates approaching capacity [26]).

In the above expression, equality holds if the effective noise (given in (3.7)) is AWGN with respect to the transmitted signal. If the effective noise is not AWGN, then the above rates represent the worst-case scenario, i.e., the lower bound [10]. In achieving the above rates, the receiver assumes that the effective noise is interference which is independent of the transmitted data with Gaussian distribution and spatial covariance matrix $\boldsymbol{\Upsilon}$. In addition, in the above expression R represents an achievable rate for reliable transmission (error-free) for the specific estimation procedure assumed. Knowing the channel response perfectly or using a better channel estimation scheme (e.g., decision driven schemes) may result in higher achievable rates.

In the following we compare the above result to an information theoretical result presented in [1] (page 2641, expression (3.3.55)). The result is presented for the conventional SISO case, introducing capacity lower bound for mismatched decoding as

$$C \geq R^* = \mathrm{E}_{\hat{h}} \left[\log_2 \left(1 + \frac{\hat{h}^2 P}{\mathrm{E}_{h|\hat{h}}(|h - \hat{h}|^2) P + N_0} \right) \right] \tag{4.2}$$

where h and $\hat{h}$ are the SISO channel response and its estimate, respectively. The above result is general, not specifying the channel response estimation procedure. The bound in (4.1) is an extension of the information theoretical bound in (4.2), capturing the more practical pilot assisted channel response estimation scheme and generalizing it to the MIMO case. Consequently,

Proposition 1. For the SISO case ($M = 1, N = 1$), the rate R in (4.1) and R^* in (4.2), are related as

$$R = \frac{K - L}{K} R^*, \quad for \ P = \frac{K}{(K - L) + \alpha^2 L} P_{av} \tag{4.3}$$

where $\hat{h}$ is obtained using the pilot assisted estimation.

5. Examples and Numerical Results

5.1. SISO Systems. To illustrate the above analysis we start with SISO systems. In the SISO case, all previously defined spatial vectors and related matrices are now single dimensional (e.g., $\mathbf{d}_i$, $\mathbf{H}$, $\widehat{\mathbf{H}}$ and $\boldsymbol{\Upsilon}$ are now scalars d_i, h, $\hat{h}$ and v, respectively). We assume that $E[|h|^2] = \Gamma$. We define the average symbol signal-to-noise (SNR) ratio as

$$SNR = 10 \log_{10} \left[\frac{\Gamma P_{av}}{N_0} \right] \tag{5.1}$$

and the effective SNR (in dB) as

$$SNR_e(A) = 10 \log_{10} \left(\frac{\epsilon}{v} \right) \tag{5.2}$$

where $\epsilon = E[|d_i \hat{h}|^2]$. In the following, we denote $SNR_{e1} = SNR_e(A_1)$ (where A_1 is defined in (2.11)) and $SNR_{e2} = SNR_e(A_2)$ (where A_2 is defined in (2.12)). In Figure 1, we plot the effective SNR (i.e., SNR_e) as a function of the allocated power to the pilot (equation (2.10)), for different values of SNR (corresponding to the channel background noise). In this example, a pilot period K is 10 symbols and coincides with the coherence time. A frequency-flat Rayleigh fading channel is assumed. The results are shown for the pilot arrangements corresponding to both case 1 and case 2. In the case of the ideal knowledge of the channel response, the effective SNR does correspond to the conventional definition of the SNR (i.e., where the channel background noise is the only noise in the system, as defined in the equation (5.1)).

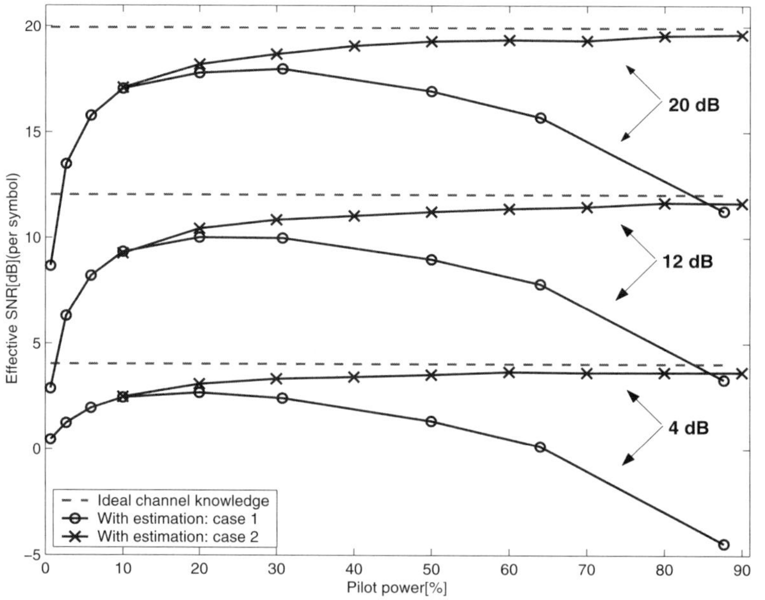

FIGURE 1. Effective SNR vs. power allocated to the pilot, SISO system, $SNR = 4, 12, 20$dB, coherence time $K = 10$ symbols, Rayleigh channel.

Further, in Figure 2, we present the rate R in (4.1) as a function of the power allocated to the pilot (equation (2.10)). The assumptions are identical to ones related to Figure 1. For the ideal knowledge of the channel response we apply the ergodic capacity formula [1]. In Figure 3 and 4, for $SNR = 20$dB, we present the achievable rates vs. power allocated to the pilot for different channel coherence

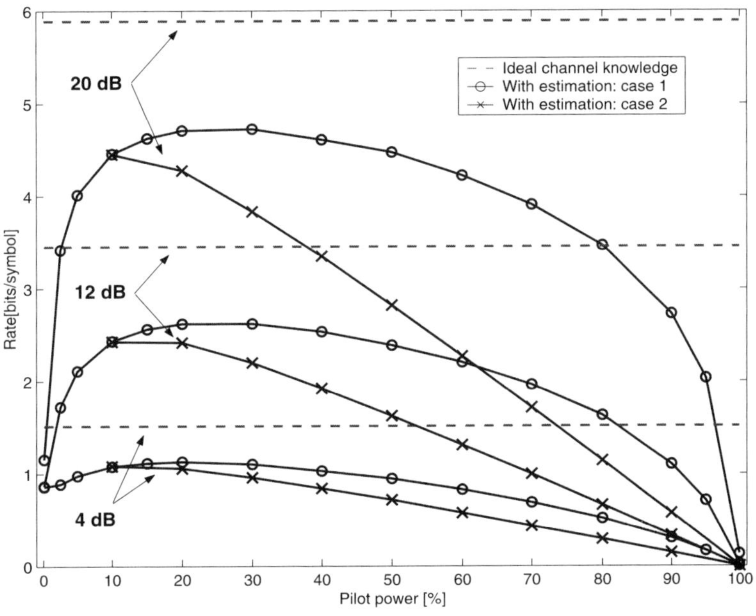

FIGURE 2. Achievable open-loop rates vs. power allocated to the pilot, SISO system, $SNR = 4, 12, 20$dB, coherence time $K = 10$ symbols, Rayleigh channel.

time lengths $K = 10, 20, 40, 100$. In case 1 and case 2, by increasing the channel coherence time length we observe that the maximum rate is reached for lower percentage of the power allocated to the pilot.

Regarding the achievable rates, from the above results we observe that case 1 is less sensitive to the pilot power allocation than case 2 (i.e., in case 2, R is dropping faster if the allocated power is different than the one that results in the maximum value). Further, for the same channel coherence time lengths, case 1 is reaching the maximum rate at a lower power allocated to the pilot than for case 2 (see Figure 3 and 4).

For the given SNR, we define the capacity efficiency ratio η as the ratio between the maximum rate R (with respect to the pilot power) and the ergodic capacity $C_{m\times n}$ in the case of the ideal knowledge of the channel response, i.e.,

$$\eta_{m\times n} = \frac{\max_\mu R}{C_{m\times n}}. \tag{5.3}$$

The index m and n correspond to number of transmit and receive antennas, respectively. In Figure 5, we show that the capacity efficiency ratio $\eta_{1\times 1}$ increases as the channel coherence time is getting longer.

5.2. MIMO Systems. In Figure 6, we present the rate R in (4.1) as a function of the power allocated to the pilot (equation (2.10)), for different number of transmit and receive antennas. In this and following cases we observe just the pilot arrangement case 1 (treating case 2 impractical for MIMO systems). We observe the rates for the Rayleigh channel, $SNR = 12$dB and the channel coherence time

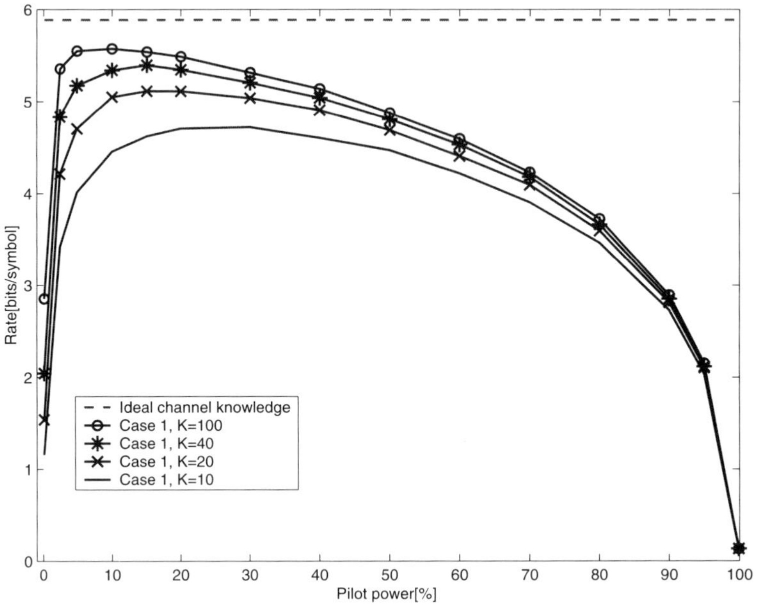

FIGURE 3. Achievable open-loop rates vs. power allocated to the pilot, case 1, SISO system, SNR = 20dB, for different channel coherence time lengths $K = 10, 20, 40, 100$ symbols, Rayleigh channel.

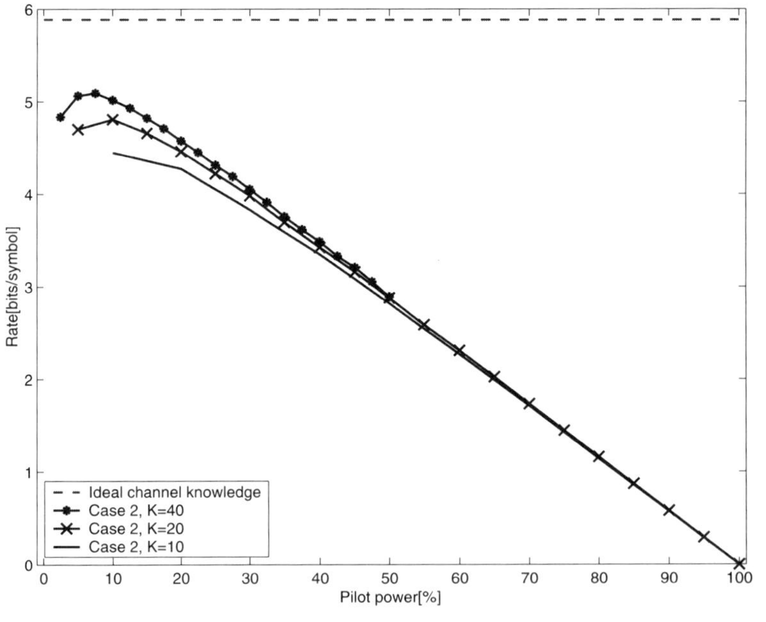

FIGURE 4. Achievable open-loop rates vs. power allocated to the pilot, case 2, SISO system, SNR = 20dB, for different channel coherence time lengths $K = 10, 20, 40$ symbols, Rayleigh channel.

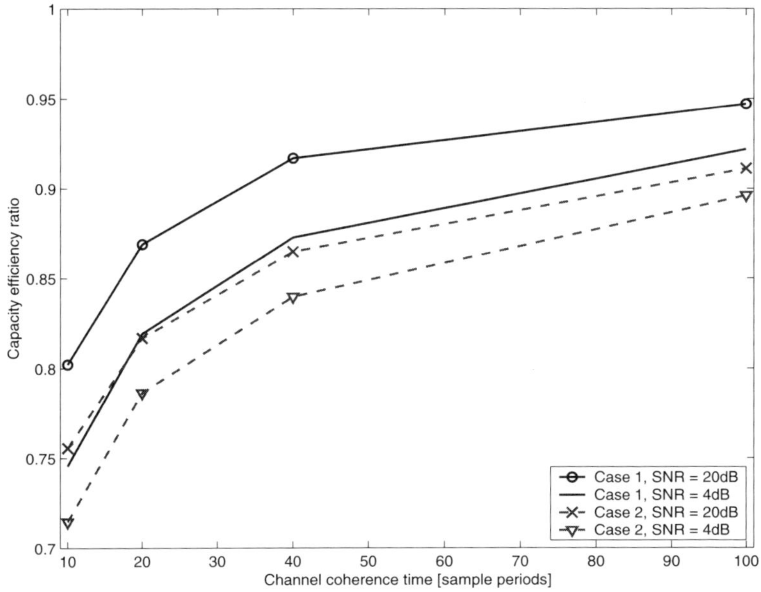

FIGURE 5. Capacity efficiency ratio vs. channel coherence time length ($K = 10, 20, 40, 100$ symbols), SISO system, $SNR = 4, 20$dB, Rayleigh channel.

length $K = 40$. Solid lines correspond to a system with the channel response estimation, and dashed lines to a system with the ideal knowledge of the channel response. Further, in Figure 7 we show the capacity efficiency ratio η for different number of transmit and receive antennas vs. different channel coherence time lengths. We observe that as the number of transmit antenna increases, the sensitivity to the channel response estimation error is more pronounced (while keeping the same number of receive antennas). For example, for the same channel coherence time length, the capacity efficiency ratio of the 4×4 system is lower than in the case of the 3×4 system.

6. Conclusion

In this paper we have studied how the estimation error of the frequency-flat time-varying channel response affects the performance of a communication system. We have considered multiple-input multiple-output (MIMO) wireless systems, where we have approximated the channel as a block-fading channel. The presented analysis may be viewed as a study of mismatched receiver algorithms in MIMO systems. We have considered two pilot based schemes for the estimation. The first scheme uses a single pilot symbol per block with the different power than the data symbol power. The second scheme uses more than one pilot symbol per block, whose power is the same as the data symbol power. We have studied the performance in terms of the effective SNR and achievable data rates (i.e., capacity lower bounds). We have presented how the performance depends on the percentage of the total power allocated to the pilot, background noise level and the channel coherence time length. Our results have shown that the first pilot-based approach

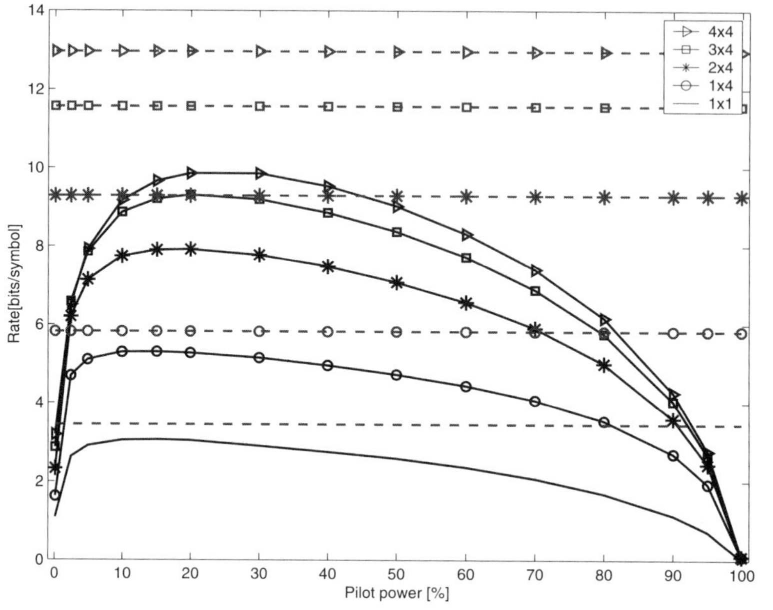

FIGURE 6. Achievable open-loop rates vs. power allocated to the pilot, MIMO system, $SNR = 12\text{dB}$, coherence time $K = 40$ symbols, Rayleigh channel, solid line corresponds to a system with the channel response estimation, and dashed line to the case of the ideal channel response knowledge.

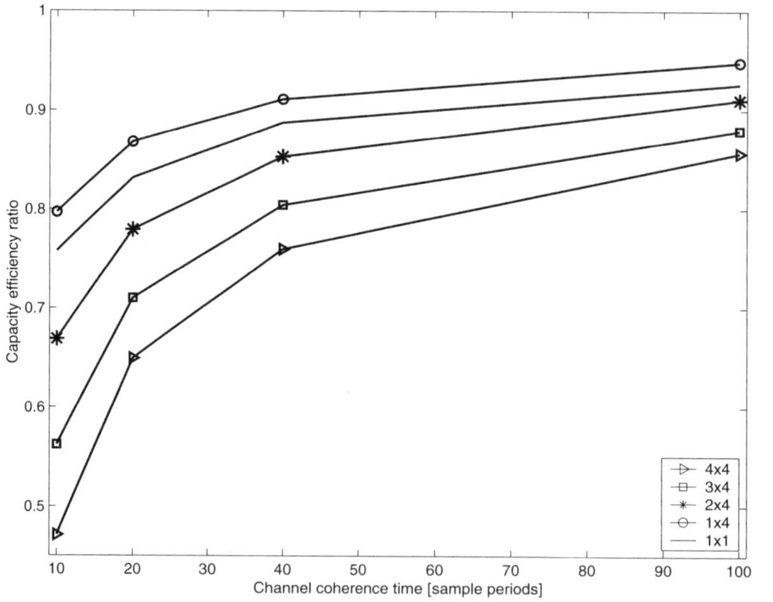

FIGURE 7. Capacity efficiency ratio vs. channel coherence time length ($K = 10, 20, 40, 100$ symbols), MIMO system, $SNR = 12\text{dB}$, Rayleigh channel.

is less sensitive to the fraction of power allocated to the pilot. Furthermore, we have observed that as the number of transmit antennas increase, the sensitivity to the channel response estimation error is more pronounced (while keeping the same number of receive antennas). The effects of the estimation error are evaluated in the case of the estimates being available at the receiver only (open loop), and the future work will address the case when the estimates are fed back to the transmitter (closed loop) allowing water pouring transmitter optimization.

Acknowledgments

The authors would like to thank Dr. Gerard Foschini and Dr. Dimitry Chizhik for their constructive comments and valuable discussions.

References

1. E. Biglieri, J. Proakis, and S. Shamai, *Fading channels: Information-Theoretic and Communications Aspects*, IEEE Transactions on Information Theory **44** (1998), no. 6, 2619 -2692.
2. A. Burg, E. Beck, D. Samardzija, M. Rupp, and et al., *Prototype Experience for MIMO BLAST over Third Generation Wireless System*, IEEE JSAC Special Issue on MIMO Systems and Applications **21** (2003), no. 3, 440–451.
3. G. Caire, G. Taricco, and E. Biglieri, *Optimum Power Control Over Fading Channels*, IEEE Transactions on Information Theory **45** (1999), no. 5, 1468 -1489.
4. D. Chizhik, G. J. Foschini, M. J. Gans, and R. Valenzuela, *Keyholes, Correlations, and Capacities of Multielement Transmit and Receive Antennas*, IEEE Transactions on Wireless Communications **1** (2002), no. 2, 361–368.
5. D. Chizhik, J. Ling, P. W. Wolniansky, R. A. Valenzuela, N. Costa, and K. Huber, *Multiple Input Multiple Output Measurements and Modeling in Manhattan*, VTC Fall **1** (2002), 107–110.
6. G. J. Foschini, *Layered Space-Time Architecture for Wireless Communication in a Fading Environment When Using Multiple Antennas*, Bell Labs Technical Journal **1** (1996), no. 2, 41–59.
7. G. J. Foschini and M. J. Gans, *On Limits of Wireless Communications in a Fading Environment when Using Multiple Antennas*, Wireless Personal Communications (1998), no. 6, 315–335.
8. G. J. Foschini, G. D. Golden, R. A. Valenzuela, and P. W. Wolniansky, *Simplified Processing for Wireless Communications at High Spectral Efficiency*, IEEE JSAC **17** (1999), no. 11, 1841–1852.
9. G. J. Foschini and J. Salz, *Digital Communications over Fading Radio Channels*, Bell Labs Technical Journal (1983), 429–456.
10. R. G. Gallager, *Information theory and reliable communications*, John Wiley and Sons, New York, 1968.
11. G. D. Golden, G. J. Foschini, R. A. Valenzuela, and P. W. Wolniansky, *Detection Algorithm and Initial Laboratory Results using V-BLAST Space-Time Communication Architecture*, Electronics Letters **35** (1999), no. 1, 14–16.
12. A. Goldsmith and P. Varaiya, *Increasing Spectral Efficiency Through Power Control*, ICC **1** (1993), 600–604.
13. J. C. Guey, M. P. Fitz, M. R. Bell, and W. Y. Kuo, *Signal Design for Transmitter Diversity Wireless Communication Systems over Rayleigh Fading Channels*, IEEE Transactions on Communications **47** (1999), 527–537.
14. J. C. Guey, M. P. Fitz, M. R. Bell, and W.Y. Kuo., *Signal Design for Transmitter Diversity Wireless Communication Systems over Rayleigh Fading Channels*, Proc. of IEEE Vehicular Technology Conference **1** (1996), 136–140, Atlanta.
15. H. Holma and A. Toskala (eds.), *Wcdma for umts*, first ed., Wiley, 2000.
16. A. Lapidoth and S. Shamai, *Fading Channels: How Perfect Need "Perfect Side Information" Be?*, IEEE Transaction on Information Theory **48** (2002), no. 5, 1118 -1134.

17. L. Li and A.J. Goldsmith, *Capacity and Optimal Resource Allocation for Fading Broadcast Channels - Part I: Ergodic Capacity*, Information Theory, IEEE Transactions **47** (2001), no. 3, 1083–1102.
18. H. V. Poor, *An introduction to signal detection and estimation*, second ed., Springer-Verlag, 1994.
19. J. G. Proakis, *Digital communications*, 3rd ed., McGraw-Hill, New York, 1995.
20. D. Samardzija, C. Papadias, and R. A. Valenzuela, *Experimental Evaluation of Unsupervised Channel Deconvolution for Wireless Multiple-Transmitter/Multiple-Receiver Systems*, Electronic Letters (2002), 1214–1215.
21. D. Samardzija, Peter Wolniansky, and Jonathan Ling, *Performance Evaluation of the VBLAST Algorithm in W-CDMA Systems*, The IEEE Vehicular Technology Conference (VTC) **2** (2001), 723–727, Atlantic City.
22. S. Shamai and T. Marzetta, *Multiuser Capacity in Block Fading With No Channel State Information*, IEEE Transaction on Information Theory **48** (2002), no. 4, 938 –942.
23. P. Soma, D. S. Baum, V. Erceg, R. Krishnamoorthy, and A. Paulraj, *Analysis and Modeling of Multiple-Input Multiple-Output (MIMO) Radio Channel Based on Outdoor Measurements Conducted at 2.5 GHz for Fixed BWA Applications*, IEEE Conference ICC 2002 (2002), 272 – 276.
24. L. Song and N. Mandayam, *Hierarchical SIR and Rate Control on the Forward Link for CDMA Data Users Under Delay and Error Constraints*, IEEE JSAC **19** (2001), no. 10, 1871 –1882.
25. G. L. Stuber, *Principles of mobile communications*, first ed., Kluwer Academic Publishers, 1996.
26. J. M. Wozencraft and I. M. Jacobs, *Principles of communication engineering*, Wiley, New York, 1965.
27. H. Zheng and D. Samardzija, *Performance Evaluation of Indoor Wireless System Using BLAST Testbed*, The IEEE Vehicular Technology Conference (VTC) **2** (2001), 905–909, Atlantic City.

Bell Labs, Lucent Technologies, 791 Holmdel-Keyport Road, Holmdel, NJ 07733, USA
E-mail address: dragan@bell-labs.com

WINLAB, Rutgers University, 73 Brett Road, Piscataway NJ 08854, USA
E-mail address: narayan@winlab.rutgers.edu

Part IV

Networking Capacity

DIMACS Series in Discrete Mathematics
and Theoretical Computer Science
Volume **62**, 2003

Challenges in UWB Signaling for Adhoc Networking

Younes Souilmi and Raymond Knopp

ABSTRACT. In this work we study the achievable rates of memoryless signaling strategies adapted to UltraWideBand (UWB) multipath fading channels. We focus on strategies which do not have explicit knowledge of the instantaneous channel realization, but may have knowledge of the channel statistics. We evaluate the average mutual information of general binary flash-signaling and achievable rates for m-PPM as a function of the channel statistics. Finally, we briefly examine the robustness of flash-signaling for interference-limited systems.

Notations

Throughout the paper, small letters $'a'$ will be used for scalars, capital letters $'A'$ for vectors, and bold capital letters $'\mathbf{A}'$ for matrices.

1. Introduction

In this work, we consider achievable rates for transmission strategies suited to *Ultra-wideband (UWB)* systems and focus non-coherent receivers (i.e. those which do not perform channel estimation, but may have prior knowledge of the second-order channel statistics). Here we take a UWB system to be loosely defined as any wireless transmission scheme that occupies a bandwidth between 1 and 10 GHz and more than 25% of it's carrier frequency in the case of a passband system.

The most common UWB transmission scheme is based on transmitting information through the use of short-term impulses, whose positions are modulated by a binary information source [**1**]. This can be seen as a special case of *flash signaling* coined by Verdu in [**2**]. Similar to direct-sequence spread-spectrum, the positions can further be modulated by a non-binary sequence (known as a *time-hopping sequence*) for mitigating inter-user interference in a multiuser setting [**3**]. This type of UWB signaling is a promising candidate for military imaging systems as well as other non-commercial sensor network applications because of its robustness to interference from signals (potentially from other non-UWB systems) occupying the same bandwidth. Based on recent documentation from the FCC it is also being considered for commercial adhoc networking applications based on peer-to-peer communications, with the goal being to provide low-cost high-bandwidth connections to the internet from small handheld terminals in both indoor and outdoor

Key words and phrases. Ultrawideband Communications, Mutual Information, Channel Coding, Interference Channels.

Eurecom's research is partially supported by its industrial partners: Hasler Stiftung, Swisscom, France Telecom, La Fondation Cegetel, Bouygues Telecom, Thales, ST Microelectronics, Hitachi Europe and Texas Instruments. The work reported here was also partially supported by the French National (RNRT) project ERABLE settings. Proposals for indoor wireless personal area networks (WPAN) in the 3-5 GHz band (802.15.3) are also considering this type of transmission scheme.

In this work, we focus on the case of non-coherent detection since it is well known [**7**][**8**] that coherent detection is not required to achieve the *wideband* AWGN channel capacity, $C_\infty = \frac{P_R}{N_0 \ln 2}$ bits/s, where P_R is the received signal power in watts, and N_0 is the noise power spectral density. In [**8**] Telatar and Tse showed this to be the case for arbitrary channel statistics in the limit of infinite bandwidth and infinite carrier frequency. Their transmission model was based on frequency-shift keying (FSK) and it was shown that channel capacity is achieved using very impulsive signals.

In [**2**] Verdu addresses the spectral efficiency of signaling strategies in the wideband regime under different assumptions regarding channel knowledge at the transmitter and receiver. The characterization is in terms of the minimum energy-per-bit to noise spectral density ratio $(E_b/N_0)_{\min}$ and the wideband slope S_0. The latter quantity is measured in bits/s/Hz/3dB and represents growth of spectral efficiency at the origin as a function of E_b/N_0. Verdu's work is fundamental to our problem since is shows that approaching C_∞ with non-coherent detection is impossible for practical data rates (>100 kbit/s) even for the vanishing spectral efficiency of UWB systems. This is due to the fact that S_0 is zero at the origin for non-coherent detection. To get an idea of the loss incurred, consider a system with a 2GHz bandwidth and data rate of 20 Mbit/s (this would correspond to a memoryless transmission strategy for channels with a 50ns delay-spread) yielding a spectral-efficiency of .01 bits/s/Hz. For Rayleigh statistics the loss in energy efficiency is on the order of 3dB, which translates into a factor 2 loss in data rate compared to a system with perfect channel state information at the receiver. The loss becomes less significant for lower data rates and/or higher bandwidths.

The main goal of this work is to examine under what conditions different non-coherent signaling strategies can approach the wideband channel capacity with perfect channel knowledge at the receiver subject to a large but finite bandwidth constraint and different propagation conditions. Section II deals with the underlying system model for transmission and reception as well as the channel model. In section III we evaluate expressions for the achievable information rates of different signaling schemes based on reasonably simple analog filter receivers. In section IV we present some numerical evaluations of the expressions from section III. Finally in section V we examine some issues related to multiple-access interference.

2. System Models

We restrict our study to strictly time-limited memoryless real-valued signals, both at the transmitter and receiver. The time-limited and memoryless assumptions are made possible due to the virtually unlimited bandwidth of UWB signals. The transmitted pulse, of duration T_p, is passed through a linear channel, $h(t,u)$, representing the response of the channel at time t to an impulse at time u. We assume that the impulse response of the channel is of duration $T_d \gg T_p$. The channel is further assumed to be a zero-mean process.

The received signal bandwidth W is roughly $1/T_p$, in the sense that the majority of the signal energy is contained in this finite bandwidth. The received signal is given by

$$r(t) = \int_0^{T_p} x(u)h(t,u)du + z(t) \tag{2.1}$$

where $z(t)$ is white Gaussian noise with power spectral density $N_0/2$. The channel is further assumed to satisfy

$$\int_0^{T_d+T_p} \int_0^{T_p} h^2(t,u)dtdu < \infty$$

which rules out impulsive channels and reflects the bandlimiting nature of analog transmit and receive chains (antennas and radio-frequency components.) The transmitted signal is written as

$$x(t) = \sum_{k=0}^{N} s(u_k)\sqrt{E_s}p\left(t - kT_s\right) \tag{2.2}$$

where k is the symbol index, T_s the symbol duration, $E_s = PT_s$ the transmitted symbol energy, $u_k \in \{1, \ldots, m\}$ is the transmitted symbol at time k, $p(t)$ and $s(u_k)$ are the assigned pulse and amplitude for symbol u_k. For all k, $p(t)$ is a unit-energy pulse of duration T_p. The considered model encompasses modulation schemes such as flash signaling, m-ary PPM, amplitude, and differential modulations. A guard interval of length T_d is left at the end of each symbol (from our memoryless assumption) so that $T_s \geq T_p + T_d$. From the point of view of spectral efficiency, we have that $\frac{E_b}{N_0} = \frac{P}{N_0 C(P/N_0)}$, where $C(P/N_0)$ is the average mutual information of the underlying signaling scheme as a function of the SNR.

The large bandwidths considered here (>1GHz) provide a high temporal resolution and enable the receiver to resolve a large number of paths of the impinging wavefront. Providing that the channel has a high diversity order (i.e. in rich multipath environments), the total channel gain is slowly varying compared to its constituent components. It has been shown [**4, 5, 6**] through measurements that in indoor environments, the UWB channel can contain hundreds of paths of significant strength. We may assume, therefore, that for all practical purposes, the total received energy should remain constant at its average path strength, irrespective of the particular channel realization. Variations in the received signal power will typically be caused by shadowing rather than fast fading.

The finite-energy random channel may be decomposed as

$$h(t,u) = \sum_{i=1}^{\infty}\sum_{j=1}^{\infty} h_{i,j}\theta_j(u)\phi_i(t)$$

where $h_{i,j}$ are the projections of the channel on the the input and output eigenspaces, $\{\theta_j(u)\}$ is the set of eigenfunctions (for $L^2(0,T_p)$) of the transmit pulse and $\{\phi_i(t)\}$ is the set of eigenfunctions (for $L^2(0,T_p+T_d)$) of the received signal. Since the input in (2.2) is one-dimensional, the most appropriate choice for $p(t)$ is the one which maximizes the expected energy of the channel output

$$p(t) = \operatorname*{argmax}_{f(t)} \mathrm{E}\left(\int_0^{T_p} h(t,u)f(u)du\right)^2 = \theta_1(t)$$

where $\theta_1(t)$ is the eigenfunction corresponding to the maximum eigenvalue, μ_1, of the input cross-correlation function

$$R_{\mathrm{i}}(u,u') = \mathrm{E}\int_0^{T_s} h(t,u)h(t,u')dt = \int_0^{T_s} R_h(t,t;u,u')dt$$

and $R_h(t,t';u,u') = \mathrm{E}h(t,u)h(t',u')$. The use of this input filter is conditioned on the emission requirements of UWB systems, and thus it may not be possible to satisfy the maximal energy solution in practice.

The above decomposition allows us to write (2.1), for each symbol k, as the equivalent channel

$$r_{k,i} = \sqrt{\mu_1}h_i s(u_k) + z_i, i = 1, ..., \infty$$

where z_i is $N(0, N_0/2)$. For notational convenience we have dropped the index corresponding to the input projection from h_{ij} since we are constrained one-dimensional inputs. Furthermore, if we choose the output eigenfunctions to be the solution to

$$\lambda_i\phi_i(t) = \int_0^{T_d+T_p}\int_0^{T_p}\int_0^{T_p} R_h(t,u;t',u')\theta_1(u)\theta_1(u')\phi_i(t')dudu'dt'$$

we have that the h_i are uncorrelated and have variance λ_i.

Because of the bandlimiting nature of the channels in this study, the channel will be characterized by a finite number, D, of significant eigenvalues which for rich environments will be roughly equal to $1 + 2WT_d$, in the sense that a certain proportion of the total channel energy will be contained in these D components. Under our rich scattering assumption D is limited by bandwidth and not insufficient scattering and we may in some cases make the following approximation

$$\mu_1 \sum_{i=1}^{D} h_i^2 \approx 1$$

for all channel realizations. This assumption essentially says that the received signal energy is not impaired by signal fading due to the rich scattering environment.

For notational convenience, we will assume that the eigenvalues are ordered by decreasing amplitude. An example of an eigenvalue distribution is shown in Fig. 1. This corresponds to an exponentially decaying multipath intensity profile with delay-spread 50ns filtered by a window function of width 1ns, resulting in a system bandwidth of approximately 1 GHz.

3. Non-Coherent Detection

In this section we consider non-coherent receivers, that may or may not have access to the second-order channel statistics. The motivation for such a study is to derive receivers that are reasonable from an implementation point of view. We particularly focus on solutions whose front-end can be implemented with analog technology, as shown in Fig. 2. We assume that the transmitter does not have side information about the instantaneous channel realization and that is constrained to the use of flash-like signaling. We first numerically compute the average mutual information for such a system, then we derive a lower bound on the achievable rates for three receivers, based on energy-detection with m-ary PPM modulation, using different front-end filters. This modulation can be seen as a specially-designed channel code for flash-signaling.

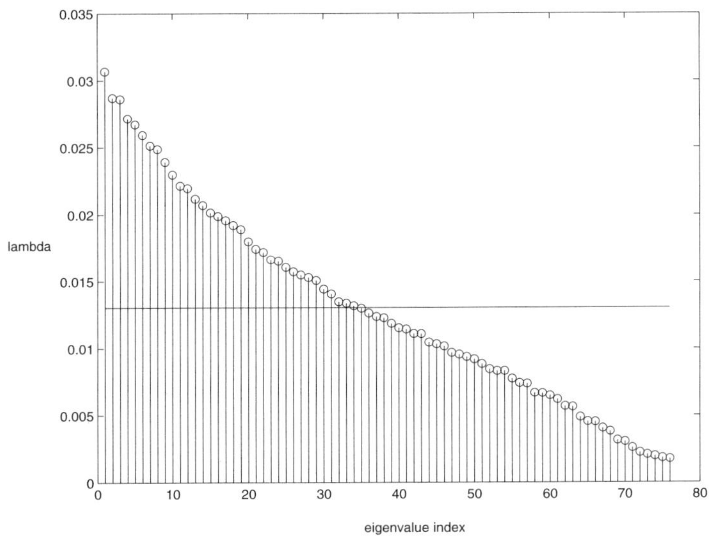

FIGURE 1. Example Eigenvalue Distribution: $T_d = 50$ ns, $W \approx 1$ GHz.

3.1. Average Mutual Information. From the results of Verdu in [**2**] we have in our rich scattering case that $(E_b/N_0)_{\min}$ is $\frac{\ln 2}{\mu_1 \mathrm{E} \sum_{i=1}^{D} h_i^2} \approx \ln 2$. In the case of vanishing spectral efficiency, binary flash-signaling is first-order optimal, and achieves this minimal E_b/N_0. Using the notation from the previous section, we express the binary flash signaling scheme as

$$u_k = \begin{cases} 1 & \text{with probability } \eta \\ 0 & \text{with probability } (1-\eta) \end{cases}$$

$s(0) = 0$, $s(1) = \sqrt{\frac{E_s}{\eta}}$, and $T_s = T_d + T_p$. We assume that the h_i are Gaussian ergodic sequences, which implies that the system's temporal resolution is not fine enough to resolve all the degrees of freedom of the considered channel and that the projection of $h(t,u)$, on each of the kernel's directions, is the combination of a relatively large number of independent multipath components. Measurements of UWB channels[**4**] have shown that channel components can be considered to fade according to Rayleigh statistics in non-line-of-sight conditions, indicating that this assumption is quite reasonable. Conditioned on u_k, R_k is a zero-mean Gaussian vector with covariance matrix $\mathrm{E}\left[R_k R_k^T\right] = \mathrm{diag}(s(u_k)E_s\lambda_i + \frac{N_0}{2})$. It is easily shown that

$$\begin{aligned} I(U;R) &= -\tfrac{1}{T_s}\mathop{\mathrm{E}}_{Y}\Big[\eta \log\Big(\eta + (1-\eta)\sqrt{\det(\mathbf{I}+\mathbf{A}^{-1})}\exp\left(-\tfrac{1}{2}Y^T\mathbf{A}^{-1}Y\right)\Big) \\ (3.1) \quad &+ (1-\eta)\log\Big((1-\eta) + \tfrac{\eta}{\sqrt{\det(\mathbf{I}+\mathbf{A}^{-1})}}\exp\left(\tfrac{1}{2}Y^T(\mathbf{A}+\mathbf{I}))^{-1}Y\right)\Big)\Big] \quad \text{bits/s} \end{aligned}$$

with Y a zero-mean gaussian random vector with covariance matrix $\mathbf{I}$ and $\mathbf{A} = \mathrm{diag}(\frac{N_0\eta}{2E_s\lambda_i}))$. This is easily computed numerically.

3.1.1. m-**ary PPM with Energy Detection.** In this section we consider two non-coherent detectors for m-ary PPM: the optimal non-coherent detector, for

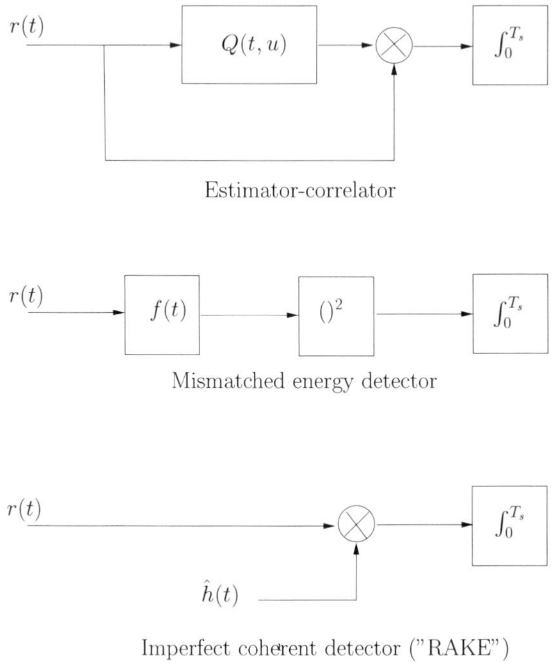

FIGURE 2. Receiver Structures

known second order channel statistics and a suboptimal mismatched version. We use base-band m-ary PPM signals to transmit the information bits. Each PPM symbol corresponds to choosing one out of m symbol times in which to emit the transmit pulse $p(t)$, which is a special case of the flash signaling system described above with $\eta = 1/m$. In practice, we would consider a coded modulation scheme with a binary code mapped to m-PPM symbols. Here, $T_s = T_p + T_d$, so that the channel can be considered to be memoryless.

The data is encoded using a randomly generated codebook $C = \{C_1, \ldots, C_M\}$ of cardinality M and codeword length N. Each codeword C_l is a sequence $C_l = (c_{1,l}, c_{2,l}, \ldots, c_{N,l})$ of m-PPM symbols. Let C_w be the transmitted codeword, using the notations of model (2.2) we have $u_k = c_{k,w}$, and $s(u_k) = 1$.

For all $n \in [1, N]$ and $k \in [1, m]$ let

$$R_{n,k} = [i \in [0, D]\,, \quad < r(t), \phi_{i,n}(t - (k - c_{n,w})T_s) >] = S_{n,k} + Z_{n,k}$$

where $S_{n,k}$ and $Z_{n,k}$ are the signal and noise components of $R_{n,k}$, and w denotes the index of the transmitted codeword.

$$Z_{n,k} = [i \in [0, D]\,, \quad < z(t), \phi_{i,n}(t - (k - c_{n,w})T_s) >]$$

$Z_{n,k}$ is a Gaussian random vector with mean zero and covariance matrix $\mathbf{K_z} = \frac{N_0}{2}\mathbf{I}$. Using the same reasoning as in the previous section, we will assume that $S_{n,k}$ has a Gaussian distribution.

3.1.2. **Optimal detector.** The Maximum likelihood non-coherent detector can be written as

$$\hat{k} = \underset{k}{\operatorname{argmax}} \ \Pr\left(r(t)/w = k\right) = \underset{k}{\operatorname{argmax}} \ \frac{1}{N}\sum_{n=1}^{N} q_{n,k}$$

with $q_{n,k} = R_{n,k}Q^{-1}R^T_{n,k}$ and $Q = \mathbf{diag}(\frac{N_0}{2}\left(1 + \frac{N_0}{2E_s\lambda_i}\right))$ [1].

We use an equivalent receiver, the decoder forms the decision variables

$$q_k = \frac{1}{N}\sum_{n=1}^{N} q_{k,n}$$

and uses the following asymptotically optimal (for infinite length codewords) threshold rule to decide on a message: if q_k exceeds a certain threshold ρ for exactly one value of k, say $\hat{k}$, then it will declare that $\hat{k}$ was transmitted. Otherwise, it will declare a decoding error. This is the same sub-optimal decoding scheme considered in [**8**].

An upper bound of the decoding error probability is then given by the following theorem

THEOREM 3.1. *The probability of codeword error is upper bounded by*

$$\Pr[\text{error}] \leq M \min_{t>0}$$

$$\exp -N\left[t\rho - \ln\left((1-\eta)\prod_{i=1}^{D}\frac{1}{\sqrt{1-\frac{2t}{1+\frac{N_0}{2E_s\lambda_i}}}} + \eta\prod_{i=1}^{D}\frac{1}{\sqrt{1-\frac{4E_s\lambda_i t}{N_0}}}\right)\right] \tag{3.2}$$

with $\rho = (1-\epsilon)\frac{2E_s}{N_0}$, *and* $\eta = 1/m$.

PROOF. The decision variable for the transmitted codeword C_w is given by

$$\frac{1}{N}\sum_{n=1}^{N}\left(S_{n,w}+Z_{n,w}\right)Q^{-1}\left(S_{n,w}+Z_{n,w}\right)^T$$

by the ergodicity of the noise process, this time-average will exceed the threshold with probability arbitrarily close to 1 for any $\epsilon > 0$ as N gets large. For all $k \neq w$ We bound the probability $\Pr\left[q_k \geq \rho\right]$ using a Chernoff bound

$$\begin{aligned}
\Pr\left[q_k \geq \rho\right] &= \Pr\left[Nq_k \geq N\rho\right] \\
&= \underset{C}{E}\left[\Pr\left[Nq_k \geq N\rho/C\right]\right] \\
&\leq \underset{C}{E}\left[\min_{t>0} e^{-tN\rho}\prod_{n=1}^{N} E\left[e^{tq_{n,k}}/C\right]\right]
\end{aligned}$$

[1]this detector is equivalent to the classical estimator-correlator [**11**]

We have that for all $c_{n,k} = c_{n,w}$

$$\mathrm{E}\left[e^{tq_{n,k}}\right] = \prod_{i=1}^{D}\left(1 - \frac{4E_s\lambda_i t}{N_0}\right)^{-\frac{1}{2}}$$

and for all $c_{n,k} \neq c_{n,w}$

$$\mathrm{E}\left[e^{tq_{n,k}}\right] = \prod_{i=1}^{D}\left(1 - \frac{2t}{1+\frac{N_0}{2E_s\lambda_i}}\right)^{-\frac{1}{2}}$$

Let l be the number of collisions between codewords C_w and C_k

$$l = \mathrm{card}\left(n = 1 \ldots N / c_{n,w} = c_{n,k}\right)$$

then we have that

$$\begin{aligned} \Pr\left[q_k \geq \rho/C\right] &= \Pr\left[q_k \geq \rho/l\right] \\ &\leq \min_{t>0} e^{-Nt\rho}\left(\prod_{i=1}^{D}\left(1 - \frac{4E_s\lambda_i t}{N_0}\right)^{-\frac{1}{2}}\right)^{l}\left(\prod_{i=1}^{D}\left(1 - \frac{2t}{1+\frac{N_0}{2E_s\lambda_i}}\right)^{-\frac{1}{2}}\right)^{N-l} \end{aligned}$$

Averaging over all the realizations of the randomly generated codebook we obtain

$$\begin{aligned} \underset{C}{E}\ \left[\Pr\left[q_k \geq \rho/C\right]\right] &\leq \underset{l}{E}\left[\min_{t>0} e^{-Nt\rho}\prod_{i=1}^{D}\frac{1}{\left(1-\frac{4E_s\lambda_i t}{N_0}\right)^{\frac{l}{2}}\left(1-\frac{2t}{1+\frac{N_0}{2E_s\lambda_i}}\right)^{\frac{N-l}{2}}}\right] \\ &\overset{(a)}{\leq} \min_{t>0}\underset{l}{E}\left[e^{-Nt\rho}\prod_{i=1}^{D}\frac{1}{\left(1-\frac{2t}{1+\frac{N_0}{2E_s\lambda_i}}\right)^{\frac{(N-l)}{2}}\left(1-\frac{4E_s\lambda_i t}{N_0}\right)^{\frac{l}{2}}}\right] \\ &= \min_{t>0} e^{-Nt\rho} \\ &\quad \sum_{l=0}^{N}\begin{pmatrix} l \\ N \end{pmatrix}\eta^{l}(1-\eta)^{N-l}\prod_{i=1}^{D}\left(1-\frac{2t}{1+\frac{N_0}{2E_s\lambda_i}}\right)^{-\frac{N-l}{2}}\left(1-\frac{4E_s\lambda_i t}{N_0}\right)^{-\frac{l}{2}} \\ &= \min_{t>0} e^{-Nt\rho}\left((1-\eta)\prod_{i=1}^{D}\frac{1}{\sqrt{1-\frac{2t}{1+\frac{N_0}{2E_s\lambda_i}}}} + \eta\prod_{i=1}^{D}\frac{1}{\sqrt{1-\frac{4E_s\lambda_i t}{N_0}}}\right)^{N} \end{aligned}$$

note that in (a) we perform a looser minimization operation for the sake of feasibility of the analytical developments. Using a union bound we obtain the desired result. □

The decoding error probability in equation (3.4) decays to zero exponentially in N as long as the transmission rate R satisfies

$$\begin{aligned} R &= \frac{1}{mNT_s}\log(M) \\ &\leq \max_{t>0}\frac{1}{T_s}\left(t\rho - \ln\left((1-p)\prod_{i=1}^{D}\left(1-\frac{2t}{1+\frac{N_0}{2E_s\lambda_i}}\right)^{-\frac{1}{2}} + p\prod_{i=1}^{D}\left(1-\frac{4E_s\lambda_i t}{N_0}\right)^{-\frac{1}{2}}\right)\right) \end{aligned}$$

Due to the finite cardinality of the symbol alphabet our information rate is bounded by

$$R \leq \frac{1}{T_s} \log_2(m) \quad \text{bits/s}$$

3.1.3. **Mismatched non-coherent detector.** We now consider the case where the receiver does not have access to channel statistics and/or is constrained to use a time-invariant front-end filter because of implementation considerations. The received signal is first filtered by the time-limited unit-energy filter $f(t)$ of duration T_f,this filtering aims to reduce the amount of collected noise while capturing the majority of its information bearing part. The optimal joint choice of such a filter $f(t)$ and transmission pulse shape $p(t)$, in the sense of maximizing the received energy of the information bearing signal, is $p = f = f_{\text{opt}}$ such that $f_{\text{opt}} * f_{\text{opt}} = v(t)$ with $v(t)$ the maximum eigenvalue eigenvector of the autocorrelation matrix of the channel.

$$\begin{aligned} r_f(t) = (r * f)(t) &= ((s + z) * f)(t) \\ &= s_f(t) + z_f(t) \end{aligned}$$

then for each potential emission position $t_{n,k} = (n-1)mT_s + kT_s$ we capture the received energy on the interval from $t_{n,k}$ to $t_{n,k} + T_d + T_p$

$$q_{n,k} = \int_{t_{n,k}}^{t_{n,k}+T_d+T_p} r_f^2(t)\, dt$$

$$q_{n,k} = \begin{cases} \sum_{i=1}^{D} \lambda_i r_{f,i}^2 & c_{n,w} = k \\ \sum_{i=1}^{D} \mu_i r_{f,i}^2 & c_{n,w} \neq k \end{cases}$$

λ_i and μ_i are defined similarly to in (2)

$$\begin{aligned} \lambda_i \phi_i(t) &= \int_0^{T_d+T_p} \mathbf{K}_{s_f+z_f}(t,u)\phi_i(u)\, du \\ \mu_i \theta_i(t) &= \int_0^{T_d+T_p} \mathbf{K}_{z_f}(t,u)\theta_i(u)\, du \end{aligned}$$

where $\mathbf{K}_{z_f}(t,u)$ and $\mathbf{K}_{s_f+z_f}(t,u)$ are the autocorrelation matrices of the filtered received signal respectively in the presence and not of transmitted signal, $\{\phi_{n,1}, \ldots, \phi_{n,D}\}$ and $\{\theta_{n,1}, \ldots, \theta_{n,D}\}$ are the the Karhunen-Loeve expansion kernels of the filtered received signal $r_f(t)$ on the interval from t_n to $t_n + T_d + T_p$ respectively in presence and not of transmitted signal, and $r_{f,i}$ are zero mean unit variance random variables.

Using the same decoding rule as in the previous section, we derive an upper bound on the decoding error probability

THEOREM 3.2. *The probability of codeword error is upper bounded by*

(3.3)

$$\Pr[\text{error}] \leq M \min_{t>0} \exp -N \left[t\rho - \ln \left((1-\eta) \prod_{i=1}^{D} \frac{1}{\sqrt{1-\mu_i t}} + \eta \prod_{i=1}^{D} \frac{1}{\sqrt{1-\lambda_i t}} \right) \right]$$

with $\rho = (1-\epsilon)\sum_{i=1}^{D} \lambda_i$, *and* $\eta = 1/m$.

The proof is similar to the proof of theorem (3.1).

3.2. Quasi-Coherent Detection. We now consider the case where a noisy estimate of the channel $\tilde{h}(t,u) = h(t,u) + n(t,u)$ is available at the receiver, where $n(t,u)$ is a white Gaussian zero-mean random process with variance σ_h^2. Projecting $\tilde{S}$ over the same set of basis functions as in (2) we obtain

$$\begin{aligned} \tilde{S_n} &= [i \in [0,D]\,, \quad < \tilde{s}(t), \phi_{i,n}(t)) >] \\ &= S_{n,w} + N_{n,w} \end{aligned}$$

The received signal is then correlated against the channel estimate for all the possible pulse emission positions at each signal frame

$$\begin{aligned} q_{n,k} &= R_{n,k}\tilde{S_n}^T \\ &= (S_{n,k} + Z_{n,k})\,(S_{n,w} + N_{n,w})^T \end{aligned}$$

An upper bound of the decoding error probability is given by the following theorem

THEOREM 3.3. *The probability of codeword error is upper bounded by*

$$\Pr[\text{error}] \quad \leq M \min_{t>0} \exp -N \left[t\rho - \ln\left((1-\eta) \prod_{i=1}^{D} \frac{1}{\sqrt{1 - \frac{N_0}{2}\left(E_s\lambda_i + \sigma_h^2\right) t^2}} + \eta \prod_{i=1}^{D} \frac{1}{\sqrt{1 - \left(E_s\lambda_i + \frac{N_0}{2}\right)\left(E_s\lambda_i + \sigma_h^2\right) t^2}} \right) \right]$$

with $\rho = (1-\epsilon)E_s$, *and* $\eta = 1/m$.

The proof uses the same technique as in (3.1.2).

4. Numerical Results

We show the numerical evaluation of the bounds in the previous sections in Figs. 3,4, where the symbol alphabet size (i.e. m) has been optimized for each SNR. The effective bandwidth of the system is 1 GHz and the channel is an exponentially decaying multipath channel with a delay spread of 50ns. Two different decaying factors, respectively corresponding to a flat eigenvalues distribution and a fast decaying eigenvalues distribution, are considered. We see that the typical information rate losses with respect to optimal flash signaling are less than a factor 2 with m-PPM and resonably simple linear filter analog receivers.

5. Multi-access Interference

The networks which will likely employ UWB signaling, for example *Wireless Personal Area Networks(WPAN)* and *sensor networks*, are characterized not only by a rich scattering propagation environment but also by requirements for adhoc and peer-to-peer communications. This latter requirement has a significant impact on systems design, since the signaling schemes must be robust to strong impulsive interference (from nearby interferers) as shown in Fig. 5. Here we show a small network consisting of 2 transmitter-receiver pairs. The receiving nodes are both far from their respective transmitters and suffer from strong interference. In contrast

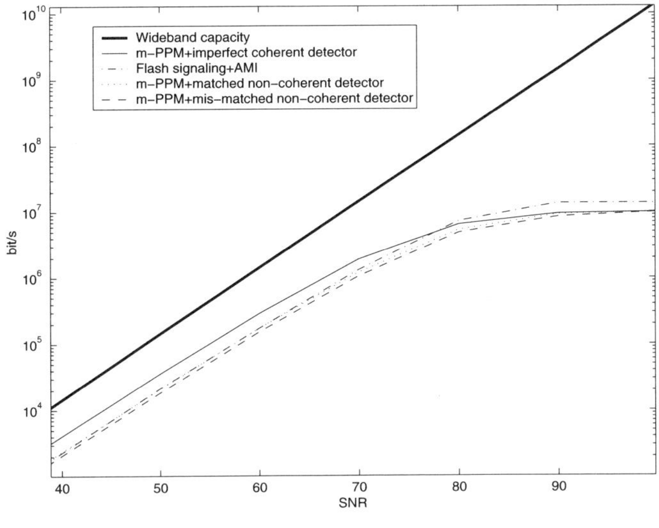

FIGURE 3. Achievable rates of energy detection based receivers+"flat" eigenvalues distribution: Td=50 ns, W=1GHz

to CDMA networks with a basestation/mobile topology, UWB adhoc networks will likely not benefit significantly from centralized or distributed power control resulting in extreme near-far interference. Even the substantial processing gains (with coherent detection) of these systems will do little in such an environment. The purpose of this section is to analyze the robustness of flash-signaling with respect to near-far interference.

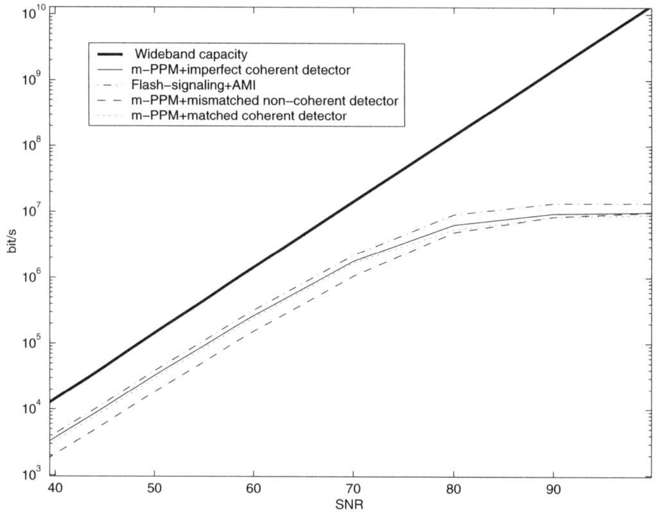

FIGURE 4. Achievable rates of energy detection based receivers+non-flat eigenvalues distribution: Td=50 ns, W=1GHz

Generalizing the model of the previous sections by adding a single interferer we have

$$r_{k,i}^{(1)} = \sqrt{\frac{E_{s,1}\mu_1}{\eta}} h_{k,i}^{(1)} s^{(1)}(u_k^{(1)}) + \underbrace{\sqrt{\frac{E_{s,2}\mu_1}{\eta}} h_{k,i}^{(2)} s^{(2)}(u_k^{(2)})}_{\text{interference}} + z_{k,i}^{(1)}, \quad i = 1, ..., \infty$$

We have assumed for simplicity that the interferer is synchronous to the desired signal. Consider an upper-bound on the achievable rate where we assume a genie-aided decoder which passes the symbol positions of the interferer to the receiver of the desired signal. For this rate to be achievable without a genie, we must have that $R_2 \leq I(U^{(2)}; R^{(1)}))$, and use a multiuser receiver to decode the desired signal.

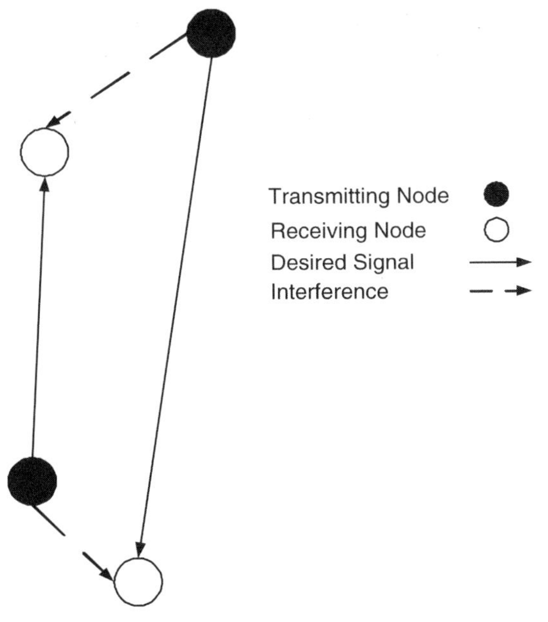

FIGURE 5. Simple Peer-to-Peer Network Example

In the genie-aided case, the rate of the desired user's signal is

$$\begin{aligned} R_1 &\leq I(U^{(1)}; R^{(1)}, U^{(2)}) \\ &= I(U^{(1)}; R^{(1)}|U^{(2)}) \\ &= (1-\eta)I(U^{(1)}; R^{(1)}|U^{(2)}=0) + \eta I(U^{(1)}; R^{(1)}|U^{(2)}=1) \end{aligned} \tag{5.1}$$

Each of the mutual information functionals in (5.1) can be expressed as in (3.1), and in the case of very strong interference, the second term will be negligible. We see, therefore, that the influence of the interference is a reduction in throughput by a factor $1-\eta$. The fact that channel knowledge is unavailable at the receiver makes the problem similar to that of the erasure channel, where erasures are generated by the presence of an interfering symbol. In practice, the genie-aided scheme could be implemented using a threshold rule on the front-end filter output, which is chosen so that the probability of detecting the presence of strong interference (and thus declaring an erasure) is very close to 1 when an interferer is transmitting. Random coding bounds similar to those of Section 3 for m-PPM modulation and a modified decoding rule taking into account interference are readily obtained.

6. Conclusions

In this work we studied the achievable rates of memoryless signaling strategies over ultrawideband (UWB) multipath fading channels. In particular we focused on strategies which do not have explicit knowledge of the instantaneous channel realization, but may have knowledge of the channel statistics. We evaluated the mutual information of general binary flash-signaling and achievable rates for m-PPM as a function of the channel statistics, which can be seen as a practical coded-modulation strategy for implementing flash-signaling. We can conclude that

m-PPM combined with simple analog front-end receivers can provide virtually the same data rates as general binary flash-signaling. Finally, we briefly examine the robustness of flash-signaling for interference-limited systems, where it is shown that flash-signaling seems to be an efficient means for combating severe near-far interference.

References

[1] Win, M. Z. , Scholtz, R. A. , "Impulse radio: how it works " *IEEE Communications Letters*, Volume: 2 Issue: 2, Pages: 36 -38, Feb. 1998.
[2] Verdu, S. ," Spectral efficiency in the wideband regime " *IEEE Transactions on Information Theory*, Vol. 48 Issue: 6 , June 2002 pp. 1319 -1343.
[3] Win, M. Z. , Scholtz, R. A. , " Ultra-wide bandwidth time-hopping spread-spectrum impulse radio for wireless multiple-access communications " *IEEE Transactions on Communications* Volume: 48 Issue:4, Pages: 679 -689, April 2000.
[4] Pendergrass, M. , IEEE P802.15 Working group, "Empirically Based Statistical Ultra-Wideband Channel Model," April 2002.
[5] Win, M. Z. ; Scholtz, R. A,"On the energy capture of ultrawide bandwidth signals in dense multipath environments " *IEEE Communications Letters*, Vol. 2, Issue: 9 pp. 245-247, 1998.
[6] Cramer, J. M. ; Scholtz, R. A. ; Win, M. Z," Spatio-temporal diversity in ultra-wideband radio" *Wireless Communications and Networking Conference, IEEE*, Vol. 2, pp. 888-892, 1999.
[7] R. S. Kennedy, *Fading Dispersive Communication Channels.* John Wiley & sons, 1969.
[8] Telatar, I. E.; Tse, D. N. C. ,"Capacity and mutual information of wideband multipath fading channels " *IEEE Transactions on Information Theory*, Vol. 46, Issue:4, pp. 1384-1400, 1400.
[9] Gallager, R.G., *Information Theory and Reliable Communication*, Wiley and Sons, 1968.
[10] R. Papoulis, *Probability, Random Variables, and Stochastic Processes.* McGraw-Hill, 1984.
[11] C. W. Helstrom, *Elements of Signal Detection and Estimation.* PTR Prentice Hall, 1995.

Institut Eurécom,2229 route des Crêtes,06904 Sophia Antipolis Cedex,France
E-mail address: `souilmi@eurecom.fr,knopp@eurecom.fr`

DIMACS Series in Discrete Mathematics
and Theoretical Computer Science
Volume **62**, 2003

Efficient Multihop Broadcast for Wideband Systems

Ivana Maric and Roy Yates

ABSTRACT. In this paper we address the minimum-energy broadcast problem. To increase the energy efficiency, we allow nodes that are out of the transmission range of a transmitter to collect the energy of unreliably received overheard signals. As a message is forwarded through the network, a node will have multiple opportunities to reliably receive the message by collecting energy during each retransmission. We refer to this strategy as accumulative broadcast. Under the assumption that the nodes reliably forward messages, we formulate the minimum-energy accumulative broadcast problem. We present a solution employing two subproblems. First, we identify the ordering in which nodes should transmit. Second, we determine the optimum power levels for that ordering. While the second subproblem can be solved by means of linear programming, the ordering subproblem is found to be NP-complete. We devise a heuristic algorithm to find a good ordering and evaluate the performance of the algorithm. Preliminary results show the performance of the heuristic algorithm is generally close to the optimum solution. Results also show a significant improvement compared to the well known BIP algorithm for constructing an energy-efficient broadcast tree.

1. Introduction.

In the problem of minimum-energy broadcasting in a wireless network, the objective is to broadcast data reliably to all network nodes at a fixed rate with minimum transmitted power. The problem of broadcasting in a wireless network was introduced in [13] and formulated as a minimum-cost broadcast tree problem. In a wired network, the minimum-cost broadcast tree can be found in $O(n^2)$ operations [8]. However, in the wireless network, this problem was shown in [6] to be NP-hard and later on, in [1, 2, 7] to be NP-complete. The greater difficulty of the wireless broadcast problem tree stems from the "wireless multicast advantage" [13], the fact that a wireless transmission can be received by all nodes in the transmission range. Several heuristics for constructing energy-efficient broadcast trees have been proposed in literature and evaluated through simulations. The quantitative characterization of three such algorithms is given in [12].

The wireless formulation of the minimum energy broadcast problem assumes that a node can benefit from a certain transmission only if the received power is above a threshold required for reliable communication. This is a pessimistic assumption. A node for which the received power is below the required threshold,

1991 *Mathematics Subject Classification.* Primary 68M10; Secondary 90B18.

This work was supported in part by the NJ Center for Wireless Technology.

but above the receiver noise floor, can collect energy from the unreliable reception of the sent information. For example, in a Bluetooth system [9], the nominal transmitted power is 1 mW resulting in a transmission range of 10 meters. However, for a typical path exponent of $\alpha = 3$, the received signal at a node within 90 meters of the transmitter is likely to be above the receiver noise floor.

Moreover, it was observed in the relay channel [3] that utilizing unreliable overheard information was essential to achieving capacity. We borrow this idea and re-examine the minimum energy broadcast problem under the assumption that nodes can exploit the energy of an unreliable reception. For the broadcast problem, a node will have multiple opportunities to receive a message as it is forwarded through the network. We refer to this strategy as accumulative broadcast.

Even in the simplest case of a single relay node, finding the maximum achievable common rate for a given set of transmit powers is, in general, an unsolved open problem. Even in the special case of the physically degraded single relay channel, key techniques employed in [3] to enable coordination of the transmissions of the source and the relay are not easily extensible to multiple node networks. In this work, we seek to employ overheard information in a large scale network. To do this, we make certain simplifying assumptions. First, our focus will be on networks in which bandwidth is plentiful relative to the transmit power available to the nodes. Second, we impose a *reliable forwarding* constraint that a node can forward a message only after reliably decoding that message. Although reliable forwarding is generally a suboptimal strategy, we will see that it greatly simplifies the system architecture while still allowing us to benefit from unreliable overheard information.

In the following section, we give the network model. In Section 3, we describe our approach for finding the best broadcasting solution and show that the problem is NP-complete. A heuristic algorithm and its performance is presented in Section 4. The proofs for the theorems are given in the Appendix.

2. System model.

We consider a wireless network of N nodes such that from each transmitting node k to each receiving node m, there exists an AWGN channel of bandwidth W characterized by a frequency non-selective link gain h_{mk}. We further assume large enough bandwidth resources to enable each transmission to occur in an orthogonal channel, thus causing no interference to other transmissions. Each node has both transmitter and receiver capable of operating over all channels.

A receiver node j is said to be in the transmission range of transmitter i if the received power at j is above a threshold that ensures the capacity of the channel from i to j is above the code rate of node i. We assume that nodes can use different power levels, which will determine transmission ranges. The nodes beyond the transmission range can exploit the fact that a message is sent through multiple hops on its way to all the nodes. Repeated transmissions act as a repetition code for all the nodes beyond the transmission range.

We view each orthogonal channel as a discrete-time Gaussian channel by representing a waveform of duration T as a vector in the $2WT$ dimensional space [4]. Then, during the ith slot, a source node, labeled node 1, transmits a codeword (vector) $X^n(i)$ from a $(2^{nR}, n)$ Gaussian code that is generated according to the distribution $p(X^n) = \prod_{l=1}^{n} p(x_l)$ where $p(x) \sim N(0,1)$. Under the reliable forwarding constraint, a node j is permitted to retransmit (forward) codeword $X^n(i)$

only after reliably decoding $X^n(i)$. With an appropriate set of retransmissions at appropriate power levels, eventually every node will have reliably decoded X^n. Henceforth, we drop the index i and focus on the broadcast of a single codeword X^n. We will say a node is *reliable* once it has reliably decoded X^n.

The constraint of reliable forwarding imposes an ordering on the network nodes. In particular, a node m will decode X^n from the transmissions of a specific set of transmitting nodes that became reliable prior to node m. Starting with node 1, the source, as the first reliable node, a solution to the accumulative broadcast problem will be characterized by a *reliability schedule*, which specifies the order in which the nodes become reliable.

A reliability schedule $[n_1, n_2, n_2, \dots, n_N]$ is simply a permutation of $[1, 2, \dots, N]$ that always starts with the source node $n_1 = 1$. Given a reliability schedule, it will be convenient to relabel the nodes such that the schedule is simply $[1, 2, \dots, N]$. After each node $k \in \{1, \dots, m-1\}$ transmits codeword X^n with average energy per symbol P_k, the received signal at node m for each symbol x in the codeword is

$$\mathbf{y} = \mathbf{h}x + \mathbf{n}, \tag{1}$$

where $\mathbf{h} = [h_1, \dots, h_{m-1}]^T$ has kth element $h_k = (\sqrt{h_{mk}P_k})$ equal to the received energy corresponding to the transmission of node k and $\mathbf{n}$ is is a random noise vector with covariance matrix $\mathbf{K_n} = \sigma^2 \mathbf{I}_K$. It follows that the mutual information is

$$I(x; \mathbf{y}) = \frac{1}{2} \log_2 \left(1 + \frac{\sum_{k=1}^{m-1} h_{mk} P_k}{\sigma^2} \right) \quad \text{bits.} \tag{2}$$

This is precisely the capacity of a multi-antenna system derived in [10] with $m-1$ transmitting antennas and one receiving antenna. It follows from Equation (2) that the maximal number of bits per second that can be transmitted in the system given by Equation (1) is

$$r_m = W \log_2 \left(1 + \frac{\sum_{k=1}^{m-1} h_{mk} p_k}{N_0 W} \right) \quad \text{bits/s,} \tag{3}$$

where p_k is the transmit power at node k and N_0 is the one-sided power spectral density of the noise.

Achieving $r_m = \overline{r}$ implies $\sum_{k=1}^{m-1} h_{mk} p_k \geq \overline{P}$ where

$$\overline{r} = W \log_2 \left(1 + \frac{\overline{P}}{N_0 W} \right) \quad \text{bits/s} \tag{4}$$

is the fixed data rate used for broadcasting.

When the nodes are power limited and the data rate $\overline{r}$ is small relative to the channel bandwidth W, the system operates in the wideband regime where energy per bit is close to its minimum value and the increase in rate with power is linear:

$$\overline{r}_\infty = \lim_{W \to \infty} \overline{r} = \lim_{W \to \infty} W \log_2 \left(1 + \frac{\overline{P}}{W N_0} \right) = \frac{\overline{P}}{N_0 \ln 2} \quad \text{bits/s.} \tag{5}$$

From the above equation, transmission of $\overline{r}_\infty = E_T/(N_0 \ln 2)$ bits requires received energy of $E_T = \overline{P}T$, revealing the well known fact that for the reliable reception of one information bit, a receiver has to collect at least $E_T = N_0 \ln 2$ Joules/bit. [11]. This energy can be collected at a node m during one transmission interval $[0, T]$ when a transmitter j is signaling with power $p = (N_0 \ln 2)/(h_{mj}T)$. However, in the system given by Equation (1), the required energy E_T is collected in $m-1$ repeated transmissions. In the wideband regime, Equation (3) is

$$\lim_{W\to\infty} r_m = \frac{1}{N_0 \ln 2}\sum_{k=1}^{m-1} h_{mk}p_k. \tag{6}$$

Next we show that for the same transmitted powers, using a coding strategy other than the repetition code described above, cannot increase the rate achieved at the receiving node m.

THEOREM 1. *For the wideband regime, fixed transmitted powers* $\{p_1,\dots,p_N\}$ *and a reliability schedule* $[1,2,\dots,N]$, *the maximum rate achievable from the source to node m is given by Equation (2) and is achieved by the repetition coding strategy.*

Proof. An upper bound to the achievable rate between the source and the destination is the maximum conditional mutual information across a minimum cut [4]. Consider the multiaccess cut in the given network that separates the destination node from the rest of the network. Let X_j denote a symbol transmitted at node j and Y denote the received signal at the destination. The maximum mutual information across this cut is given by

$$C_{\text{MAC}} = I(X_1,\dots X_{m-1};Y). \tag{7}$$

In the considered network, each orthogonal channel is assigned bandwidth W and hence the mutual information above is given by the sum of rates achieved in each of the channels. For Gaussian channels,

$$C_{\text{MAC}} = W\sum_{k=1}^{m-1}\log_2\left(1+\frac{h_{mk}p_k}{N_0 W}\right). \tag{8}$$

In the wideband regime, Equation (8) becomes

$$C_{\text{MAC}} = \lim_{W\to\infty} W\sum_{k=1}^{m-1}\log_2\left(1+\frac{h_{mk}p_k}{N_0 W}\right) = \frac{1}{N_0\ln 2}\sum_{k=1}^{m-1} h_{mk}p_k, \tag{9}$$

which is precisely the rate given by (6) achieved using the repetition strategy. Since this rate is achievable, this cut is the minimum cut. No better rate can be achieved since it would violate the condition for the upper bound. □

Not allowing the noisy observations to be amplified and sent at relays makes achieving the broadcast cut-set bound impossible. As shown in [5], this bound can be asymptotically achieved in a network with one source-destination pair as the number of relays goes to infinity. Again, this situation is equivalent to the multi-antenna system [10], but in this case, with one transmitting antenna and K receiving antennas.

3. Approach

Under the constraint of reliable forwarding, an optimal solution to the minimum power accumulative broadcasting problem must specify the reliability schedule as well as the transmitter power levels used at each node. An optimal choice of the reliability schedule will result in minimum total transmitted power over the set of nodes.

From the above arguments, it follows that our problem of finding the minimum-energy accumulative broadcast differs from the minimum-energy broadcast tree

[1,2,6,7,12–14] in the sense that no received energy is wasted at the nodes while they collect energy unreliably. The two problems are similar in that finding the optimum solution involves the right ordering of relay nodes and determination of transmitter power levels. Defining the broadcast tree determines all the transmission levels and thus solves the problem completely; a relay that is the parent of a group of nodes in the broadcast tree transmits with the power needed to reliably reach the most disadvantaged node in the group. Hence, the arcs in the broadcast tree uniquely determine the power levels for each transmission.

In our scenario, however, the broadcast tree loses its meaning since nodes collect energy from the transmissions of many nodes. The optimum solution may require that a relay transmits with a power level different from the level precisely needed to reach a group of nodes reliably; the nodes can collect the rest of the needed energy from the future transmissions of other nodes. In fact, the optimum solution favors such situations because all nodes beyond the range of a certain transmission are collecting energy while they are unreliable; the more such nodes, the more efficiently the transmitted energy is being used. Therefore, if during one transmission, an unreliable node is collecting energy faster due to a good channel gain with a current transmitting node, it may be desirable for the transmitting node to stop transmitting before the collected energy at the node in favor reaches the threshold of reliable communication; that node will collect the rest of the energy needed during future transmissions of other nodes.

The next theorem shows that this approach can never do worse than the solution to the minimum-energy broadcast tree approach. In fact, accumulative broadcasting will always result in a more energy-efficient solution.

THEOREM 2. *The sum of transmitted powers of the minimum-energy broadcast tree is an upper bound to the sum of transmitted powers for accumulative broadcasting.*

The differences in our problem from the minimum-energy broadcast tree dictate a new approach. The crucial step is finding the best reliability schedule. Given a schedule, we can formulate a linear program (LP) that will find the optimum solution for that schedule. Such a solution will identify those nodes that should transmit and their transmission power levels. Solving the LP for all possible schedules and taking the minimum-energy solution among all the LP solutions will result in the optimum schedule, and optimum transmission power levels. This divides the problem into two subproblems.

To define the LP for a certain schedule, we use the observation that every node selected to transmit by the optimum solution, needs to transmit only once. This fact is given by the next theorem.

THEOREM 3. *Given a solution to the accumulative broadcast problem consisting of a sequence of transmissions where a node j is assigned to transmit K times with power levels $P_j^1, \dots P_j^K$ then there is a feasible optimum solution in which node j transmits only once with power level $\sum_{k=1}^{K} P_j^k$.*

A reliability schedule can be represented by a matrix $\mathbf{X}$ where

$$x_{ij} = \begin{cases} 1 & \text{if node } i \text{ is scheduled to transmit after node } j \\ 0 & \text{otherwise} \end{cases} \tag{10}$$

Each x_{ij} is an indicator that a node i collects energy from a transmission by node j. Note that $x_{ii} = 0$, for all i and $x_{ji} = 1 - x_{ij}$. Given $\mathbf{X}$ we define a gain matrix

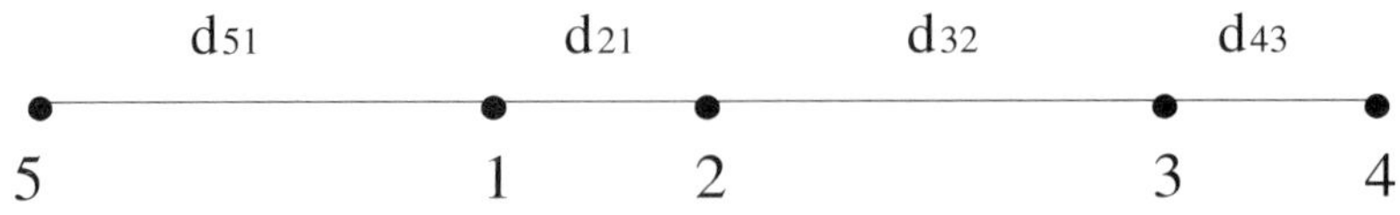

FIGURE 1. A five node network example.

for that schedule $\mathbf{H}(\mathbf{X})$. Each element (i, j) in $\mathbf{H}(\mathbf{X})$ is given as $h_{ij}x_{ij}$. In terms of the vector $\mathbf{p}$ of transmitted powers, the LP for that schedule is

$$\begin{aligned} \rho(\mathbf{X}) = \min\ & \mathbf{1}^T\mathbf{p} \\ \text{subject to } & \mathbf{H}(\mathbf{X})\mathbf{p} \geq \mathbf{1}\overline{P}, \\ & \mathbf{p} \geq \mathbf{0}. \end{aligned} \tag{11}$$

The inequality $\mathbf{H}(\mathbf{X})\mathbf{p} \geq \mathbf{1}\overline{P}$ contains $N-1$ constraints requiring that the received power at all the nodes but the source is above the required threshold $\overline{P}$. Given a schedule $\mathbf{X}$, we will use $\mathbf{p}^*(\mathbf{X})$ to denote a power vector $\mathbf{p}$ that achieves total transmitted power $\rho(\mathbf{X})$.

In a schedule, all N nodes are given a chance to transmit the data (since p_j can be greater than 0 for every node) and only the order of transmission is different for different schedules. Every schedule is a different ordered sequence of N nodes. Since the source always transmits first, there are $(N-1)!$ schedules corresponding to the number of permutations of $N-1$ elements. Thus, out of $N^{(N-2)}$ broadcast trees, we consider a subset of $(N-1)!$ schedules. The reason is that, if the best solution is that only a subset of nodes should be transmitting, the appropriate LP will find that solution by setting appropriate powers to zero. In general, however, the problem of finding a best schedule is intractable.

THEOREM 4. *The existence of a schedule* $\mathbf{X}$ *such that* $\rho(\mathbf{X}) \leq B$ *is an NP-complete problem.*

In the following, we describe scheduling heuristics.

4. Scheduling Heuristics

We observe that we can restrict ourselves to scheduling nodes in an order in which they can become reliable one at a time. When a node j is scheduled to be the next node in a schedule after a set of nodes S, then a transmission from that set has to make node j reliable. If the power that is needed to reach node j is enough to reach another unreliable node i as well, than we could have done better by assigning node i for transmission before node j. This is because i cannot benefit from a transmission from node j (since it is made reliable before j) but j might benefit from a transmission from i. This reasoning will be used in the proposed heuristic algorithm.

The next theorem shows how the solution of a LP for a certain schedule gives information about the subsequent schedule to be considered.

THEOREM 5. *If the solution* $\mathbf{p}^* = \mathbf{p}^*(\mathbf{X})$ *of the LP (11) has a subset of nodes* I*, such that* $p_i^* = 0$ *for all* $i \in I$*, then* $\rho(\hat{\mathbf{X}}) \leq \rho(\mathbf{X})$ *for a new schedule* $\hat{\mathbf{X}}$ *that assigns nodes in* I *to transmit last.*

```
s = [1]; p = 0
while (|S| < N) do
    k = arg max_{i∈S} Σ_{j∈U} h_{ji};
    j = arg min_{m∈U} (P̄ − Σ_{i∈S} h_{mi} p_i)/h_{mk};
    n = I_k(s);
    if (n ≠ |S| and p_k ≠ 0)
        p_k ← p_k + (P̄ − Σ_{i∈S} h_{ji} p_i)/h_{jk};
        for i = n + 1 : |S| do p_{s_i} = 0 end;
        s ← [s_1, ... , s_n];
        do
            l = arg max_{m∈U} Σ_{i∈S} h_{mi} p_i;
            if (Σ_{i∈S} h_{li} p_i ≥ P̄) s ← [s, l];
        while (Σ_{i∈S} h_{li} p_i ≥ P̄)
    else
        p_k ← p_k + (P̄ − Σ_{i∈S} h_{ji} p_i)/h_{jk};
        s ← [s, j]
    end;
end
```

Given a partial schedule $\mathbf{s}$*,* S *is the unordered set of nodes in* $\mathbf{s}$ *and its complement* U *is the set of unreliable nodes. The cardinality of* S *is given by* $|S|$*. In addition,* $I_k(\mathbf{s})$ *returns the position of node* k *in the schedule* $\mathbf{s}$*. That is, if* $s_n = k$*, then* $I_k(\mathbf{s}) = n$*.*

FIGURE 2. Greedy Filling Heuristic Algorithm - The Offline Version.

```
s = [1]; p = 0
while (|S| < N) do
    k = arg max_{i∈S} Σ_{j∈U} h_{ji};
    j = arg min_{m∈U} (P̄ − Σ_{i∈S} h_{mi} p_i)/h_{mk};
        p_k ← p_k + (P̄ − Σ_{i∈S} h_{ji} p_i)/h_{jk};
        s ← [s, j]
end
```

FIGURE 3. Simplified Greedy Filling Heuristic Algorithm.

We illustrate our approach on an example network with 5 nodes on a line as shown in Figure 1. For every two nodes j and k, the link gain is calculated from $h_{jk} = d_{jk}^{-3}$ for the distances given in Figure 1. Let node 1 be a source. The number of possible spanning trees is 125. By Theorem 3, $(N-1)! = 24$ schedules need to be considered given by all the possible permutations of all the nodes except the source who always transmits first. Two observations help reduce the number of schedules: Since the closest neighbor to the source is node 2, this node will become reliable first regardless of a schedule. Following the reasoning from above, all the considered schedules should assign this node for a possible transmission after the source. For the same reason node 3 should always be scheduled to transmit before node 4, thus giving node 4 the chance to collect energy from the transmission of node 3. This leaves only three possible schedules to be considered:

$$\mathbf{s}_1 = [1,2,3,4,5], \quad \mathbf{s}_2 = [1,2,3,5,4], \quad \mathbf{s}_3 = [1,2,5,3,4].$$

We consider schedule $\mathbf{S_1}$ first. The LP for that schedule is

$$\min\ p_1 + p_2 + p_3 + p_4$$
$$\begin{bmatrix} h_{21} & & & \\ h_{31} & h_{32} & & \\ h_{41} & h_{42} & h_{43} & \\ h_{51} & h_{52} & h_{53} & h_{54} \end{bmatrix} \begin{bmatrix} p_1 \\ p_2 \\ p_3 \\ p_4 \end{bmatrix} \geq \mathbf{1}\overline{P}$$
$$p_1, p_2, p_3, p_4 \geq 0$$

Solving the LP we get the optimum power levels for this schedule. In particular, the solution finds that $p_4 = 0$. Therefore, although node 4 was given a chance to transmit as soon as possible, its transmission is not needed; by theorem 5 we can move node 4 to the end of the schedule. In both schedules $\mathbf{s}_2$ and $\mathbf{s}_3$ node 4 is scheduled to be the last node in the schedule. Hence, we consider both $\mathbf{s}_2$ and $\mathbf{s}_3$. By solving the LP for those two schedules, and comparing the minimum energy solution for three schedules, we get the optimum solution. The solution gives nodes 1, 2 and 3 as the transmitting nodes. The power level at node 1 is higher than needed to reach node 2 but less than needed to reach node 5; the rest node 5 collects from later transmission at nodes 2 and 3. The transmitted power at node 2 is enough to reach node 3 given what it collected from p_1. Node 3 transmits to 4 such that enough energy from three transmissions is collected at node 4 to meet the threshold.

Because of the complexity of the problem of finding the best schedule, we now propose a heuristic algorithm that finds a good schedule. Once the schedule is determined, the LP for that schedule is solved to find the optimum power levels. We propose two versions of the algorithm: *The offline version* that requires a schedule adjustment, and its simplified counterpart where no such adjustment is needed. We evaluate the performance of the algorithm through simulation and compare its power efficiency to the optimum solution as well as to the performance of BIP, the heuristic algorithm proposed in [13].

We present the offline version of the algorithm first. The algorithm pseudocode is given in Figure 2. The algorithm starts with a partial reliability schedule $\mathbf{s} = [1]$ that contains only the source. In general, given $\mathbf{s} = [s_1, \dots, s_n]$, if $s_l = k$, then node k is the lth node in the schedule. Given a partial schedule $\mathbf{s}$, we find the reliable node k that maximizes the fill rate of the unreliable set. By Theorem 3, node k should transmit only once. Therefore, we check to make sure that k has not yet transmitted, or if it did, k was the last reliable node to transmit. In this case, we increase p_k by the power needed to add one more node, node j, to the reliable set and we simply append node j to the schedule. Otherwise, if k is not the last reliable node to transmit and $p_k \neq 0$, then k has been chosen to transmit earlier in the schedule. By Theorem 3, k should have transmitted at that earlier time with a higher power level, making more of its neighbors reliable at the same step. Thus we reset the schedule after node k in the following way. All transmitting nodes that followed node k in the schedule have their powers set to zero and we truncate the schedule after node k. However, those nodes that are made reliable by transmissions of the truncated schedule are appended to the schedule in order

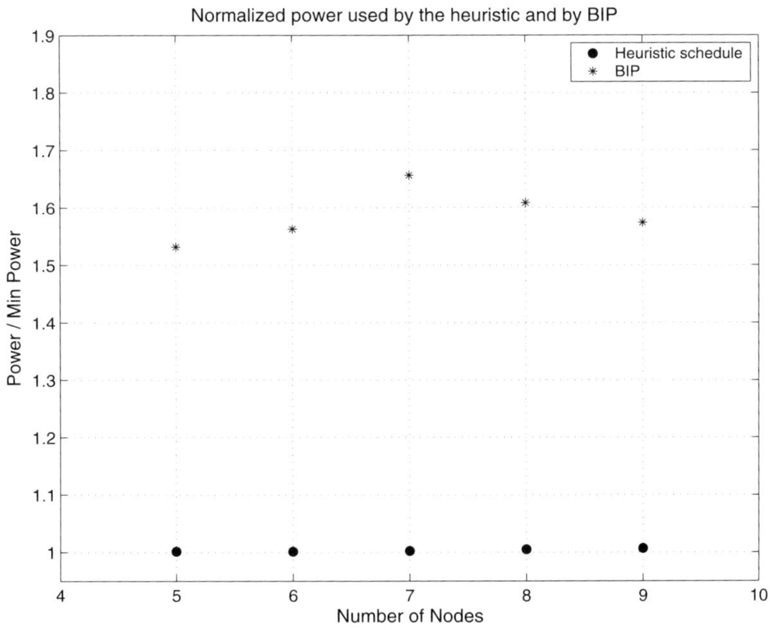

FIGURE 4. Normalized power used for broadcasting.

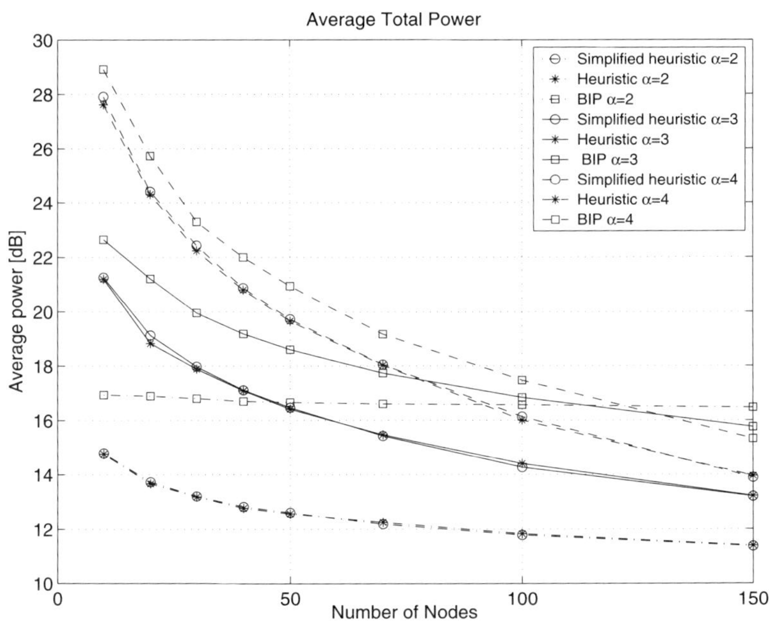

FIGURE 5. Average total power used for broadcasting.

of decreasing total received power. Once the schedule is complete, the LP is solved to find the optimum power levels for that schedule.

To avoid the offline schedule adjustment we consider a simplification of the above algorithm where we drop the requirement on a single transmission at a node. Then, in each step of the algorithm we only need to choose a reliable node k for transmission and to append a single node j to the existing schedule. As the simulation results will show, the loss in the performance due to the simplification is negligible. The corresponding simplified pseudocode is given in Figure 3.

We evaluated the performance of the algorithms and compared them to the optimal solution as well as to the performance of BIP. Networks with a small number $(5-10)$ of randomly positioned nodes were generated. We also compared the performance of all three heuristics for more dense networks. The transmitted power was attenuated as r^α for three different values of propagation exponent $\alpha = 2, 3, 4$. The received power threshold was chosen to be $\overline{P} = 1$. The results were based on the performance of 100 randomly chosen networks. When comparison with the optimum solution was performed, the performance metric used was the normalized total power in the network. In each simulation run, the power used when a heuristic algorithm was employed was normalized by the power used in the optimum solution.

Results are shown in Figure 4 as a function of the number of network nodes for $\alpha = 2$. In this case, the offline version of the algorithm was used. Results show the heuristic algorithm performance very close to the optimum. This is a desirable and important characteristic, given the complexity of finding the optimum solution. The simulation result also show the noticeable improvement in the power saving in our approach compared to power used when the minimum-energy broadcast tree approach is used and the broadcast tree is found using BIP.

For the networks with a larger number of nodes, performance results for the accumulative broadcast heuristics and BIP are shown in Figure 5. The metric used was the average total power used for broadcasting. We observe that total power decreases with the number of nodes due to the increased number of shorter hops. The decrease in the case of the accumulative broadcast is steeper since the increased number of transmissions allows for more energy to be collected. Hence the relative improvement over BIP increases with the number of network nodes. For smaller values of propagation exponent α, the smaller path loss allows for the higher gains from the accumulative broadcast. Results also show that, for a larger number of nodes the total power required is smaller for larger values of α. This counterintuitive result is due to the fact that, as the networks becomes more dense, the most of the distances become less than 1. Results show no significant performance loss when the simplified heuristic algorithm is used.

5. Conclusion

In this paper we address the minimum-energy broadcast problem. To increase the energy efficiency, we realize that a lot of energy is radiated into the network in each transmission and thus, we allow nodes outside of the transmission range to collect the energy of the unreliably received signal. A node will have multiple opportunities to receive reliably the message by collecting energy while the message is forwarded through a network. We demonstrate that finding the optimum solution to this problem is complex and therefore, propose a heuristic algorithm that has energy-efficiency close to optimal and can still provide energy-saving compared to broadcasting that doesn't collect energy from unreliable transmission.

6. Appendix: Additional Proofs.

Proof (Theorem 2): Let the optimum solution to the minimum-energy broadcast problem be a vector of transmitted powers, $\hat{\mathbf{p}}^*$ and a matrix $\hat{\mathbf{X}}$ determined by the minimum-energy broadcast tree (MEBT) such that $\hat{x}_{ij} = 1$ if j is a parent of i in MEBT and zero otherwise. Again, $\hat{x}_{ij}$ is an indicator that node i collects energy from a transmission at node j. We show that $\hat{\mathbf{p}}^*$ belongs to the feasibility region of the LP (11) for some schedule $\mathbf{X}$ and can therefore only be worse than the optimum solution for accumulative broadcast.

As in accumulative broadcast, the solution to the MEBT problem has to satisfy the constraint that the received power at all nodes except the source is above the required threshold. Using matrix $\hat{\mathbf{X}}$, we can represent these constraints as in (11) in the form

$$\mathbf{H}(\hat{\mathbf{X}})\hat{\mathbf{p}}^* \geq \mathbf{1}\overline{P}. \tag{12}$$

Given the MEBT, we can form an ordered sequence of all the nodes in which each parent node appears before its children. The nodes that are at the same depth from the source appear in any order relative to each other. The sequence defines one permutation of N elements with the source being the first element and thus defines a possible schedule for accumulative broadcast. For convenience, we relabel the nodes in the order they have been scheduled such that the schedule is $[1, 2, \dots, N]$. Let $\mathbf{X}$ be a corresponding matrix for that schedule. Each element of $\mathbf{X}$ can be represented as $x_{ij} = \hat{x}_{ij} + \mathbf{1}(j < i)(1 - \hat{x}_{ij})$ where $\mathbf{1}(\cdot)$ is an indicator function. Thus, $\mathbf{X} = \hat{\mathbf{X}} + \tilde{\mathbf{X}}$ where $\tilde{x}_{ij} = \mathbf{1}(j < i)(1 - \hat{x}_{ij})$. The feasibility constraints (11) become

$$\mathbf{H}(\hat{\mathbf{X}})\mathbf{p} + \mathbf{H}(\tilde{\mathbf{X}})\mathbf{p} \geq \mathbf{1}\overline{P}. \tag{13}$$

From (12) and nonnegativity of $H(\tilde{X})$, it follows that $\hat{\mathbf{p}}^*$ satisfies (13). Therefore, $\hat{\mathbf{p}}^*$ belongs to the feasibility region of the LP for schedule $\mathbf{X}$ and the solution of that LP can only be better than $\hat{\mathbf{p}}^*$. □

Proof (Theorem 3): For the purpose of this proof we represent a solution to the accumulative broadcast problem by a vector with each entry i containing the ith transmitting node n_i and the ith transmitted power level P_i. Non-transmitting nodes are appended in an arbitrary order at the end and assigned zero power levels. A solution $\mathbf{S}$ is represented as

$$\mathbf{S} = \left[\ (n_1, P_1) \quad (n_2, P_2), \quad \dots \quad (n_M, P_M)\ \right]^T \tag{14}$$

for some $M \geq N$.

Let $\mathbf{S}^*$ be the optimum solution. Given $\mathbf{S}^*$, we can represent a feasibility condition at every node j as follows. Assume that a first transmission of a node j occurs at the kth step for some $k \in [1, M]$. Then, $n_k = j$ and $n_i \neq j$ for all $i < k$. From the feasibility of $\mathbf{S}^*$ it follows that every node becomes reliable before it transmits. Therefore, a feasibility condition at node j

$$\sum_{i=1}^{k-1} h_{n_k n_i} P_i \geq \overline{P} \tag{15}$$

is satisfied.

Assume that $\mathbf{S}^*$ schedules the same node for a transmission more than once. Let l denote the smallest integer such that there exists an integer $m > l$ with

$n_l = n_m$. Consider the policy $\hat{\mathbf{S}}^*$, a vector of length $M-1$ with elements $(\hat{n}_i, \hat{P}_i)$ such that

$$(\hat{n}_i, \hat{P}_i) = \begin{cases} (n_i, P_i) & \text{if } i < l, \\ (n_l, P_l + P_m) & \text{if } i = l, \\ (n_i, P_i) & \text{if } l < i < m, \\ (n_{i+1}, P_{i+1}) & \text{if } i \geq m. \end{cases} \tag{16}$$

The solution $\hat{\mathbf{S}}^*$ combines transmissions at steps l and m into a single transmission with power $P_l + P_m$ at step l. The rest of the nodes are scheduled as in $\mathbf{S}^*$.

From (16),

$$\sum_{i=1}^{k-1} h_{\hat{n}_k \hat{n}_i} \hat{P}_i = \begin{cases} \sum_{i=1}^{k-1} h_{n_k n_i} P_i & \text{if } k \leq l, \\ \sum_{i=1}^{k-1} h_{n_k n_i} P_i + h_{n_k n_l} P_m & \text{if } l < k < m, \\ \sum_{i=1}^{k-1} h_{n_k n_i} P_i & \text{if } m \leq k \leq M-1. \end{cases} \tag{17}$$

From (15) and (17) it follows that the feasibility condition is satisfied for every node j in $\hat{\mathbf{S}}^*$. Optimality of the solution $\hat{\mathbf{S}}^*$ follows from the fact that the total power transmitted in both solutions is the same. Since the solution $\mathbf{S}^*$ minimizes the total power, the solution $\hat{\mathbf{S}}^*$ does too. The above combining technique can be repeatedly applied until each node transmits just once. □

Proof (Theorem 4): Let Π_i denote the set of all vectors $\pi = [\pi_0, \ldots, \pi_i]$ that are permutations of $[0, 1, \ldots, i]$. A formal statement of the ACCUMULATIVE BROADCAST (AB) problem is

AB: Given a nonnegative matrix specified by $\{h_{j,k} | 1 \leq j \leq m, 0 \leq k \leq m\}$, and a constant c, does there exist a permutation $\pi \in \Pi_k$ with $\pi_0 = 0$ and a non-negative vector $\mathbf{p} = [p_0, p_1, \ldots, p_m]$ such that $\sum_{k=0}^{m} p_k \leq c$ and $\sum_{k=0}^{j-1} h_{\pi_j, \pi_k} p_{\pi_k} \geq 1, \quad j = 1, \ldots, m.$

Thus an instance of AB is specified by the pair $(\{h_{j,k}\}, c)$. Note that we set the reliability threshold to unit power since any scaling can be specified by the constant c. We observe that AB is in NP since given a permutation π and vector $\mathbf{p}$, it is easy to check whether the AB constraints are met.

We will show that the ACCUMULATIVE BROADCAST problem is NP complete by a polynomial time reduction of the DIRECTED HAMILTON PATH (DHP) [8] problem. Formally the DHP problem is

DHP: Given a directed graph $G = (V, A)$ with nodes $V = \{0, \ldots, n\}$, does there exist a permutation $\pi \in \Pi_n$ such that $\pi_0 = 0$ and $(\pi_i, \pi_{i+1}) \in A$ for $i = 0, \ldots, n-1$.

We now describe the transformation of DHP into an instance of AB. Without loss of generality, we assume that the instance of DHP is such that node 0 has a single outgoing arc $(0, 1)$ and that node n is a sink node reachable by an arc (i, n) from each node $i \in \{1, \ldots, n-1\}$. Note that if this condition does not hold, we can add such source and sink nodes and solve an equivalent DHP. Thus, for each such graph, the Hamilton path, if it exists, will start at node 0 and terminate at node n.

Given $G = (V, A)$ for DHP, we construct a set of nodes G' and matrix $\{h_{j,k}\}$ for an instance of AB. In particular, for each node $k \in G$, we construct a cluster of nodes $C_k \subset G'$. In particular, the cluster C_k includes a node $i_{j,k}$ for each incident arc $(j, k) \in A$ and a node $o_{k,l}$ for each outgoing arc $(k, l) \in A$. That is, in terms of each arc $(j, k) \in A$, we have created an *incident node* $i_{j,k} \in C_k$ and an *outgoing node* $o_{j,k} \in C_j$. Note that cluster C_0 contains only the single node $o_{0,1}$ and that the sink node n has the cluster $C_n = \{i_{j,n} | 1 \leq j < n\}$ of only incoming nodes.

To avoid an explicit enumeration of the nodes in G', we describe the matrix $\{h_{j,k}\}$ in terms of a function $h(a, b)$ that gives the channel gain from node b to node a. Similarly, we will use the notation $p(a)$ to denote the transmitted power of the node a. Corresponding to each arc $(j, k) \in A$, we have $h(i_{j,k}, o_{j,k}) = 1$. Within each cluster C_k, we have that for any pair of incident nodes $i_{j,k}$ and $i_{j',k}$, $h(i_{j,k}, i_{j',k}) = 1$. In addition, for each outgoing node $o_{k,l} \in C_k$, and each incoming node $i_{j,k} \in C_k$, $h(o_{k,l}, i_{j,k}) = 1$. For all other pairs of nodes $a, b \in G'$, we set $h(a, b) = 0$. Keep in mind that if $h(a, b) = 1$, then $p(b) = 1$ yields received power $h(a, b)p(b) = 1$ at node a. We will see in our AB construction, each node a will use power $p(a) \in \{0, 1\}$.

To prove that AB is NP-complete, we show that the graph G has a Hamilton path if and only if the resulting instance $(h(\cdot, \cdot), c = 2n)$ of AB is feasible. Consider a Hamilton path that starts at node zero and proceeds through all nodes to node n. Suppose the Hamilton path uses arc (j, k), then for the AB problem, we set $p(o_{j,k}) = 1$, $p(i_{j,k}) = 1$, $p(o_{j,k'}) = 0$ for all $k' \neq k$, and $p(i_{j',k}) = 0$ for all $j' \neq j$. In the context of AB, node $o_{j,k}$ transmits to make node $i_{j,k}$ reliable and then node $i_{j,k}$ transmits to make all nodes in cluster C_k reliable. If the next arc in the Hamilton path is (k, l), then in the AB, $o_{k,l}$, which has already been made reliable by the transmission of $i_{j,k}$, will transmit to make $i_{k,l}$ reliable. We call the event that an incoming node $i_{j,k}$ is made reliable a *visit* to cluster C_k. The sequence of nodes in the Hamilton path corresponds exactly to the sequence of cluster visits. To calculate the total transmitted power, note that in cluster C_0, node $o_{0,1}$ will transmit. In clusters $1, \ldots, n-1$, one incoming node and one outgoing node will transmit. Lastly, in cluster C_n, one incoming node will transmit to make the other incoming nodes in C_n reliable. The total transmitted power will be exactly $2n$. We note that the node ordering required by the formal statement of AB will not be uniquely specified. If cluster C_k is visited before cluster C_l, then all nodes in C_k must be ordered ahead of nodes in C_l. In a cluster C_k, if incoming node $i_{j,k}$ is made reliable then $i_{j,k}$ must be first in the cluster but other nodes in the cluster can be ordered arbitrarily.

To complete the proof, suppose we have a solution to the AB problem. This AB solution must make every node in the graph G' reliable. For each cluster C_k, $1 \leq k \leq n$, at least one incoming node $i_{j,k}$ must be made reliable by the transmission of the corresponding outgoing node $o_{j,k}$. However, since this transmission of $o_{j,k}$ makes only $i_{j,k}$ reliable, one such transmission is needed for each cluster C_k. Over all clusters C_k, $1 \leq k \leq n$, we require n such transmissions. Further, within each cluster, the outgoing nodes can be made reliable only by the transmission of an incoming node in the cluster. Thus for each cluster C_k, $1 \leq k \leq n$, at least one incoming node $i_{j,k}$ must transmit to make all other nodes in the cluster reliable; this requires n additional transmissions. Thus $2n$ is a lower bound to the number of transmissions for the AB problem. Moreover, if the solution to AB achieves the minimum $2n$, then each outgoing node transmission must be to an incoming node in a cluster that has had no other incoming nodes receive a transmission from its corresponding outgoing node. That is, each cluster can be visited only once for the $2n$ lower bound to be met. Starting with node 0 and cluster C_0, node $o_{0,1}$ will transmit to make node $i_{0,1}$ reliable. Node $i_{0,1}$ must then transmit to make all other nodes in cluster C_1 reliable. An outgoing node $o_{1,k}$ will then transmit to make a node $i_{1,k}$ reliable, constituting a visit to cluster k. To achieve the $2n$ lower bound, each cluster will be visited precisely once, with termination at cluster C_n. Since moving from cluster C_j to visit C_k can occur only if (j,k) is an arc in G, the AB solution corresponds to a Hamilton path in the graph G. □

Proof (Theorem 5): Let $LP_{\mathbf{X}}$ and $LP_{\hat{\mathbf{X}}}$ denote the LPs for schedules $\mathbf{X}$ and $\hat{\mathbf{X}}$, respectively. We show that $\mathbf{p}^* = \mathbf{p}^*(\mathbf{X})$ belongs to the feasibility region for $LP_{\hat{\mathbf{X}}}$ given by (11) and therefore the optimum solution of $LP_{\hat{\mathbf{X}}}$ can only be better than $\mathbf{p}^*$.

Consider the schedule $\mathbf{X}$. From the assumption that $p_i^* = 0$ for all $i \in I$, it follows that the feasibility condition (11) for $LP_{\mathbf{X}}$ satisfied for every node k is

$$\sum_j x_{kj} h_{kj} p_j^* = \sum_{j \notin I} x_{kj} h_{kj} p_j^* \geq \overline{P}. \tag{18}$$

Next consider the schedule $\hat{\mathbf{X}}$. Since all the nodes k such that $k \notin I$ are scheduled in the same order relative to each other in both schedules $\mathbf{X}$ and $\hat{\mathbf{X}}$, it follows that $\hat{x}_{kj} = x_{kj}$ for all $k, j \notin I$. All the nodes k such that $k \in I$ are scheduled last in $\hat{\mathbf{X}}$, therefore $\hat{x}_{kj} = 1$ for all $k \in I, j \notin I$. Then,

$$\sum_j \hat{x}_{kj} h_{kj} p_j^* = \sum_{j \notin I} \hat{x}_{kj} h_{kj} p_j^* = \begin{cases} \sum_{j \notin I} x_{kj} h_{kj} p_j^* & k \notin I, \\ \sum_{j \notin I} h_{kj} p_j^* & k \in I. \end{cases} \tag{19}$$

From (18) and (19) it follows that the $\mathbf{p}^*$ satisfies the feasibility conditions for $\hat{\mathbf{X}}$. □

References

[1] A. Ahluwalia, E. Modiano, and L. Shu. On the complexity and distributed construction of energy-efficient broadcast trees in static ad hoc wireless networks. In *Proc. of Conf. on Information Science and Systems*, March 2002.

[2] M. Cagalj, J. Hubaux, and C. Enz. Minimum-energy broadcast in all-wireless networks: Np-completeness and distribution issues. In *Proc. of Mobicom'02*, 2002.

[3] T. Cover and A. El Gamal. Capacity theorems for the relay channel. *IEEE Trans. on Information Theory*, 25(5):572–584, September 1979.

[4] T. Cover and J. Thomas. *Elements of Information Theory*. John Wiley Sons, Inc., 1991.

[5] M. Gastpar and M. Vetterli. On the capacity of wireless networks: The relay case. In *Proc. of INFOCOM'02*, June 2002.

[6] F. Li and I. Nikolaidis. On minimum-energy broadcasting in all-wireless networks. In *Proc. of Local Computer Networks (LCN'01)*, 2001.

[7] W. Liang. Constructing minimum-energy broadcast trees in wireless ad hoc networks. In *Proc. of International Symposium on Mobile Ad Hoc Networking and Computing (MobiHoc'02)*, June 2002.

[8] C.H. Papadimitriou and K. Steiglitz. *Combinatorial Optimization: Algorithms and Complexity*. Prentice Hall, Englewood Cliffs, NJ, 1982.

[9] SIG. Bluetooth specification version 1.0b. *Tech. Spec.*, 1999.

[10] E. Telatar. Capacity of multi-antenna gaussian channels. In *Europ. Trans. Telecommunications*, November 1997.

[11] S. Verdú. Spectral efficiency in the wideband regime. *IEEE Trans. on Information Theory*, 48(6):1319–1343, June 2002.

[12] P-J. Wan, G. Calinescu, X-Y. Li, and O. Frieder. Minimum-energy broadcast routing in static ad hoc wireless networks. In *Proc. of INFOCOM'02*, April 2001.

[13] J.E. Wieselthier, G.D. Nguyen, and A. Ephremides. On the construction of energy-efficient broadcast and multicast trees in wireless networks. In *Proc. of Infocom 2000*, 2000.

[14] J.E. Wieselthier, G.D. Nguyen, and A. Ephremides. Algorithms for energy-efficient multicasting in static ad-hoc wireless network. *Mobile Networks and Applications*, pages 251–263, 2001.

WINLAB, Rutgers University, Piscataway, NJ 08854
E-mail address: ivanam@winlab.rutgers.edu
E-mail address: ryates@winlab.rutgers.edu

Titles in This Series

62 **Gerard J. Foschini and Sergio Verdú, Editors,** Multiantenna Channels: Capacity, Coding and Signal Processing

61 **M. F. Janowitz, F.-J. Lapointe, F. R. McMorris, B. Mirkin, and F. S. Roberts, Editors,** Bioconsensus

60 **Saugata Basu and Laureano Gonzalez-Vega, Editors,** Algorithmic and Quantitative Real Algebraic Geometry

59 **Michael H. Goldwasser, David S. Johnson, and Catherine C. McGeoch, Editors,** Data Structures, Near Neighbor Searches, and Methodology: Fifth and Sixth DIMACS Implementation Challenges

58 **Simon Thomas, Editor,** Set Theory: The Hajnal Conference

57 **Eugene C. Freuder and Richard J. Wallace, Editors,** Constraint Programming and Large Scale Discrete Optimization

56 **Alexander Barg and Simon Litsyn, Editors,** Codes and Association Schemes

55 **Ding-Zhu Du, Panos M. Pardalos, and Jie Wang, Editors,** Discrete Mathematical Problems with Medical Applications

54 **Erik Winfree and David K. Gifford, Editors,** DNA Based Computers V

53 **Nathaniel Dean, D. Frank Hsu, and R. Ravi, Editors,** Robust Communication Networks: Interconnection and Survivability

52 **Sanguthevar Rajasekaran, Panos Pardalos, and D. Frank Hsu, Editors,** Mobile Networks and Computing

51 **Pierre Hansen, Patrick Fowler, and Maolin Zheng, Editors,** Discrete Mathematical Chemistry

50 **James M. Abello and Jeffrey Scott Vitter, Editors,** External Memory Algorithms

49 **Ronald L. Graham, Jan Kratochvíl, Jaroslav Nešetřil, and Fred S. Roberts, Editors,** Contemporary Trends in Discrete Mathematics

48 **Harvey Rubin and David Harlan Wood, Editors,** DNA Based Computers III

47 **Martin Farach-Colton, Fred S. Roberts, Martin Vingron, and Michael Waterman, Editors,** Mathematical Support for Molecular Biology

46 **Peng-Jun Wan, Ding-Zhu Du, and Panos M. Pardalos, Editors,** Multichannel Optical Networks: Theory and Practice

45 **Marios Mavronicolas, Michael Merritt, and Nir Shavit, Editors,** Networks in Distributed Computing

44 **Laura F. Landweber and Eric B. Baum, Editors,** DNA Based Computers II

43 **Panos Pardalos, Sanguthevar Rajasekaran, and José Rolim, Editors,** Randomization Methods in Algorithm Design

42 **Ding-Zhu Du and Frank K. Hwang, Editors,** Advances in Switching Networks

41 **David Aldous and James Propp, Editors,** Microsurveys in Discrete Probability

40 **Panos M. Pardalos and Dingzhu Du, Editors,** Network Design: Connectivity and Facilities Location

39 **Paul W. Beame and Samuel R Buss, Editors,** Proof Complexity and Feasible Arithmetics

38 **Rebecca N. Wright and Peter G. Neumann, Editors,** Network Threats

37 **Boris Mirkin, F. R. McMorris, Fred S. Roberts, and Andrey Rzhetsky, Editors,** Mathematical Hierarchies and Biology

36 **Joseph G. Rosenstein, Deborah S. Franzblau, and Fred S. Roberts, Editors,** Discrete Mathematics in the Schools

35 **Dingzhu Du, Jun Gu, and Panos M. Pardalos, Editors,** Satisfiability Problem: Theory and Applications

34 **Nathaniel Dean, Editor,** African Americans in Mathematics

33 **Ravi B. Boppana and James F. Lynch, Editors,** Logic and Random Structures

32 **Jean-Charles Grégoire, Gerard J. Holzmann, and Doron A. Peled, Editors,** The SPIN Verification System

31 **Neil Immerman and Phokion G. Kolaitis, Editors,** Descriptive Complexity and Finite Models

30 **Sandeep N. Bhatt, Editor,** Parallel Algorithms: Third DIMACS Implementation Challenge

29 **Doron A. Peled, Vaughan R. Pratt, and Gerard J. Holzmann, Editors,** Partial Order Methods in Verification

28 **Larry Finkelstein and William M. Kantor, Editors,** Groups and Computation II

27 **Richard J. Lipton and Eric B. Baum, Editors,** DNA Based Computers

26 **David S. Johnson and Michael A. Trick, Editors,** Cliques, Coloring, and Satisfiability: Second DIMACS Implementation Challenge

25 **Gilbert Baumslag, David Epstein, Robert Gilman, Hamish Short, and Charles Sims, Editors,** Geometric and Computational Perspectives on Infinite Groups

24 **Louis J. Billera, Curtis Greene, Rodica Simion, and Richard P. Stanley, Editors,** Formal Power Series and Algebraic Combinatorics/Séries Formelles et Combinatoire Algébrique, 1994

23 **Panos M. Pardalos, David I. Shalloway, and Guoliang Xue, Editors,** Global Minimization of Nonconvex Energy Functions: Molecular Conformation and Protein Folding

22 **Panos M. Pardalos, Mauricio G. C. Resende, and K. G. Ramakrishnan, Editors,** Parallel Processing of Discrete Optimization Problems

21 **D. Frank Hsu, Arnold L. Rosenberg, and Dominique Sotteau, Editors,** Interconnection Networks and Mapping and Scheduling Parallel Computations

20 **William Cook, László Lovász, and Paul Seymour, Editors,** Combinatorial Optimization

19 **Ingemar J. Cox, Pierre Hansen, and Bela Julesz, Editors,** Partitioning Data Sets

18 **Guy E. Blelloch, K. Mani Chandy, and Suresh Jagannathan, Editors,** Specification of Parallel Algorithms

17 **Eric Sven Ristad, Editor,** Language Computations

16 **Panos M. Pardalos and Henry Wolkowicz, Editors,** Quadratic Assignment and Related Problems

15 **Nathaniel Dean and Gregory E. Shannon, Editors,** Computational Support for Discrete Mathematics

14 **Robert Calderbank, G. David Forney, Jr., and Nader Moayeri, Editors,** Coding and Quantization: DIMACS/IEEE Workshop

13 **Jin-Yi Cai, Editor,** Advances in Computational Complexity Theory

12 **David S. Johnson and Catherine C. McGeoch, Editors,** Network Flows and Matching: First DIMACS Implementation Challenge

11 **Larry Finkelstein and William M. Kantor, Editors,** Groups and Computation

10 **Joel Friedman, Editor,** Expanding Graphs

9 **William T. Trotter, Editor,** Planar Graphs

8 **Simon Gindikin, Editor,** Mathematical Methods of Analysis of Biopolymer Sequences

7 **Lyle A. McGeoch and Daniel D. Sleator, Editors,** On-Line Algorithms